Contents

PREFACE iv

PART 1. OVERVIEW 1

   1. Introduction 1
   2. Overview of Community Matrices 3
   3. Community Sampling and Measurements 13
   4. Species Diversity 25
   5. Species on Environmental Gradients 35
   6. Distance Measures 45

PART 2. DATA ADJUSTMENTS 58

   7. Data Screening 58
   8. Documenting the Flow of Analyses 62
   9. Data Transformations 67

PART 3. DEFINING GROUPS WITH MULTIVARIATE DATA 80

   10. Overview of Methods for Finding Groups 80
   11. Hierarchical Clustering 86
   12. Two-Way Indicator Species Analysis 97

PART 4. ORDINATION 102

   13. Introduction to Ordination 102
   14. Principal Components Analysis 114
   15. Rotating Ordination Axes 122
   16. Nonmetric Multidimensional Scaling 125
   17. Bray-Curtis (Polar) Ordination 143
   18. Weighted Averaging 149
   19. Correspondence Analysis 152
   20. Detrended Correspondence Analysis 159
   21. Canonical Correspondence Analysis 164
   22. Reliability of Ordination Results 178

PART 5. COMPARING GROUPS 182

   23. Multivariate Experiments 182
   24. MRPP (Multi-response Permutation Procedures) 188
   25. Indicator Species Analysis 198
   26. Discriminant Analysis 205
   27. Mantel Test 211
   28. Nested Designs 218

PART 6. STRUCTURAL MODELS 222

   29. Classification and Regression Trees 222
   30. Structural Equation Modeling 233

APPENDIX 1. ELEMENTARY MATRIX ALGEBRA 257

REFERENCES 260

INDEX 284

Preface

We wanted to write a book on community analysis with practical utility for ecologists. We chose to emphasize multivariate techniques for community ecology, because there is a great demand for accessible, practical, and current information in this area.

We hope this book will be useful for researchers, academicians, and students of community ecology. We envision its use as both a reference book and a textbook. By publishing the book as a paperback through MjM Software, rather than a major publisher, we hope to keep the price reasonable.

This book shares some of the first author's experience gained from many years of teaching a course in community analysis and his participation in developing the software package PC-ORD. This book is not a manual for PC-ORD — that already exists. The PC-ORD manual describes the mechanics for using the software, along with a very brief statement of the purpose of each analytical tool. The current book, in contrast, provides a foundation for better understanding community analysis. This book also expands the logic behind choosing one technique over the other and explains the assumptions implicit in that decision. We also illustrate many of the methods with examples.

We have tried to write a book for all community analysts, not just PC-ORD users. Many of the techniques described in this book are not currently available in PC-ORD. The reader will find, however, numerous references to PC-ORD, simply because so many readers will want to immediately try out ideas generated from reading this book.

Because of the need to relate community properties to environmental factors, Part 6 of the book deals with newly emerging methods that are well suited for this purpose. Dean Urban contributed an overview of the methods of classification and regression trees. The final chapter in Part 6 introduces structural equations, a constantly evolving body of methods for multivariate hypothesis testing that is widely used in many fields outside of the environmental sciences. The second author has spent a number of years appraising the value of structural equations and introducing these methods to the uninitiated. Our treatment of this topic is brief, but we hope to suggest some of the power of structural equation modeling for understanding ecological communities.

Bruce McCune dedicates his efforts on this book to Edward W. Beals, for his insight and teaching on many of these techniques, for his geometric view of community analysis, and for his willingness to look standard practice in the eye. Bruce also feels forever indebted to Paul L. Farber and Stella M. Coakley for graciously giving space in their gardens for this project to grow.

Jim Grace acknowledges his appreciation to his major professor, Bob Wetzel, for giving him an example of the power of perseverance and dedication, and to his mother and sister for their unwavering support.

We thank the numerous graduate students in McCune's Community Analysis class. Their collective contribution to the book is tremendous. We apologize for subjecting them to variously half-baked versions of this document, and we thank them for their patience, corrections, and suggestions.

We also thank the many people who have provided comments, insights, and encouragement to Jim Grace during his journey through the world of complex data.

The manuscript greatly benefited from careful readings by Michelle Bolda, Michael Mefford, JeriLynn Peck, and especially Patricia Muir. Selected chapters were improved by Mark V. Wilson. Portions of the chapter on cluster analysis were derived from unpublished written materials by W. M. Post and J. Sheperd. We thank Amy Charron for the cover design and its central graphic.

CHAPTER 1

Part 1. Overview

Introduction

Who lives with whom and why? In one form or another this is a common question of naturalists, farmers, natural resource managers, academics, and anyone who is just curious about nature. This book describes statistical tools to help answer that question.

Species come and go on their own but interact with each other and their environments. Not only do they interact, but there is limited space to fill. If space is occupied by one species, it is usually unavailable to another. So, if we take abundance of species as our basic response variable in community ecology, then we must work from the understanding that species responses are not independent and that a cogent analysis of community data must consider this lack of independence.

We confront this interdependence among response variables by studying their correlation structure. We also summarize how our sample units are related to each other in terms of this correlation structure. This is one form of "data reduction." Data reduction takes various forms, but it has two basic parts: (1) summarizing a large number of observations into a few numbers and (2) expressing many interrelated response variables in a more compact way.

Many people realize the need for multivariate data reduction after collecting masses of community data. They become frustrated with analyzing the data one species at a time. Although this is practical for very simple communities, it is inefficient, awkward, and unsatisfying for even moderate-sized data sets, which may easily contain 100 or more species.

We can approach data reduction by categorization (or classification), a natural human approach to organizing complex systems. Or we can approach it by summarizing continuous change in a large number of variables as a synthetic continuous variable (ordination). The synthetic variable represents the combined variation in a group of response variables. Data reduction by categorization or classification is perhaps the most intuitive, natural approach. It is the first solution to which the human mind will gravitate when faced with a complex problem, especially when we are trying to elucidate relationships among objects, and those objects have many relevant characteristics.

For example, consider a community data set consisting of 100 sample units and the 80 species found in those sample units. This can be organized as a table with 100 objects (rows) and 80 variables (columns). Faced with the problem of summarizing the information in such a data set, our first reaction might be to construct some kind of classification of the sample units. Such a classification boils down to assigning a category to each of the sample units. In so doing, we have taken a data matrix with 80 variables and reduced it to a single variable with one value for each of the objects (sample units).

The other fundamental method of data reduction is to construct a small number of continuous variables representing a large number of the original variables. This is possible and effective only if the original response variables covary. It is not as intuitive as classification, because we must abandon the comfortable typological model. But what we get is the capacity to represent continuous change as a quantitative synthetic variable, rather than forcing continuous change into a set of pigeonholes.

So data reduction is summarization, and summarization can result in categories or quantitative variables. It is obvious that the need for data reduction is not unique to community ecology. It shows up in many disciplines including sociology, psychology, medicine, economics, market analysis, meteorology, etc. Given this broad need, it is no surprise that many of the basic tools of data reduction — multivariate analysis — have been widely written about and are available in all major statistical software packages.

In community ecology, our response variables usually have distinct and unwelcome properties compared with the variables expected by traditional multivariate analyses. These are not just minor violations. These are fundamental problems with the data that seriously weaken the effectiveness of traditional multivariate tools.

This book is about how species abundance as a response variable differs from the ideal, how this creates problems, and how to deal effectively with those problems. This book is also about how to relate species abundance to environmental conditions, the various challenges to analysis, and ways to extract the most information from a set of correlated predictors.

Definition of community

What is a "community" in ecology? The word has been used many different ways and it is unlikely that it will ever be used consistently. Some use "community" as an abstract group of organisms that recurs on the landscape. This can be called the **abstract community** concept, and it usually carries with it an implication of a level of integration among its parts that could be called organismal or quasi-organismal. Others, including us, use the **concrete community** concept, meaning simply the collection of organisms found at a specific place and time. The concrete community is formalized by a sample unit which arbitrarily bounds and compartmentalizes variation in species composition in space and time. The content of a sample unit is the operational definition of a community.

The word "assemblage" has often been used in the sense of a concrete community. Not only is this an awkward word for a simple concept, but the word also carries unwanted connotations. It implies to some that species are independent and noninteracting. In this book, we use the term "community" in the concrete sense, without any conceptual or theoretical implications in itself.

Why study biological communities?

People have been interested in natural communities of organisms for a long time. Prehistoric people (and many animals, perhaps) can be considered community ecologists, since their ability to survive depended in part on their ability to recognize habitats and to understand some of the environmental implications of species they encountered. What differentiates community ecology as a scientific endeavor is that we systematically collect data to answer the question "why" in the "who lives with whom and why."

Another fundamental question of community ecology is "What controls species diversity?" This springs from the more basic question, "What species are here?" We keep backyard bird lists. We note which species of fish occur in each place where we go fishing. We have mental inventories of our gardens. Inventorying species is perhaps the most fundamental activity in community ecology. Few ecologists can resist, however, going beyond that to try to understand which species associate with which other species and why, how they respond to environmental changes, how they respond to disturbance, and how they respond to our attempts to manipulate species.

It is not possible now, nor is it ever likely to be possible to make reliable, specific, long-term predictions of community dynamics for specific sites based on general ecological theory. This is not to say we should not try. But, we face the same problems as long-term weather forecasters. Most of our predictive success will come from short-term predictions applying local knowledge of species and environment to specific sites and questions.

Purpose and structure of this book

The primary purpose of this book is to describe the most important tools for data analysis in community ecology. Most of the tools described in this book can be used either in the description of communities or the analysis of manipulative experiments. The topics of community sampling and measuring diversity each deserve a book in themselves. Rather than completely ignoring those topics, we briefly present some of the most important issues relevant to community ecology. Explicitly spatial statistics as applied to community ecology likewise deserve a whole book. We excluded this topic here, except for a few tangential references.

Each analytical method in this book is described with a standard format: Background, When to use it, How it works, What to report, Examples, and Variations. The **Background** section briefly describes the development of the technique, with emphasis on the development of its use in community ecology. It also describes the general purpose of the method. **When to use it** describes more explicitly the conditions and assumptions needed to apply the method. Knowing **How it works** will also help most readers appreciate when to use a particular method. Depending on the utility of the method to ecologists, the level of detail varies from an overview to a full step-by-step description of the method. **What to report** lists the methodological options and key portions of the numerical results that should be given to a reader. It does not include items that should be reported from any analysis, such as data transformations (if any) and detection and handling of outliers. **Examples** provide further guidance on how to use the methods and what to report. **Variations** are available for most techniques. Describing all of them would result in a much more expensive book. Instead, we emphasize the most useful and basic techniques. The references in each section provide additional information about the variants.

CHAPTER 2

Overview of Community Matrices

Kinds of variables

Individual observations of *ordinal* variables can be ranked or placed on a measurement scale. On the other hand, *categorical* or *nominal* variables are qualitative statements that have no inherent rank or measure. Sokal and Rohlf (1995) used the term "attributes" for categorical variables. Ecologists, however, sometimes use "attribute" as any observed characteristic of a sample unit, essentially synonymous with "variable." Ordinal and categorical variables must be handled quite differently in an analysis. Categorical variables are often used to define groups, for example indicating several different treatments and a control. In other cases they are observations from the sample units, such as recording whether the bedrock of a sample unit is granite, limestone, or basalt.

A categorical variable can sometimes be analyzed with methods that demand ordinal variables, by transforming the categories into a series of binary (0/1) variables. If a categorical variable has n possible states, it is recoded into a series of n-1 binary variables (sometimes called *dummy* or *indicator* variables). One category, the omitted category, is implicit when all of the other categories receive a zero. Interpretation of the results is most straightforward if the category omitted is somehow the norm or mode, and the other categories are deviations from that condition.

Ordinal variables can be *measurement* variables or *ranked* variables. Many nonparametric statistics use ranked variables. In most cases, the ranks are derived from measurement variables, rather than being recorded in the data as ranks. Converting measurements to ranks usually results in a loss of power (ability to detect a difference), but sometimes this loss is outweighed by the need to avoid distributional assumptions.

Measurement variables can be *continuous* or *discontinuous*, the latter having only fixed numerical values, such as cover classes representing ranges of percent cover. In contrast, a continuous variable theoretically has infinite possible real numbers. In practice the difference is fuzzy, because at some scale, all measurements are discontinuous. The finite number of possibilities is determined by the graduations on a measuring device (e.g., on a thermometer) or the number of significant digits recorded.

Kinds of multivariate data sets

Multivariate data sets may include many objects, each with many measurable attributes or variables. Table 2.1 lists some biological examples.

Most community analysis studies have species data from a set of sample units (e.g., plots × species). You can also use other taxonomic levels, such as genera or families, as is common for insects and occasionally for plants (e.g., del Moral & Denton 1977). Molecular markers have increasingly been used in lieu of species when sampling minute or otherwise cryptic species (e.g., bacteria, soil fungi, or samples of fine roots; Table 2.2). Using these markers has opened a new field for community analysis by revealing a huge, previously unknown component of biodiversity (Stahl 1995).

A niche-space analysis might use data on food or other resources used (e.g., individuals × resource used). Analyses of guilds or functional groups typically seek relationships among species in characteristics of ecological importance (e.g., species × life history characteristics). Multivariate behavioral analyses are heavily used in the social sciences: psychology and sociology (e.g., individuals × frequency of specific behaviors). Taxonomic analyses seek phenetic relationships among operational taxonomic units (OTUs) using a variety of quantitative or qualitative characters.

Goal of data reduction

Averages, frequency distributions, ordinations, and classifications are all frequently used for data reduction, depending on the nature of the data. A simple example of data reduction is finding the average diameter of trees in a plot. Reduction of multivariate data involves extracting a small number of composite variables (or dimensions or axes or principal components or groups) that express much of the information in the original multidimensional data.

Table 2.1. Examples of objects and attributes in ecological matrices.

Type of Study	Objects	Attributes
Community analysis	Sample plots Stands Community types	Species Molecular markers Structures or functions Environmental factors Time of sample
Niche-space analysis	Individuals Populations Species Guilds	Resources used or provided Environmental optima, limits, or responses Physicochemical characteristics of resources Habitats
Behavioral analysis	Individuals Populations Species	Activities Response to stimuli Test scores
Taxonomic analysis	Individuals (specimens) Populations Species	Morphological characters Nucleotide positions Isozyme presence Secondary chemicals
Functional or guild analysis	Individuals Populations Species Higher taxa	Life history characteristics Morphology Ecological functions Ecological preferences

Table 2.2. Examples of molecular markers used in lieu of species in community ecology. BIOLOG = carbon source utilization profiles; cpDNA = chloroplast DNA; FAME = fatty acid methyl esters; LH-PCR = length heterogeneity polymerase chain reaction; T-RFLP = terminal restriction fragment length polymorphisms.

Organisms	Kind of marker	Reference
microbial communities (fungal or bacterial)	BIOLOG microplates	Ellis et al. 1995, Garland & Mills 1991, Myers et al. 2001, Zak et al. 1994
fine roots of trees	cpDNA, restriction fragments	Brunner et al. 2001
soil microbes	FAME	Cavigelli et al. 1995; Schutter & Dick 2000, 2001
soil bacteria	FAME and LH-PCR of 16S rDNA	Ritchie et al. 2000
biosolids in wastewater treatment	FAME	Werker & Hall 2001
aquatic microbes	LH-PCR of 16S rDNA	Bernhard et al. 2002
nitrogen-fixing microbes	nitrogen fixing gene sequences (nifH)	Affourtit et al. 2001
mycorrhizal fine roots and soil microbes	phospholipid fatty acids	Myers et al. 2001, Wiemken et al. 2001
microbes in soil	T-RFLP of 16S rDNA and 16S rRNA-targeted oligonucleotide probes	Buckley & Schmidt 2001

Analyzing the data matrix

See Appendix 1 for conventions of simple matrix notation and matrix operations. In the data matrix, in its **normal** orientation (matrices **A** and **E**, Table 2.3), **rows** represent sample units, objects, entities, or cases. **Columns** represent variables or attributes of the objects. The analysis can take several general forms, defined below:

Normal analysis: The grouping or ordering of objects (Tables 2.1 and 2.4).

Transpose analysis: The grouping or ordering of attributes (Table 2.1). The matrix can be transposed, then analyzed (Table 2.4).

Q route: Arriving at a grouping or ordering of either objects or attributes by analyzing a matrix of relationships among **objects**. For example, with a matrix of sample units by species, we would analyze a matrix of relationships among sample units.

R route: Arriving at a grouping or ordering of either objects or attributes by analyzing a matrix of relationships among **attributes**. For example, with a matrix of sample units by species, we would analyze a matrix of relationships among species.

The choice between Q route and R route is usually inherent in the choice of analytical method. For example, a normal analysis with Bray-Curtis ordination is always by the Q route. A few methods, however, can be done by either route. For example, identical results from principal components analysis can be obtained for a normal analysis either by the Q route or the R route.

Table 2.3. Example data matrices in community ecology. **A**: 15 sample units (plots) × 8 species; each species indicated by a 4-letter acronym. Each element represents the abundance of a species in a plot. **E**: 15 sample units × 3 environmental variables. For the first two variables each element represents a measured value, while the third variable, "Group" represents assignments to two treatments and a control group. **S**: 3 ecological traits × 8 species. Each element represents the characteristic value for an ecological trait for a given species.

A	\	\	\	Species	\	\	\	\	**E** Environmental Variables	\	\
	ALSA	CAHU	CECH	HYDU	HYEN	HYIN	HYVI	PLHE	Elev	Moisture	Group
Plot01	1.51	0.11	0	0.35	0.21	0	0	0.24	311	9	1
Plot02	1.73	0	0	0.25	0.23	0	0	0.53	323	17	1
Plot03	1.20	0	0.02	0.03	0.05	0	0	0.05	12	10	1
Plot04	1.42	0.05	0	0.99	0.13	0	0.09	0.08	15	8	1
Plot05	1.14	0	0	0.14	0.17	0	0	0	183	12	1
Plot06	1.39	0.07	0.07	0	0	0	0	0.04	12	26	2
Plot07	2.26	0.11	0.03	0.02	0	0	0	0.07	46	29	2
Plot08	1.01	0.32	0.03	0	0	0	0	0.48	220	19	2
Plot09	1.09	0.09	0	0	0	0	0	0	61	22	2
Plot10	2.90	0	0.04	0.19	0.71	0.30	0.15	0.39	43	34	2
Plot11	3.22	0.03	0.06	0	0.56	0.41	0	0.63	256	21	3
Plot12	3.42	0.12	0	0.38	0.26	0	0	0.06	46	17	3
Plot13	2.55	0.08	0	0	0.27	0.43	0.08	0.15	76	22	3
Plot14	2.72	0.13	0	0.28	1.11	0	0.11	0	488	36	3
Plot15	3.01	0	0	0.28	0.67	0.86	0	0.25	274	23	3
S											
MaxAge	15	1	9	22	15	7	8	5			
RootDpth	1.73	0.08	0.52	0.25	0.23	0.35	2.20	0.53			
Fecundity	12	40	66	5	32	43	56	52			

Table 2.4. Characteristics of normal vs. transpose analysis of the community matrix (**A**).

	Normal	Transpose
Sample units (SUs)	SUs are points. A normal analysis begins from SUs as points in species space. Each axis of the space represents a species. A SU is a snapshot in space and time that can logically and accurately be represented as a point in species space.	SUs are axes. Each SU is an axis of a multidimensional space. A species is represented by a point. Each species is positioned along a SU axis according to its abundance in that SU.
Species	Species are axes.	Species are points.
Ordination (Ch. 13)	Points are SUs. Axes are combinations of correlated species. You can overlay species as areas or volumes in an ordination space. Because species occur in multiple sample units, a species conceptually occupies a region of the ordination space rather than a single point.	Points are species. Axes are combinations of correlated SUs. Two problems arise in interpreting the analysis: (1) Conceptually species recur in a range of points in space and time and in a range of environmental space, thus species are misrepresented (oversimplified) as points. (2) Data for infrequent species are noisy and undersampled in contrast to complete representation of a sample unit in the normal analysis.
Dimensionality of data reduction	Usually achievable in a few dimensions.	Often a higher dimensionality is required to describe correlation structure.
Response of analysis to common vs. rare species	Both common and rare species can contribute to measures of distance between SUs.	Common species separate to some extent from rare species.
Double zero problem? (Ch. 5 & 6)	Yes. Some resemblance measures use joint absence (0,0) as indication of similarity, but the information is ecologically ambiguous.	Yes. Same problem as with normal analysis.
Clustering (Ch. 10 & 11)	Frequent and infrequent species are blended in describing clusters of sample units.	Frequent and infrequent species fall into different clusters, even if they share optima. Dominant species and species that are widespread but in low abundance fall into different clusters. Deletion of rare species and relativization by species maxima or totals tend to reduce those differences such that species are grouped more by optima than by rarity.

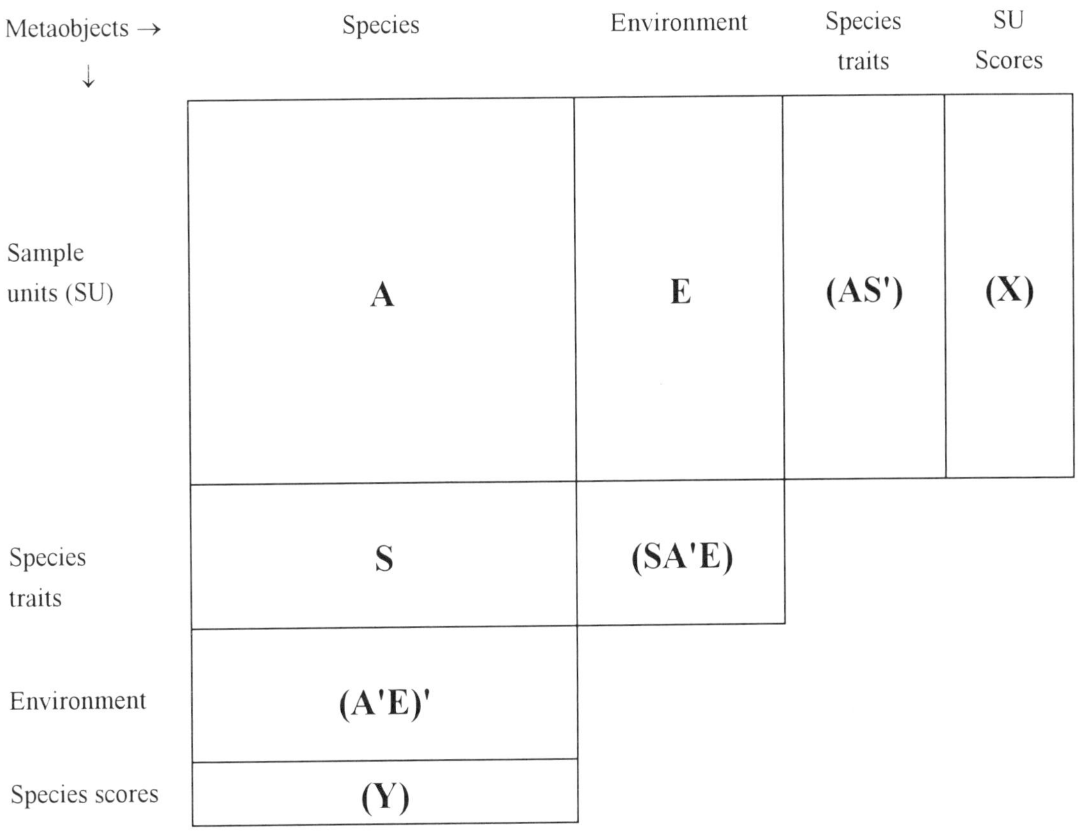

	Species	Environment	Species traits	SU Scores
Sample units (SU)	**A**	**E**	**(AS')**	**(X)**
Species traits	**S**	**(SA'E)**		
Environment	**(A'E)'**			
Species scores	**(Y)**			

Metaobjects → (top), ↓ (left)

Figure 2.1. The complete community data set. Bold uppercase letters represent different matrices. Calculated matrices are in parentheses. Sample unit (SU) scores and species scores are based on ordination or classification of sample units or species. The prime mark (') indicates a transposed matrix (see Appendix 1).

The difference between the Q route vs. R route is one of mechanics. There are, however, very important conceptual differences between the normal analysis and transpose analysis. Those differences have not been thoroughly compared and explained in the literature. Table 2.4 contrasts normal and transpose analyses, based on the literature (e.g., Clarke 1993, p. 118) and our own experience. Understanding these differences requires some experience with community analysis, so we recommend that beginners return to this table as a reference.

Community data sets

Community data sets take many forms, but most of them can be fit into a concept of basic matrices and their relationships (Fig. 2.1, Tables 2.5 & 2.6). One can think of the sample, species, environment, and species traits as the basic "metaobjects" in community ecology. In other words, a sample is a metaobject composed of the objects that are sample units. The environment is a metaobject composed of environmental variables.

You can measure or calculate a matrix for each pair of metaobjects (Fig. 2.1, Table 2.5). Furthermore, you can represent each metaobject as points in a space defined by another metaobject (Table 2.6). In other words, sample units can be represented as points in species space and vice-versa; if this concept is not immediately clear, we hope it will become clear when distance measures are introduced (Ch. 6). Similarly, species can be represented as points in environmental space, environmental variables can be represented as points in species space, etc. Although not all of these combinations are conceptually appealing, all are mathematically possible.

Table 2.5. Sources and types of multivariate ecological data used in studies of species composition. Categorical (= nominal) variables are qualitative rather than quantitative, indicating membership in one of two or more categories.

	Matrix	Matrix Fullness	Categorical* vs. ordinal	Usual data source
A	Species composition	sparse**	O	field
E	Environment	full	C, O	field
S	Species traits	full	C, O	field/literature
X	Sample ordination or group membership	full	O	calculated
Y	Species ordination or group membership	full	O	calculated
AS'	samples × traits	full	O	calculated***
A'E	species × environment	full	O	calculated***
SA'E	traits × environment	full	O	calculated***

* Categorical (nominal) variables cannot be used in calculating new matrices, unless they are converted to a series of binary (0/1) variables.
** A sparse matrix contains many zero values.
*** Rare in the literature.

Table 2.6. Matrices required for representing ecological entities in spatial coordinates defined by other sets of ecological variables. You can represent any of these entities in any of the coordinate systems. All are appropriate for multivariate analysis. The prime mark (') indicates a transposed matrix (see Appendix 1).

	Coordinate system			
Entities	Species	Samples	Environment	Species traits
Species	-	A'	A'E	S'
Samples	A	-	E	AS'
Environment	(A'E)'	E'	-	(SA'E)'
Species traits	S	(AS')'	SA'E	-

The most common data sets are either the matrix **A** (sample units in rows, species in columns) or the combination of matrices **A** and **E** (sample units in rows, environmental variables in columns). The matrix **S** is rarely used, but may be of primary interest where the focus is on variation in life history strategies or traits as it relates to environments or communities. Interest has increased in using **S** as a basis for assigning species to "functional groups."

The matrices **X** and **Y** are very commonly used. Typically **X** has only a few columns, each column representing placement on an ordination axis (the sample unit "score"). Alternatively, **X** may contain a categorization of the sample units into community types. Similarly, **Y** may contain species scores on ordination axes and/or a categorization of species into species types.

The other matrices are rarely used, but can be produced by matrix multiplication. For example, suppose you are doing a standard community analysis on species composition and environmental data. Your main matrix contains the abundance of each species in each plot and your secondary matrix contains the environmental parameters for each plot. The most common analytical approach is to examine the relationships among plots in species space. That is, plots are grouped, ordered, or otherwise arranged by their similarities in species composition, **A**. Relationships between the environmental variables, **E** and the community patterns are then sought. The purpose of this procedure is to see how species distributions are related to environmental factors.

The species × environment matrix

Why not examine species directly in environmental space? First you would need a matrix of species scores for each environmental variable. One way to produce such a matrix is to multiply your main matrix, **A**, by your environmental matrix, **E**. (See Appendix 1 for matrix multiplication.) Actually, to make the matrix algebra come out right, you must postmultiply the transpose of the main matrix (rows and columns switched) by the environmental matrix:

species × environment matrix = **A'E**

(The transpose of **A** is indicated by **A'**; see Appendix 1 for elementary matrix algebra.) The resulting matrix contains scores for each species on each environmental variable. This matrix can now be summarized further by analyzing its correlation structure (using correlation in the broad sense). A good example of this approach is in Brazner and Beals (1997).

Serious potential pitfalls in calculating **A'E** must be carefully avoided. For example, all of the environmental variables are included, whether or not they are related to species distribution and abundance. By multiplying the two matrices you implicitly assume that all of the environmental variables are important. Brazner and Beals reduced this problem by first restricting the environmental variables to those shown to be important in relation to an ordination of sample units in species space.

It is important to give some thought to prior data transformations (Ch. 9). Because matrix multiplication involves adding the products of many numbers, the resulting numbers can become very small or very large.

A worse problem is that the resulting matrix can be nearly meaningless if careful attention is not paid to scaling your variables. For example, if one of the environmental variables ranges from 1000 to 3000 and the other environmental variables range from 0 to 1, the large numbers will completely obscure the small numbers, resulting in a matrix dominated by a single environmental variable. This problem could be avoided by standardizing or relativizing the variables in the two matrices, so that each variable is given equal weight (see Chapter 9). Brazner and Beals (1997) relativized all of their environmental variables, expressing each data point as the number of standard deviations away from the mean for that variable.

If you do this kind of matrix manipulation, it is important that you examine the resulting matrices critically to be sure they contain what you think they contain. See Greig-Smith (1983 pp. 229 and 278) for further cautions about ordinations based on environmental data.

The sample unit × species traits matrix

Given the standard community matrix, **A**, and a matrix of species traits, **S**, one can multiply these to yield a matrix of sample units by species traits:

sample units × species traits = **AS'**

Analysis of this matrix (e.g., Feoli & Scimone 1984, Diaz et al. 1992, Diaz & Carbido 1997, Diaz et al. 1999, Diaz Barradas et al. 1999, Lavorel et al. 1999, Landsberg et al. 1999) would reveal how sample units are related to each other in terms of species traits. One might wish to contrast the blend of species traits in different groups of sample units (e.g., treatment vs.

control). For example, Lavorel et al. (1999) analyzed traits from **AS'**, one trait at a time, comparing experimental treatments with univariate ANOVA. Or one might wish to study how species traits covary in a sample along a gradient. For example, Diaz et al. (1999) used ordination to extract the main gradients (**X**) from the **AS'** matrix, then analyzed the relationship of those gradients to climate and disturbance (**E**).

If **A** and **S** are binary (**A** contains presence-absence data and **S** specifies 1=yes or 0=no for each trait for each species), then **AS'** gives the total number of species in a sample unit with a given trait. If either **A** or **S** or both are quantitative, the elements of the **AS'** matrix can be considered to represent the abundance or magnitude of a particular trait in a particular sample unit. If **S** is quantitative and different traits are measured on different scales (the usual case), then the traits must be relativized in some way (for example, as a proportion of the maximum value or as number of standard deviations away from the mean; see Chapter 9).

Functional groups

We can add another matrix, **G**, that assigns species to functional groups. The groups may be different states of a single categorical variable, or the groups may be multiple variables that are not necessarily mutually exclusive. For example, the Common Loon could be assigned to both the groups "divers" and "predators."

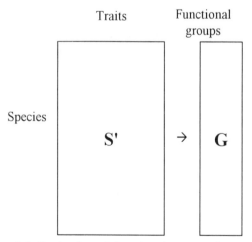

Figure 2.2. Derivation of functional groups, **G**, from a species trait matrix, **S**.

Functional groups can be assigned a priori or based on an analysis of the trait matrix, **S**. **G** can be thought of as a summary of **S**, where **G** must contain categorical or binary variables rather than measurement variables. If **G** is derived from **S** by multivariate analysis (Pillar 1999a), and different traits are measured on different scales (the usual case), then the traits must be relativized in some way. This can happen as part of the analysis itself (e.g., the choice of correlation coefficients in the cross-products matrix for PCA; see Chapter 14) or as a relativization prior to analysis (see Chapter 9).

The functional group matrix, **G**, can be analyzed in relation to a community sample, **A**, and environmental variables, **E**. Kleyer (1999), for example, first derived **G** by a cluster analysis of **S**. In this case, **G** contained a single variable, representing groups of species with similar traits. The number of individuals in each species group in each sample unit is conveniently calculated by first separating the group membership variable in **G** into a series of mutually exclusive binary variables representing membership or not (1/0) in each species group, then calculating **AG'** (analogous to **AS'**). This yields the representation of each species group in each sample unit. Because **A** consisted of counts of each species in Kleyer's case, each element of **AG'** specified the number of individuals of each functional group in each sample unit. Multivariate analysis of the **AG'** matrix is possible, but in this case, Kleyer used Gaussian logistic regression to fit the probability of co-occurrence of all species in a group along a gradient plane of resource supply and disturbance intensity.

The traits × environment matrix

Calculating the traits × environment matrix, **SA'E**, is one solution to the "fourth-corner problem" (Legendre et al. 1997). This refers to the fourth corner (lower right) in the square arrangement of matrices in Figure 2.1. This matrix can be used to represent how species traits (morphological, physiological, phylogenetic, and behavioral) are related to habitat or site characteristics. Three data matrices are required: **S**, **A**, and **E**.

With presence-absence species data and nominal variables in **S** and **E**, then **SA'E** is a contingency table. Legendre et al. (1997) pointed out that one cannot use a G statistic to test for independence of traits and environment in this contingency table, because several species are observed in each sample unit. This makes the observations nonindependent. They propose a permutation method for assessing statistical significance.

With quantitative data, all of the cautions about calculating **A'E** apply here as well. Carefully done, the resulting matrix is essentially an ordination of each species trait on each environmental variable. If species traits are intercorrelated or environmental variables are intercorrelated, then further data reduction of the **SA'E** matrix may be desirable. For example, one could ordinate species traits in environmental space.

Analysis of a focal species and its habitat

Assume that one has sampled across a variety of habitats and that the sample units have variable use or performance by a focal species. Very often, the focal species is a rare species or some other species of conservation or management interest. Performance of the species in each sample unit, represented by **F** (Fig. 2.3), might be measured by presence-absence, abundance, reproductive success, presence of a nest, etc. A matrix of habitat variables, **H**, have been measured in the same set of sample units. The habitat variables may be environmental factors, other species, or experimental treatments.

There are two fundamentally different views of a focal species in relation to its habitat, and each view carries with it a different approach to data analysis (Fig. 2.3, Table 2.7). One can take either or both approaches with a particular data set. The focal species-centered approach attempts to predict performance of the focal species from the habitat variables. The general habitat-centered approach describes the variation in the available habitats, then evaluates the focal species performance within that variation. The former approach is much more common.

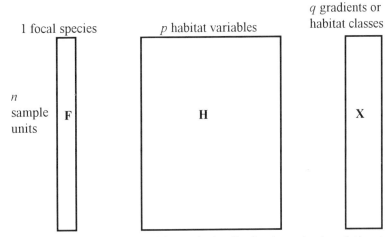

Figure 2.3. Basic matrices used in analysis of a focal species in relation to its habitat. Presence, abundance, or other measure of performance of the focal species in each sample unit is in matrix **F**. The p habitat variables for each sample unit are in **H**. The matrix **X** summarizes the habitat variables along q synthetic gradients (such as ordination axes) or habitat classes. See Table 2.7 for further explanation.

Table 2.7. Comparison of a focal species-centered approach to a general habitat-centered approach for analysis of a focal species in relation to its habitat. **F**, **H**, and **X** are defined in Figure 2.3.

	Focal species-centered approach	General habitat-centered approach
Goal	Predict focal species (i.e., predict presence or abundance or performance of focal species based on habitat variables)	Describe variation in available habitats, then describe position of the focal species within that variation.
Matrix concepts	**F** = f(subset of **H**)	Derive **X** from **H**, then see how **F** is related to **X**
Example tools	Logistic regression Multiple regression Discriminant analysis	Extract primary gradients **X** in habitat by ordination or classification of habitat matrix **H**. Overlay **F** on this ordination and calculate correlations between **F** and **X**.
Advantages	Better predictive power for focal species	Better description of habitat variation in general. Potentially applicable to a wider range of species.
Disadvantages	Little or no description of habitat variation in general. Not applicable to other focal species.	Worse predictive power for focal species.
Cautions	Need to evaluate reliability of predictions, preferably using an independent data set, an unused subset of the data set, or a resampling procedure.	Need to carefully consider (1) relative weights of variables that are on different scales and (2) which variables to include.

CHAPTER 3

Community Sampling and Measurements

Sampling is the process of selecting objects of study from a larger number of those objects (the population). Each object is then subjected to one or more measurements or observations. Although the word "sampling" is often used in a broad sense to include a discussion of the measurements, the two concepts are distinct.

This book does not include sampling theory, nor does it contain a comprehensive survey of sampling methods. It does include a few sampling basics and a discussion of some recurrent issues in community sampling and measurement.

To be perfectly clear, we will use the word "sample" to refer to a collection of sampling units or sample units (SUs). In casual conversation, a "sample" is often used to mean a single sample unit or a collection of sample units.

Measures of species abundance

1. **Cover** is the percentage of some surface covered by a vertical projection of the perimeter of an organism. Note that when summed for a given sample unit, percent cover can exceed 100% because of multiple layering. Cover excels as an abundance measure in speed, repeatability, comparability between different estimation methods, and because it can be measured nondestructively. Conversion to biomass estimates is possible but requires additional data collection for calibration (e.g., Forman 1969, McCune 1990). Cover is the most commonly used abundance measure for plants.

Percent cover is often scored as **cover classes**, rather than estimates to the nearest percent (although these too are classes, just much narrower). Using cover classes rather than attempting to estimate cover to the nearest one percent tends to speed sampling and data entry. Cover classes do not pretend to achieve more accuracy than is realistic. Furthermore, they yield statistical results that are similar to unclassed data, provided that the classes are not too broad (Sneath & Sokal 1973). Cover classes have been shown to be effective surrogates for direct biomass measurement (Hermy 1988, McCune 1990) and detectors of community changes through time (Mitchell et al. 1988). Although most analysts treat cover classes as if they were continuous data, some methods are available that explicitly recognize their ordered, multistate nature (e.g., Guisan & Harrell 2000). A disadvantage of cover classes is the potential for consistent differences (bias) between observers (Sykes et al. 1983).

The most useful and most commonly used cover classes are narrow at the extremes and broad in the middle. These approximate an arcsine-squareroot transformation, which is generally desirable for proportion data. The cover classes can thus be analyzed directly, improving normality and homogeneity of variances among groups without converting to percentages. Unless transformed (Chapter 9), multivariate analysis of raw percent cover data tends to emphasize the dominant species at the expense of species with medium to low abundance. Cover class data seldom have this problem. If, however, cover classes are transformed into midpoints of the ranges, then the problem reappears.

Many cover class schemes have been devised. Some of the most common and/or logical are listed in Table 3.1. Note the high degree of similarity among the systems.

Frequency is the proportion (or percentage) of sample units in which a species occurs. The best traits of frequency are that it is relatively sensitive to infrequent species and it is fast to score in the field. Frequency measures should, however, generally be avoided because frequency, unlike cover or density, is highly dependent on the size of the sample unit. Because there is little standardization in size of sample units, use of frequency measures restricts opportunities for comparison with other studies.

Frequency does, however, carry information about spatial distribution. For example, consider two populations of equal density, one highly aggregated and the other dispersed. The second population is more frequent.

If individuals are randomly located then frequency is an asymptotic function of density (Fig. 3.1). A little known fact: an average density of two individuals/SU gives about 86% frequency. Why? What underlying distribution is this based on? (Answer: the Poisson distribution, the same distribution as is used to describe the number of chocolate chips in chocolate chip

13

Table 3.1. Cutoff points for cover classes. Question marks for cutoff points represent classes that are not exactly defined as percentages. Instead, another criterion is applied, such as number of individuals. Cutoffs in parentheses are additional cutoffs points used by some authors.

Name	Cutoff points, %	Notes	References
Arcsine squareroot	0 1 5 25 50 75 95 99	Designed to approximate an arcsine squareroot transformation of percent cover.	Muir & McCune (1987, 1988)
Braun-Blanquet	0 ? ? 5 25 50 75	Uses two categories of low cover not exactly defined as percents. Commonly used in Europe.	Braun-Blanquet (1965), Mueller-Dombois & Ellenberg (1974)
Daubenmire	0 (1) 5 25 50 75 95	Widely used in western U.S. in habitat-typing efforts by U.S. Forest Service and many other studies.	Daubenmire (1959)
Domin	0 ? 1 5 10 25 33 50 75	One category of low cover not exactly defined as percent.	Krajina (1933); Mueller-Dombois & Ellenberg (1974)
Hult-Sernander (modified)	0 (.02 .05 .10 .19 .39 .78) 1.56 3.13 6.25 12.5 25 50 75 ...	Based on successive halving of the quadrat.	Oksanen (1976)

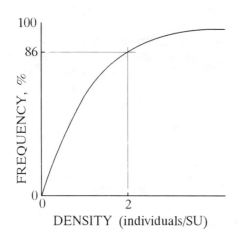

Figure 3.1 Expected percent frequency of presence in sample units (SU) as a function of density (individuals/SU).

cookies.) As SU size increases, frequency loses sensitivity and plateaus at 100%.

Counts (density). Density is the number of individuals per unit area. Density is not dependent on size of the SU. Relative density of species j is the proportion of the p species that belong to species j:

$$RD_j = \frac{Density_j}{\sum_{j=1}^{p} Density_j}$$

Density is useful if the target organisms have readily distinguishable individuals (these may be ramets of clonal plants) and the individuals do not vary much in size. Density is of questionable utility when applied to organisms that vary greatly in size, such as trees, unless applied to restricted size classes (e.g., seedling density).

Biomass. Biomass values are usually relatively destructive and tedious to obtain directly. Most often, biomass is estimated by regression equations based on more easily measured "predictor" variables. Of course, someone at sometime had to go through the tedium of obtaining data on both the biomass and the predictors. Biomass is often chosen as a measure of abundance when the functional role of the organisms is important. For example, it may be important to estimate available forage for animals.

Basal area is a measure applied to individual trees (units usually cm^2), then aggregated to the stand level, where it is sometimes referred to as **dominance**. The usual units for stand-level basal area are m^2/hectare (or ft^2/acre historically). We like basal area as a descriptor of forests because it is more closely proportional to leaf area and foliage mass than are the other common measures, such as density or frequency. Thus, we would argue that it has more functional significance than most other simple descriptors of forest structure. See Box 1 for an example description of forest composition based on individual-tree data.

Relative measures: Density, frequency, and dominance of species may all be expressed as relative proportions. For example, relative density of species j is the ratio of its density to the overall density, the sum of the densities of the p species. These relative measures are commonly expressed as percents, by multiplying the proportions by 100, for example:

$$Relative\ density_j\% = \frac{(100 \cdot Density_j)}{\sum_{j=1}^{p} Density_j}$$

Importance values are averages of two or more of the above parameters, each of which is expressed on a relative basis. For example, a measure often used for trees in eastern North American forests is:

$IV\%$ = (Relative frequency + relative dominance + relative density) / 3

where relative dominance is based on basal area. Some ecologists feel that that importance values muddle interpretation, while others appreciate the simplification of several measures into one overall measure of abundance.

The advantage of importance values is that they are not overwhelmingly influenced either by large tree size (as is relative dominance) or large numbers of small trees (as are relative frequency and relative density). The importance values add to 100 when summed across species in a stand.

The chief disadvantage of importance values is that you never know quite what a number represents. For example, consider an IV based on relative density and relative dominance. The following two species have the same IV, yet they will look very different in the stand (Table 3.2).

Table 3.2. Example of identical importance values representing different community structures.

	Species 1	Species 2
Relative Density	42	8
Relative Dominance	10	44
Sum	52	52
IV%	26	26

Presence-absence is a very useful measure in large-scale studies. It is also what is recorded in point-intercept sampling. In large regional studies or any other study in which the heterogeneity of the SUs is large, most of the information in the data will be carried in the presence-absence of species. But presence-absence is not useful in detecting more subtle differences in more homogeneous areas. For example, in a comparison of old-growth and managed second-growth forests in Montana, the lichen *Alectoria sarmentosa* was present in all stands but was consistently much more abundant in old growth than in second growth (Lesica et al. 1991).

Size-class data (or age classes or life history stages) can give useful insight into the history and future of a species. For example, an ideal "climax" species should be well represented in all size classes, indicating that the species is reproducing and replacing itself at a site. On the other hand, a species that is present in only the largest size classes may gradually be lost from a population as the large, old individuals die.

Size-structured populations are sometimes incorporated into community analysis by treating each size class (or age class or stage) as a separate "species" in the analysis. This has a desirable effect of incorporating information that may better integrate life history patterns into an analysis of community patterns. On the other hand, the goals of a study may be sufficiently broad that including this additional detail contributes little or nothing to the results.

Box 3.1. Example of stand description, based on individual tree data from fixed-area plots. The variance-to-mean ratio, V/M, is a descriptor of aggregation, values larger than one indicating aggregation and values smaller than one indicating a more even distribution than random. The variance and mean refer to the number of trees per plot. *IV* and other measures are defined in the text.

Raw data for three tree species in each of four 0.03 hectare plots. Each number represents the diameter (cm) of an individual tree.

Species	Plot 1	Plot 2	Plot 3	Plot 4
Carya glabra	23	22		31
		24		
Cornus florida		10	10	12
		10	11	
		12		
Quercus alba	13	20	11	10
	17		30	32
	44			

Frequencies, counts, total basal areas, stand densities, and stand basal areas.

Species	Freq.	No. Trees	BA (dm^2)	Freq.%	Density Trees/ha	BA dm^2/ha
Carya glabra	3	4	20.0	75	33.3	166.9
Cornus florida	3	6	5.6	75	50.0	46.4
Quercus alba	4	8	38.8	100	66.7	323.3
Totals	10	18	64.4		150.0	536.56

Relative abundances, importance values, and variance statistics.

Species	Relative Abundance			IV(%)	Variance	
	Frequency	Density	Dominance		no. trees	V/M
Carya glabra	30.0	22.2	31.1	27.8	0.67	0.67
Cornus florida	30.0	33.3	8.7	24.0	1.67	1.11
Quercus alba	40.0	44.4	60.3	48.2	0.67	0.33
				All species	1.00	0.22

Number of quadrats = 4
Empty quadrats = 0
Quadrat size = 0.030 hectares
Area sampled = 0.120 hectares
Average BA/tree = 3.577 dm^2
BA/hectare = 5.366 m^2/hectare
Trees/hectare = 150
Trees/quadrat = 4.5

Defining the population

Write down the definition of your population. It is important that the population in the statistical sense be defined in writing during the planning stages. Revise the definition in the field as you encounter unexpected situations. Perhaps the clearest practical way of defining the population is to make a list of criteria used to reject SUs. This list should be reported in the resulting publication. For example, one might include the sentence, "Plots were selected on southeast to west aspects between 1000 and 1500 m in elevation. Only stands with the dominant cohort between 70 and 120 years in age were included. Rock outcrop, talus, and riparian areas were excluded."

Homogeneity within sample units. Very often we apply a criterion of homogeneity to sample units, the idea being that SUs are internally more-or-less homogeneous. Typically, sites or "stands" are considered to be areas that, at the scale of the dominant vegetation, are essentially homogeneous in vegetation, environment, and history. This is almost always applied in a very loose way. One rule of thumb is that the leading dominant should vary no more than by chance. This could be evaluated by subsampling, but in practice, homogeneity is usually assessed by eye.

Placement of sample units

> An anonymous early ecologist: "The most important decision an ecologist makes is where to stop the car."

1. **Random sampling** requires the application of two criteria: each point has an equal probability of being included and points are chosen independently of each other.

2. **Stratified random sampling** has the additional feature that a population (or area) is divided into strata: subpopulations (or subareas) with known proportions of the whole. Sample units are selected at random within strata. Stratification allows sampling intensity to vary among different strata, yet you can still calculate overall population estimates for your parameters.

3. **Regular (systematic) sampling** has sample units that are spaced at regular intervals. In most cases, the consequences of sampling in a regular pattern are not severe unless the target organisms are patterned at a scale similar to the distance between SUs. It has repeatedly been shown, however, that the p-values emerging from hypothesis tests based on systematic sampling are NOT accurate (e.g., see Whysong and Miller 1987). In many cases, though, the practical advantages of systematic sampling (say with a grid of points) outweigh the reduction in faith in our p-values. One practical advantage is that it is usually much easier to relocate systematically placed permanent plots than randomly placed plots.

4. **Arbitrary but without preconceived bias**. We find this to be an apt phrase for what biologists usually do when they are claiming "random sampling." Unless you are strictly following a randomization scheme that is applied to your WHOLE population, you cannot really say you are sampling at random. Very often we try hard not to bias the sample but do not carefully randomize. What are the consequences of this? Surely they depend on the goal of the study. The more important it is to make a statement with known error about the population as a whole, the more important it is that sampling be truly random. When describing your sampling method, consider calling it "arbitrary but without preconceived bias" or "haphazard" instead of bending the truth by calling it random sampling if it is, in fact, not random.

5. **Subjective**. In some cases it makes sense to locate samples subjectively, but you should be very cautious about using subjective sampling — it has a long and partly unpleasant history. Sampling in Europe (especially the Braun-Blanquet school of phytosociology) and North America was often based on subjectively placed SUs, the criteria being whether or not the community was "typical" of a preconceived community type. Clearly, one cannot use such data to make objective statements about topics such as the existence of continuous vs. discrete variation in communities.

In other cases, subjectivity is a necessary and integral part of a study design. For example, if you want to find or study the most diverse spots in the landscape, then it is reasonable to use a subjective visual assessment of diversity to choose SUs. Of course, the price is that you immediately reduce the scope of inference from the study. In this example, you would obviously be remiss to use the resulting diversity estimates to make a statement about average diversity in the landscape as a whole.

Sources of random digits. Often one needs a source of random numbers in the field. Some people copy a page from the random numbers tables contained in most compilations of statistical tables. Some people have used a die in a clear container. Random digits can also be assigned in advance and entered on data forms, before going into the field. Fulton (1996) pointed out that the digital stopwatch built into most people's watches is a good source of random digits, when the digit appearing in one of the most rapidly changing positions is used.

Types of sample units

Fixed-area: Fixed-area sample units are of a set size and shape. Usually these are called quadrats or plots. Quadrats need not be four-sided. The ease of use and statistical efficiency of quadrats depend on their shape.

Circles are very fast to lay out, only one marker is needed for permanent relocation and they minimize the number of edge decisions (lowest perimeter to area ratio), but they have the poorest shape for estimation from aggregated distributions, yielding a high variance among SUs.

Squares are slow to lay out when large, two or four markers are needed for permanent relocation, and they have a poor shape for aggregated distributions.

Rectangles are slow to lay out when they are large and two or four markers are needed for permanent relocation, but they have a better shape for aggregated distributions (the narrower the better for that; i.e., lower variances among SUs). Rectangles require more edge decisions than do other shapes.

Point intercept: Percent cover is calculated as proportion of hits by (theoretically) dimensionless points. Points are usually arrayed in pin frames, ticks on tapes, etc. With multilayered vegetation more than one hit can be recorded per pin. Some details follow. Size of pin makes a difference in rarer species (Goodall 1952, 1953b). The more abundant the species, the less the pin size matters. Doubling the pin diameter from 2 mm to 4 mm makes about a 5-10% difference in % cover. Point sampling is difficult to apply to grasses or tall vegetation, and easiest to apply in low-growing vegetation such as tundra. Clustering pins in frames is convenient but reduces the quality of estimates for a given number of points, because the points are not independent.

Line intercept. Percent cover is calculated as a proportion of a line directly superimposed on a species (by vertical projection). This method was developed for use on desert shrubs. It is difficult or impossible to use if the highest plants are taller than your eye level. It is a relatively good method for both common and rare species. The line intercept method allows an estimate of cover but not density, unless individuals are of uniform size.

Distance methods. Distance-based methods have been most frequently used for sampling forest structure (Cottam & Curtis 1956) as well as for animal populations (Buckland et al. 1993). If you are sampling forests, you measure distances from randomly chosen points to the nearest trees. The diameter (or basal area) and species of those trees are then recorded. The most commonly used of these methods is the point-centered quarter method (Cottam & Curtis 1956). In the point-quarter method the observer measures the distance to the nearest individual in each of four quadrants around randomly chosen points.

Distance methods are based on the concept that the distances can be used to calculate a "mean area" occupied by the objects. The average of the distances is equal to the square root of the mean area. Mean area is the reciprocal of density. Density is converted into some measure of dominance or biomass by multiplying the density by the average size of each object. Usually this means multiplying tree density by the average basal area of the trees to arrive at a total basal area per unit land area.

Distance methods can also be used to estimate the quantity of any discrete objects in any area. For example, Peck and McCune (1998) used the point-centered quarter method to estimate the biomass of harvestable epiphytic moss mats in forests in western Oregon. Batcheler (1971) used point-to-object distances to estimate the density of animal pellet groups and introduced a correction factor for a fixed maximum search distance.

The main drawback to distance methods is that their effectiveness diminishes as the objects become increasingly aggregated. Distance methods perform well when the objects are distributed at random. Because plots need not be laid out, distance methods are usually more rapid to apply than area-based methods. Using a laser-based tool for measuring distances makes the method even more rapid and accurate.

Another criticism of distance methods is that they require judgements by the analyst that can lead to differences in estimates of density. Anderson and Southwell (1995) compared results from a panel of students and experts. All participants used the same data, but they were not restricted to particular software. The authors concluded that "the subjective aspects of the analysis of distance sampling may be overcome with some education, reading and experience with computer software."

Number of sample units

In community ecology, the number of sample units required depends on the complexity of the community and the goals of the study. With some preliminary sampling and a univariate measure of interest, one can use standard statistical techniques to determine the sample size required to achieve some specified degree of accuracy. In practice, however, such decisions are often made without good information based on preliminary sampling. Furthermore, there is not a good single variable that one can use to evaluate sample adequacy or sample size in community studies. In some cases, one can use a rule of thumb (slightly modified from Tabachnik and Fidell 1989) that for every important controlling factor or gradient, 20 additional sample units are needed. For example, to determine community differences along a transect on a gradient away from a point source of air pollution on a homogeneous plain, 20 sites would be a reasonable plan. If two substrates were involved and there were elevational differences, then 60 sites would be desirable (3 important factors $\times$ 20 sites/factor = 60 sites).

Measures of sample adequacy

As mentioned above, the "textbook" methods for evaluating sample adequacy are hardly applicable to typical sampling situations in community ecology. We are faced with multivariate, nonnormal, highly skewed, zero-rich data. We seldom have a set target for a degree of difference that we seek to detect.

It is easier to estimate the number of sample units needed for a subsample that represents a larger sample unit (e.g., the number of plots needed to represent a forest stand). Historically, ecologists have used two methods to evaluate sample adequacy: species-area curves and, where a variable can be singled out, seeking a sample size that will yield a standard error less than or equal to 10% of the mean for that variable.

Species-area curves are discussed below under diversity indices (Chapter 4). Species-area curves can be used to determine sample size by plotting number of species against number of sample units (e.g., Coddington et al. 1996). The curve will generally flatten out, meaning that increasing the sample size yields only small increases in the number of species. A sample size is chosen such that most of the readily obtained species are captured in the sample.

SE $\leq$ 10% of mean. You may recall that the standard error is the sample standard deviation divided by the square root of the sample size. The 10% rule is actually a fairly stringent criterion in field biology, where a high degree of spatial variability seems to be the rule. There is nothing magical about 10%. You can change the percentage to suit your purposes. In our experience, 20% tends to give broad, but often acceptable confidence bounds. A more strict criterion, such as 5%, might be used if you need to detect subtle differences.

Data quality assurance

In general, ecologists who are not doing methodological studies can seem quite informal about data quality. In our opinion, most ecologists deal with quality assurance in an intuitive way. There is, however, an increasing awareness of the need for explicit documentation of data quality, particularly in large-scale long-term studies. Government funding agencies in the United States, led by the Environmental Protection Agency (EPA), are increasingly insistent on explicit data quality assurance practices.

The basic procedure put forth by the EPA is that scientists should set QA goals, construct estimates of data quality, then assess the quality of the data relative to the goals. The most important components of the QA goals are accuracy, precision, and bias. There are many ways of expressing these, and some people use these terms almost interchangeably. The following definitions represent the most consistent use of the terms in the technical literature:

Precision is the degree of agreement in a series of measurements. The more significant digits in those measurements, the more precise they can be.

Accuracy describes how close measurements are to a true value. Measurements can be very precise but highly inaccurate — as with a digital pH meter reading 4 decimal places but biased by 2.0 pH units.

Bias is a systematic, directional error. Highly biased measurements can be very precise (as in the preceding example), but they can never be accurate. Unbiased measurements can be accurate or not. Bias is assessed by observing the signed differences between a set of measurements and true values.

Table 3.3. Average accuracy and bias of estimates of lichen species richness and gradient scores in the southeastern United States. Results are given separately for experts and trainees in the multiple-expert study. Extracted from McCune et al. (1997). N = sample size.

Activity	N	Species richness		Score on climatic gradient		Score on air quality gradient	
		% of expert	Bias	Acc.	Bias	Acc.	Bias
Reference plots	16	61	-39	4.4	+2.4	11.1	-10.5
Multiple-expert study, experts	3	95	-5	3.6	+3.6	4.7	-4.7
Multiple-expert study, trainees	3	54	-46	8.0	+8.0	5.0	-5.0
Certifications	7	74	-26	2.7	+2.4	2.1	-2.1
Audits	3	50	-50	10.3	+3.7	6.0	+2.7

How do we apply these to community ecology? First, we usually express the parameters of interest in a univariate way. Some univariate measures are species richness, ordination scores, and abundances of dominant species. Second, we need "true" values for comparison. This can be done with computer simulation, but in field studies, our best approximation of the "truth" can be obtained by resampling an area intensely, using multiple observers and a large number of sample units.

For an extended example, we will list some results from the Lichen Community component of the Forest Health Monitoring program (McCune et al. 1997a). Data quality was assessed for each plot-level summary statistic (air quality index, climatic index, and species richness) with several criteria: species capture, bias, and accuracy (Table 3.3). The indices were scores on major gradients, as determined by ordination methods. "Species capture" was the proportion of the "true" number of species (S_{true}) in a plot that was captured in the sampling. Accuracy was the absolute deviation from "true" gradient scores, as determined from expert data. Bias was the signed deviation from "true" gradient scores, as determined from expert data. An expert was considered to be a person with extensive experience with the local lichen flora, in most cases with two or more peer-reviewed publications in which the person contributed floristic knowledge of lichens. Percent deviation in gradient score is calculated as 100 × (observer's score - expert's score) / length of the gradient.

If S_{obs} is the observed number of species, x_{obs} is the observed value of variable x, and x_{true} is the true value of parameter x, then:

$$\text{Species capture, \%} = 100\,(S_{obs} / S_{true})$$

$$\text{Accuracy, \%} = \frac{100\,|x_{obs} - x_{true}|}{x_{true}}$$

$$\text{Bias, \%} = \frac{100\,(x_{obs} - x_{true})}{x_{true}}$$

To quote from the report (McCune et al. 1997) from which Table 3.3 was extracted:

Two results of this study seem most important. First, species richness is a very difficult parameter to estimate, being strongly dependent on the skill, experience, and training of the observer. Second, scores on compositional gradients are relatively consistent across observers, even in cases where there is considerable variation in species capture by the different observers. Each of these points is discussed further below.

With the concept of "biodiversity" becoming deeply entrenched in the management plans of government agencies and conservation organization, there comes a great need to inventory, monitor, and understand underlying factors controlling diversity. Perhaps the simplest and most readily communicated descriptor of diversity is species richness (Whittaker 1972). Some sampling characteristics of species richness, in particular species-area relationships have been studied in detail (Arrhenius 1921, Connor and McCoy 1979, MacArthur and Wilson 1967) and are widely applied to determine the number of subsample units needed to adequately characterize an area. Although many people are aware that different investigators can have different species capture rates, this has rarely been quantified or published. It has, however, been noted that changes of investigators in long-term monitoring of permanent plots can produce spurious apparent changes (Ketchledge and Leonard 1984; McCune and Menges 1986).

The preceding examples simplify community data to one or more univariate measures. Although we have not seen it in the literature, there is no reason why we cannot express accuracy and bias in a multidimensional space. This requires application of some distance measures (see Chapter 6).

What follows is perhaps the simplest multivariate example of evaluating accuracy and bias. Assume we have five people each shooting three times at a bull's eye at (0,0; Table 3.4; Fig. 3.2). We can evaluate the accuracy and bias for each person using distance measures (Table 3.5).

In this case, the "true value" is (0,0). We can calculate the inaccuracy for each person as the average Euclidean distance from the true value. Because there is no fixed outer limit for the inaccuracy, it cannot be easily rescaled to a percent accuracy value, where 100% means perfect accuracy.

The bias in this 2-D example has two components, the x dimension and the y dimension. In this case the average bias is the same as the centroid (multivariate average) for each person's shots. You can also think of the bias as a vector from the true value to the centroid of the observed values. The centroid for each person defines the 2-D coordinates of the tip of the vector.

One application of this concept is to evaluating the differences between predicted and observed community composition (McCune 1992). But, there is no reason it could not be applied to evaluate accuracy and precision of community sampling.

Table 3.4. Raw data for two-dimensional example of accuracy and bias, plotted in Figure 3.2.

Person	x	y	Person	x	y
1	3.0	5.2	4	0.9	-0.4
1	4.0	5.1	4	-0.6	1.1
1	4.0	2.6	4	-1.1	-2.0
2	0.9	-0.7	5	-2.7	-1.0
2	0.1	-2.9	5	-3.1	-1.2
2	0.4	-2.9	5	1.3	-1.9
3	0.0	0.0			
3	0.6	0.3			
3	-2.1	-1.6			

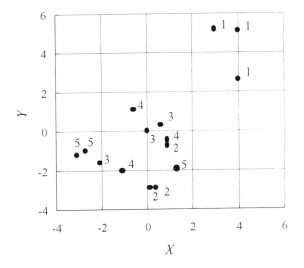

Figure 3.2. Two-dimensional example of accuracy and bias. Each person (1, 2,... ,5) aims at the center (0,0), representing the "true value." Deviations are measured in two dimensions, x and y.

Table 3.5. Inaccuracy and bias for two-dimensional example (Fig. 3.2).

Person	Inaccuracy (Ave. distance to 0,0)	Average bias	
		x	y
1	5.75	3.67	4.30
2	2.32	0.47	-2.17
3	1.10	-0.50	-0.43
4	1.51	-0.27	-0.43
5	2.84	-1.50	-1.37

Recurrent issues in community sampling

Tradeoffs between size and number of sample units. All studies with fixed-area plots face tradeoffs between size and number of sample units (Table 3.6). The "many-but-small" strategy will yield relatively accurate abundance estimates for the most common species but will yield a very incomplete species list. The "large-but-few" strategy will yield a relatively complete species list but will tend to overestimate the cover of rarer species and yields imprecise estimates of the more common species (McCune and Lesica 1992).

Kenkel and Podani (1991) recommended that a plot size somewhat larger than the mean patch size will likely provide the most efficient sampling design. They also recommended that to maximize efficiency, plots be as large as possible, given the constraints on sampling effort.

Pseudoreplication. Experimental units are often subsampled and the subsamples analyzed as independent replicates. This is pseudoreplication because subsample units are not independent. Designs that include subsampling (nested designs) are fine, but it is crucial to identify the level in the design at which treatments are applied and analyze the data accordingly.

Repeated measures. Successive samples from permanent sample units are usually correlated, so analyzing successive dates as if they were independent replicates of a treatment is invalid. Permanent sample units are excellent for detecting temporal change (Lesica and Steele 1997) because they allow you to separate spatial variation from the temporal change. However, seldom can "time" or "date" be included as main effects in a factorial ANOVA. In most cases, if you are following an experiment with permanent plots, you should be aware that you are using a repeated measures design.

Most permanent plot studies are subject to some degree of error from changes in observers and inexact relocation of sample units (Ketchledge & Leonard 1984, McCune & Menges 1986). This was called "pseudo-turnover" of species by Fischer and Stöcklin (1997). Although rarely measured directly, it is important to know the size of this error relative to the size of observed changes in community composition.

Nesting. Ecologists frequently subsample their sample units, thus creating nested designs. Nested designs are also called **multistage sampling**. Sample units are subsampled (two-staged) and the subsamples may be subsampled (three stage). From a statistical standpoint, it is desirable to randomize the sampling at each stage in the design. In forest inventories, clusters of plots are often placed randomly but with a systematic pattern of plots within clusters. According to Husch et al. (1972), "Fixed clusters of this type do not permit a valid measure of within-cluster variation... The entire cluster would have to be considered as the ultimate sampling unit."

Pairing. Paired designs are potentially very powerful ways to isolate single factors in a field setting. In theory, pairing has the potential to isolate the ever-present site-to-site variation from variation related to the factor of interest. Pairs are selected such that the members are "identical" except that one member receives (or received) a treatment or disturbance and one member does not. The most serious problem with this design in practice is that it is usually very difficult to find two adjacent spots that are identical. This is particularly true if one is studying historical disturbances. Many disturbances, such as fire, tend to leave edges at natural topographic breaks, such that the areas inside and outside the disturbance are likely to differ in some important way.

Topographic variables

Although the topic of environmental measures is not covered here, a couple of issues about topography recur so frequently that a short discussion of them is worthwhile.

Aspect of a slope (the direction or azimuth that a slope faces) is commonly measured in field studies. Untransformed, aspect is a poor variable for quantitative analysis. For example, 1° is adjacent to 360°, and although the numbers suggest a large difference, the aspect is about the same. So aspect needs to be transformed in one of several ways, depending on the precision with which it was measured and the environmental factor(s) you would like it to represent.

Heat load. Heat load is not symmetrical about the north-south axis. A slope with afternoon sun will be warmer than an equivalent slope with morning sun. A reasonable approximation of heat load for slopes in the northern hemisphere is to make the scale symmetrical about the northeast-southwest line. The following formula rescales aspect to a scale of zero to one, with zero being the coolest slope (northeast) and one being the warmest slope (southwest).

Table 3.6. Tradeoffs between few-and-large and many-and-small sample units.

	Few-and-large	Many-and-small
Bias against cryptic species	Higher. There is a hazard that some species, particularly cryptic species, are inadvertently missed by the eye.	Lower. Small sample units force the eye to specific spots, reducing inadvertent observer selectivity in detection of species.
Degree of visual integration	High. The use of visual integration over a large area is an effective tool against the normally high degree of heterogeneity, even in "homogeneous" stands.	Low. Minimal use is made of integrative capability of eye, forcing the use of very large sample size to achieve comparable level of representation of the community.
Inclusion of rare to uncommon species	High. Visual integration described above results in effective "capture" of rare species in the data.	Low. Unless sample sizes are very large, most rare to uncommon species are missed.
Accuracy of cover data on common species	Lower. Cover classes in large sample units result in broadly classed cover estimates with lower accuracy and precision than that compiled from many small sample units.	Higher. More accurate and precise cover estimates for common species.
Bias of cover estimates for rare species	High (overestimated).	Low.
Sampling time	Varies by complexity and degree of development of vegetation. No consistent difference from many-and-small.	Varies by complexity and degree of development of vegetation. No consistent difference from few-and-large.
Analysis time	Faster. With a single large plot, data entry at site level leads directly to site-level analysis.	Slower. Point data or microplot data require initial data reduction (by hand, calculator, or computer) to site-level abundance estimates.
Analysis options	Estimates of within-site variance are poor or impossible, restricting analyses to individual sites as sample units.	Within-site variance estimates are possible as long as sample units are larger than points.
Recommendations	The extreme case (single large plot) is most useful with extensive (landscape level) inventory methods. In many cases it is better to compromise with a larger number of medium-sized sample units.	The extreme case (point sampling) is most useful when rare to uncommon species are of little concern and accurate estimates are desired for common species. In most cases a compromise by using a smaller number of larger sample units is better.

$$\textit{Heat load index} = \frac{1 - \cos(\theta - 45)}{2}$$

where θ = aspect in degrees east of true north. A very similar equation but ranging from zero to two, was published by Beers et al. (1966).

While this and related equations are useful, they ignore the steepness of the slope. For example, a 1° south slope would receive the same heat load index as a 30° south slope, even though the latter will be a considerably warmer site.

Light. A very useful reference for converting slope and aspect to an estimate of direct solar radiation is provided by Buffo et al. (1972). This consists of tables for direct light with entry points by slope, latitude, and aspect. Some interpolation is required.

Slope correction of distances. Physical distances measured in the field are often transformed according to the slope on which they were taken. The purpose of such a transformation is to represent distances in a plane projection. If the slope is small, corrections are often not made. The plane-corrected distance D' for a distance D on an angle of S above the horizontal is:

$$D' = D/\cos S.$$

CHAPTER 4

Species Diversity

Background

Species richness is defined as the number of species in a sample unit or other specified area. According to Whittaker (1972). "Diversity in the strict sense is richness in species, and is appropriately measured as the number of species in a sample of standard size."

Although species richness is an intuitive measure of diversity, including inequality in relative abundance as a component of diversity can be intuitive as well. Consider, for example, two plots, each with three species (Table 4.1). Plot 1 has equal amounts of the three species, but Plot 2 has mostly one species and just a bit of the other two. Most people agree that Plot 1 seems more diverse than Plot 2. This intuitive notion of diversity incorporates the **evenness (equitability)** of abundance. A sample unit with more even abundances is, all else being equal, more diverse than a sample unit with abundant and sparse species.

Table 4.1. Which plot is more diverse?

	species 1	species 2	species 3
plot 1	10	10	10
plot 2	28	1	1

Whittaker (1960, 1965, 1972) defined three levels of diversity.

Alpha diversity: diversity in individual sample units

Beta diversity: amount of compositional variation in a sample (a collection of sample units)

Gamma diversity: overall diversity in a collection of sample units, often "landscape-level" diversity

Each of these can be measured in various ways. There have been numerous reviews of the pros and cons of various diversity measures. Some of the better known and more complete references are Auclair and Goff (1971), Hill (1973a), Hurlbert (1971), Magurran (1988), Peet (1974), Pielou (1966, 1975), Rosenzweig (1995), and Whittaker (1972). A selection of the most popular diversity measures follows.

Alpha diversity

Proportionate diversity measures

Many diversity measures are special cases of a general equation proposed by Hill (1973a) and Rényi (1961). For an observed abundance x_i, (numbers, biomass, cover, etc.) of species i in a sample unit, let

p_i = proportion of individuals belonging to species i:

$$p_i = x_i / \sum_{i=1}^{S} x_i$$

a = constant that can be assigned and alters the property of the measure

S = number of species

D_a = diversity measure based on the constant a. The units are "effective number of species"

$$D_a = \left(\sum_i^S p_i^a\right)^{\frac{1}{1-a}}$$

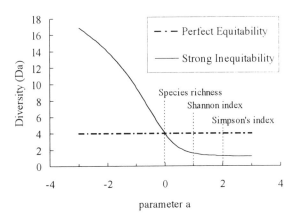

Figure 4.1. Influence of equitability on Hill's (1973a) generalized diversity index. Diversity is shown as a function of the parameter a for two cases: a sample unit with strong inequitability in abundance and a sample unit with perfect equitability in abundance (all species present have equal abundance; see Table 4.1).

This equation is one of a small handful of unifying equations (an equation that unifies otherwise disparate-seeming equations) in ecology. As a increases, the index gives more weight to dominant species, less weight to rare species, and sample size (either number or spatial size) is less important (Fig. 4.1).

Although this equation was originally based on information theory, it can also be viewed as the reciprocal of a weighted average. More abundant species are given more weight, and the intensity of this weighting is set by a. The measure D_a is in units of species. You can think of it as equal to the diversity of a community of D_a equally abundant species. It has been called the "effective number of species" (MacArthur 1965, MacArthur & MacArthur 1961).

D_0 = species richness

$$D_0 = \sum_i^S p_i^0$$

When $a = 0$, D_a is simply species richness. Species richness is discussed further below.

D_2 and Simpson's index

$$D_2 = \left(\sum_i^S p_i^2\right)^{-1} = \frac{1}{\sum_i^S p_i^2}$$

D_2 is another diversity measure, but Simpson's (1949) original index ($1/D_2$) is a measure of dominance rather than diversity. Simpson's index represents the likelihood that two randomly chosen individuals will be the same species. It varies inversely with diversity. The complement of Simpson's index of dominance is

$$Diversity = 1 - \sum_i^S p_i^2$$

and is a measure of diversity. It is the likelihood that two randomly chosen individuals will be different species. This measure is little affected by addition or loss of rare species and it emphasizes common species. Therefore it is relatively stable with sample size.

Simpson's index applies to an infinite population (i.e., once an individual is sampled, it does not change the probability of getting it as the next individual). There are other versions for finite populations.

D_1 and Shannon-Wiener index

If $a = 1$ then D_1 is a nonsense equation because the exponent is 1/0. But if we use limits to define D_1 as a approaches 1 then

$$D_1 = \lim a \to 1 (D_a)$$

$$D_1 = \log^{-1}\left(-\sum_i^S p_i \log p_i\right)$$

The logarithmic form of D_1 is the Shannon-Wiener index (H'), which measures the "information content" of a sample unit:

$$H' = \log(D_1) = -\sum_i^S p_i \log p_i$$

The units for D_1 are "number of species of equal abundance" while the units for H' are the log of the number of species of equal abundance. The minimum value for H' is zero, obtained when one species is present. H' is undefined if $S = 0$. This equation was first used in ecology by MacArthur and Macarthur (1961) and has a basis in information theory. If we put diversity into the context of information theory, then maximum diversity yields maximum uncertainty. Think of it as drawing individuals at random from a community. The higher the diversity, the more uncertainty you will have about which species you will draw next (Box 4.1). Determining the species of an individual drawn from a diverse sample unit relieves more uncertainty (provides more information) than from a sample unit with low diversity.

If you use the antilog form (D_1 or $\log^{-1}(H')$), then the choice of base of logarithms (base e or base 10) makes no difference. Both forms have been commonly used.

D_1 and H' are intermediate between species richness and Simpson's index in its sensitivity to rare species. It is widely used, but has the disadvantage that the numerical value is not as directly meaningful as some other indices, especially species richness.

Whittaker (1972) summarized the interpretation of H': "There is no particular reason to interpret diversity or equitability as information or uncertainty, but the index has distinctive and appropriate qualities. It is most strongly affected by importances of species in the middle of the sequence. For large samples the index is consequently somewhat damped against effects of differences in quantitative proportions of the first few species. Effects of the rarer species are also damped, rendering the index, like [Simpson's index], relatively independent of sample size... for samples that are not too small."

Box 4.1. How is information related to uncertainty?

> You are blindfolded next to two plots, one with equal numbers of two species and one with many individuals of one species and few of the other species. Your partner calls out species names of individuals as they are encountered. After 100 individuals have been tallied from each plot, your data are:
>
> | | Sp A | Sp B | p_A | p_B |
> |--------|------|------|-------|-------|
> | Plot 1 | 99 | 1 | 0.99 | 0.01 |
> | Plot 2 | 50 | 50 | 0.50 | 0.50 |
>
> For the 101st individual, in which plot is your uncertainty greater? In which plot does the next individual provide more information?
>
> $$\text{information content} = H' = -\sum_{i=1}^{2} p_i \log p_i$$
>
> For plot 1
>
> $$H' = -1\bigl[0.99 \cdot \log(0.99) + 0.01 \cdot \log(0.01)\bigr] = 0.024$$
>
> For plot 2,
>
> $$H' = -1\bigl[0.5 \cdot \log(0.5) + 0.5 \cdot \log(0.5)\bigr] = 0.301$$
>
> Clearly, the information content of the next individual chosen from plot 2 is much higher than for plot 1, because it resolves more uncertainty. In plot 1 you are fairly certain that the next individual chosen will be species A, but it is more uncertain in plot 2. The more uncertainty is relieved, the more information you have obtained. This much is clear. But Hurlburt (1971) and others question how the concept of information is relevant to biological diversity.

Species richness

Species richness is simply calculated as the number of species in a sample unit (SU), whether the sample unit is defined as a specific number of individuals, area, or biomass. If expressed per unit area, it is called species density (Hurlbert 1971). SUs of different sizes cannot, however, be compared directly, because the relation between species richness and SU size is nonlinear (see species-area curves below).

Species richness as a measure of diversity is very attractive to ecologists because it is simple, easily calculated, readily appreciated, and easy to communicate to policy makers and other lay people (Purvis & Hector 2000). For example, consider trying to explain $H' = 2.12$ versus $S = 33$. Peters (1991) said we should try harder to present results in a way that does not obfuscate simple underlying units.

On the other hand, species richness is very sensitive to the sample unit area and the skill of the observer. Measurement error is high for small, cryptic, mobile, or taxonomically difficult organisms (e.g., Coddington et al. 1996; McCune et al. 1997a).

Whittaker's bottom line (1972) was: "For the measurement of alpha diversity relations I suggest, first, use of a direct diversity expression, [species richness], as a basic measurement wherever possible; second, accompaniment of this by a suitable slope expression [a measure incorporating equitability] when the data permit."

Beta diversity

Beta diversity is the amount of compositional change represented in a sample (a set of sample units). There are various ways of measuring beta diversity, depending on our concepts or measurements of underlying sources of the compositional variation (Table 4.2). The phrase "beta diversity" need not invoke specific gradients. **Species turnover**, on the other hand, is a special case of beta diversity applied to changes in species composition along explicit environmental gradients (Vellend 2001). The term "beta diversity" has also recently been applied in a different way (e.g., Condit et al. 2002), as a rate of decay in species similarity with increasing distance, without respect to explicit environmental gradients.

Three applications of beta diversity in the usual sense are:

1. **Direct gradient** — beta diversity is the amount of change in species composition along a directly measured gradient in environment or time,
2. **Indirect gradients** — beta diversity is the length of a presumed environmental or temporal gradient as measured by the species, and
3. **No specific gradient** — beta diversity measures compositional heterogeneity without reference to a specific gradient.

Measures of beta diversity depend on the underlying gradient model and the data type (Table 4.2). After a brief summary of the usefulness of measuring beta diversity, we describe measures of beta diversity for each class of underlying gradient model.

Usefulness of beta diversity

Greig-Smith (1983) pointed out that beta diversity is a property of the sample, not an inherent property of the community. This is well illustrated by the results of Økland et al. (1990): "Beta diversity, measured as the length of the first DCA axis, invariably increased upon lowering of sample plot size... This is explained as a consequence of the weakening of structure in the data matrices when the fine-grained patterns of the vegetation are emphasized." Thus, beta diversity is controlled by a combination of biological and sampling processes.

The performance of multivariate methods strongly depends on beta diversity. Estimates of beta diversity help to inform us about which ordination methods might be appropriate for a particular data set and whether differences between ordination methods should be anticipated. The greater the beta diversity, the more ordination methods are challenged and the more results will differ among methods.

Beta diversity can be used to compare responsiveness of different groups of organisms to environmental differences in a sample (e.g., McCune & Antos 1981). Such comparisons are not, however, biologically meaningful between studies using different methods.

Beta diversity along a direct gradient

Beta diversity integrates the rate of change. Do not confuse the **rate** of species change with the **amount** of change. The rate of change, R, refers to steepness of species response curves along gradients (not to be confused with Minchin's R, which is a measure of the amount of change). For example, if you sample with a series of plots along a gradient and calculate the dissimilarity among adjacent plots, then you can graph dissimilarity as a measure of R against position on the gradient (Fig. 4.2). This concept of "rate of species change" has historically been used to address theoretical questions about sharpness of boundaries between communities.

Table 4.2. Some measures of beta diversity. See Wilson and Mohler (1983) and Wilson and Shmida (1984) for other published methods. "DCA" is detrended correspondence analysis. A direct gradient refers to sample units taken along an explicitly measured environmental or temporal gradient. Indirect gradients are gradients in species composition along presumed environmental gradients.

Underlying gradient model	Data type	
	Quantitative	Presence-absence
Direct gradient	HC (Whittaker's half changes) β_G (gleasons) Minchin's R Δ (total gradient length)	β_T (beta turnover) Minchin's R
Indirect gradient	Axis length in DCA	Axis length in DCA
No specific gradient	β_D Dissimilarity β (half changes)	β_w (Whittaker's beta, $\gamma/\alpha - 1$) =

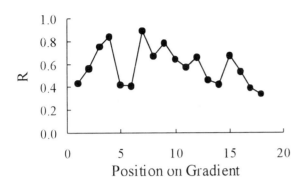

Figure 4.2. Example of rate of change, R, measured as proportional dissimilarity in species composition at different sampling positions along an environmental gradient. Peaks represent relatively abrupt change in species composition. This data set is a series of vegetation plots over a low mountain range. In more homogeneous vegetation, the curve and peaks would be lower.

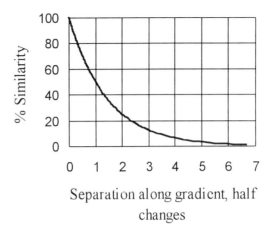

Figure 4.3. Hypothetical decline in similarity in species composition as a function of separation of sample units along an environmental gradient, measured in half changes. Sample units one half change apart have a similarity of 50%.

We seldom try to estimate R along gradients because we rarely have the appropriate quality of information. Usually the primary interest is in the total amount of change along the gradient, in other words, the integral of R. Measures of the total amount of change, beta diversity, are discussed below. One of them, Wilson and Mohler's (1983), is based on estimates of R using similarity measures. Other methods for estimating R proposed by Oksanen and Tonteri (1995) assume Gaussian (bell-shaped) response functions, summing the absolute values of the slopes of individual species' response functions.

The amount of change, β, is the integral of the rate of change:

$$\beta = \int_a^b R(x)\, dx$$

where a and b refer to the ends of an ecological gradient x.

Beta diversity can be calculated in various ways, depending on the available data. First consider species abundance data collected in sequential sample units along one directly measured gradient. Wilson and Mohler (1983) introduced "**gleasons**" as a unit of species change. This measures the steepness of species response curves. It is the sum of the slopes of individual species at each point along the gradient.

$$R(x) = \sum_{i=1}^{N} \left| \frac{dY_i(x)}{dx} \right|$$

where Y is the abundance of species i at position x along the gradient. This can be integrated into an estimate of beta diversity along a whole gradient with

$$\beta_G = 2\sum_{i=1}^{n-1} [IA - PS(i, i+1)]$$

where $PS(a,b)$ is the percentage similarity of sample units a and b and IA is the expected similarity of replicate samples (the similarity intercept on Fig. 4.3).

For simulated data, Minchin (1987) defined a measure of beta diversity as "R units." To reduce confusion with Wilson's use of R as a rate of change, we refer to Minchin's measure of the amount of change as "Minchin's R." Minchin measured beta diversity using the mean range of the species' physiological response function:

$$\text{Minchin's } R = \frac{L}{\sum_{i=1}^{n} r_i / n}$$

where r_i is the range of species i along the gradient. L is the length of the gradient, and r and L are measured in the same units. For example, Minchin's R = 2 means that the gradient is twice as long as the mean range of species occurrence. For simulated presence-absence data this method is closely related to β_T (below). Real data (i.e., responses measured by abundance) will usually give higher values than simulated data, all else being equal. This occurs because the tails of abundance curves are truncated versions of the tails of the underlying physiological response curves: species will be absent from many of the sample units in which they are barely able to survive.

Oksanen and Tonteri (1995) asserted that unless species response functions are fitted and the slopes are obtained from those functions, beta diversity measured on real data will primarily reflect the noise in the data. Furthermore, they point out that the direct measure of gleasons depends on the abundance scale used. Oksanen and Tonteri (1995) proposed the following measure of total gradient length:

$$\Delta_{ab} = \int_a^b \delta_A(x) \cdot dx$$

where δ_A is the absolute compositional turnover (rate of change) of the community between points a and b on gradient x. They recommended evaluating Δ_{ab} by summing the analytic integrals of the response functions for individual species. Oksanen and Tonteri concluded: "Direct estimation of compositional turnover can be used to assess the importance of observed gradients in different data sets, and compare the gradients to each other. In this way the comparison is independent of incidental span of sampling for a gradient in a data set. The measure can also be used for assessing successional change."

Half-changes are the amount of change resulting in a 50% percent similarity. For example, a gradient that is 3 half changes long means that the community has changed by half for 3 times. Half changes are related to the rate of change by:

$$R(a,b) = \frac{HC(a,b)}{\Delta x(a,b)}$$

Either presence-absence or quantitative data can be used.

If one constructs a graph similar to Figure 4.3, but with the vertical axis on a logarithmic scale, the series may approximate a straight line. The intercept with the vertical axis will typically not be 100% because replicate samples will not be identical. This semi-log plot is the basis for Whittaker's (1960) method of calculating the number of half changes along the gradient segment from a to b, $HC(a,b)$:

$$HC(a,b) = \frac{\log(IA) - \log(PS(a,b))}{\log 2}$$

where $PS(a,b)$ is the percentage similarity of sample units a and b and IA is the expected similarity of replicate samples (the y intercept on the figure just described).

Given **presence-absence data** collected along one direct gradient, the amount of species change can be measured as "beta turnover" (β_T, Wilson and Shmida 1984; see cautions of Vellend 2001). Beta turnover measures the amount of change as the "number of communities." Where g = the number of species gained, l = the number of species lost, and $\bar{\alpha}$ = the average species richness in the sample units:

$$\beta_T = \frac{|g + l|}{2\bar{\alpha}}$$

Beta diversity along an indirect gradient

Axis length in DCA (detrended correspondence analysis; Ch. 20) can be used as a measure of beta diversity along that axis. The length of axes in DCA are determined by Hill's scaling [see Gauch (1982a, p. 155) and a more comprehensive explanation by Oksanen and Tonteri (1995)]. With Hill's scaling, the average standard deviation of species turnover is used as a unit for scaling an ordination axis. A species comes and goes over a span of about 4 standard deviations. One half-change occurs in about 1-1.5 standard deviations. The axis length is measured in number of standard deviations. DCA expands and contracts axes such that the rate of species turnover is more-or-less constant along an ordination axis. This is achieved by attempting to equalize the average within-sample dispersion of the species scores at all points along the axis. Oksanen and Tonteri (1995) concluded: "Direct estimation [of total turnover or gradient length] is straightforward, and there is no need for indirect estimation through rescaling of ordination axes, a much less reliable alternative."

Beta diversity with no specific gradient

Most community data have more than one underlying gradient, measured or not. We may wish to measure overall beta diversity, without reference to specific gradients. We can calculate compositional change through a multidimensional hyperspace.

The simplest descriptor of beta diversity and one that can be applied to any community sample, is

$$\beta = \frac{\gamma}{\overline{\alpha}}$$

where γ is the landscape-level diversity and $\overline{\alpha}$ is the average diversity in a sample unit. Whittaker (1972) stated that a generally appropriate measure of this is

$$\beta_w = \frac{S_c}{S} - 1$$

where β_w is the beta diversity, S_c is the number of species in the composite sample (the number of species in the whole data set), and S is the average species richness in the sample units. The one is subtracted to make zero beta diversity correspond to zero variation in species presence.

While this measure does not have any formal units, one can think of the result in approximate units as the "number of distinct communities." If $\beta_w = 0$, then all sample units have all of the species. The larger the value of β_w, the more challenging the data set for ordination. As a rule of thumb in that context, values of $\beta_w < 1$ are rather low and $\beta_w > 5$ can be considered "high."

The maximum value of β_w is S_c -1 (one less than the total number of species across all sample units). The maximum value is obtained when no species are shared among sample units. Thus β_w does not have a fixed maximum value.

Wilson and Shmida (1984) compared six different measures of beta diversity using presence-absence data and found this simple measure to be one of two measures most suitable for the ecological analysis of community data. Using β_w does not assume a simple gradient structure.

The measure β_w is very useful for gauging the amount of compositional heterogeneity in a sample, which in turn informs us about the difficulty of certain multivariate analyses (especially ordination) of the data set. In general, differences between ordination methods become greater when β_w is high.

Three drawbacks to this application of β_w are (1) rare species tend to influence β_w greatly, while they may or may not influence the ordination results, (2) β_w considers only presence-absence, not abundance, and (3) experience has shown that β_w has a nonlinear (exponential) relationship to difficulty of obtaining a useful ordination.

These drawbacks do not apply to β_D, which calculates the number of *half changes corresponding to the average dissimilarity* among sample units. Using a proportion coefficient as a measure of dissimilarity, the average dissimilarity ($\overline{D}$) in a sample can be used as a descriptor of the amount of compositional change in a multidimensional hyperspace. The scaling of the dissimilarity is improved if it is converted to a beta diversity measured in half changes (β_D; Fig. 4.4):

$$\beta_D = \frac{\log(1-\overline{D})}{\log(0.5)}$$

This can be rewritten as

$$1-\overline{D} = 0.5^{\beta_D}$$

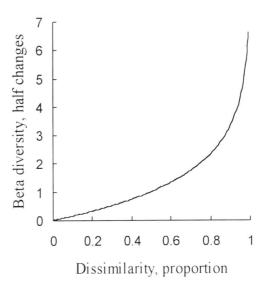

Figure 4.4. Conversion of average dissimilarity $\overline{D}$, measured with a proportion coefficient, to beta diversity measured in half changes (β_D).

Gamma diversity

"Richness in species over a range of habitats (a landscape, a geographic area, an island) is a *gamma* diversity and is consequent on the alpha diversity of the individual communities and the range of differentiation or beta diversity among them" (Whittaker 1972). Gamma diversity is calculated from a collection of sample units. The diversity measures applied to alpha diversity can be applied to gamma diversity. Perhaps the simplest and most informative is simply the number of species in the combined sample.

As operationally defined, gamma is as much a property of the sample as the system. Gamma cannot be measured in its pure theoretical form because we can never know the number of species in a landscape. It is not a useless concept, however, because we can use it in a relative sense, comparing different groups of species within a given sample. Also, it is a basic descriptor of the diversity of a sample.

Jackknife estimators of species richness

A worthy goal is to estimate the true species richness of an area (Colwell & Coddington 1994). A sample will almost always underestimate the true species richness — not all species will be encountered. Palmer (1990, 1991) compared several ways of estimating species richness of an area when it is subsampled with smaller sample units. Included in these comparisons were two jackknife estimators, nonparametric resampling procedures. The number of observed species in a subsample will typically be smaller than the true number of species. These jackknife estimators improve accuracy and reduce bias, at least when subsampling a restricted area. Palmer (1990) warned that they may not be appropriate when sampling large heterogeneous regions because the estimated number of species can never be more than twice the number of observed species. Palmer (1995) pointed out the disadvantage of these techniques as being very sensitive to the number of rare species. In the same paper he indicated other methods of extrapolating and interpolating estimates of species richness.

The first-order jackknife estimator (Heltshe & Forrester 1983, Palmer 1990) is:

$$Jack_1 = S + \frac{r1(n-1)}{n}$$

where S = the observed number of species, $r1$ = the number of species occurring in only one sample unit, and n = the number of sample units.

The second-order jackknife estimator (Burnham & Overton 1979; Palmer 1991) is:

$$Jack_2 = S + \frac{r1(2n-3)}{n} - \frac{r2(n-2)^2}{n(n-1)}$$

where $r2$ = the number of species occurring in exactly two sample units. The second-order jackknife can perform poorly with small sample sizes. For example, $Jack_2$ can be less than S when $r1 = 0$ and $r2 > 0$ (in other words, when there are no singletons but some doubletons).

Hellman and Fowler (1999) compared the bias, precision, and accuracy of jackknife and bootstrap estimators of species richness. Simple species richness was the most precise, but also the most biased of the estimators, always underestimating the true species richness considerably. For small sample sizes the second-order jackknife was least biased, followed by the first-order jackknife. However both jackknife estimators were positively biased with large sample sizes (over 25% of the total community sampled).

Evenness

Many measures, such as the Shannon index, measure evenness in part. Other measures attempt to exclude species richness and focus solely on evenness. The literature on evenness measures is disproportionately large compared to the usefulness of the concept. Measures of evenness have numerous methodological problems, such as dependence on the species count (Sheldon 1969; Hayek and Buzas 1997).

An easy-to-use measure (Pielou 1966, 1969) is "Pielou's J"

$$J = \frac{H'}{\log S}$$

where H' is the Shannon-Wiener diversity measure and S is the average species richness. If there is perfect equitability then $\log(S) = H'$ and $J = 1$. The main problem with this kind of evenness measure is that it is a ratio of a relative stable number, H', (i.e., one that is not strongly dependent on sample size) with an unstable number, S, which is strongly dependent on sample size (Sheldon 1969, Hurlbert 1971).

More recently, Hayek and Buzas (1997) partitioned H' into richness and evenness components based on the equation

$$H' = \ln(S) + \ln(E)$$

where $E = e^{H'}/S$ and e is the base of the natural logarithms. They then describe the dependence of H', $\ln(S)$, and $\ln(E)$ on the sample size to evaluate such questions as, "Is there more dominance than expected?" under a particular assumption about the distribution of abundance among species.

Species-area curves

The large literature on species-area curves extends back to early in the 20th century. Species-area curves have three distinct applications: (1) comparing separate sample units that differ in area, such as individual islands, (2) comparing increasingly large sets of contiguous sample units with a strict spatial nesting, and (3) comparing aggregates of differing numbers of sample units from the same data set. These three types of species-area curves differ in their underlying ecological processes and in sampling phenomena.

Both ecological processes and sampling phenomena alone will result in an increased number of species with area (Hill et al. 1994). Hill demonstrated this by sampling unique words in books and plotting the number of unique words against the number of words examined. But Hill et al. also showed that only ecological processes could be expected to increase the number of species *per unit area*. We should be cautious, however, about interpreting species-area curves as biological phenomena. For example, Hill et al. (1994) questioned the application of species-area curves as a theoretical basis for design of nature reserves (e.g., May 1975 and many others listed in Hill et al.), reasoning that species-area curves partly result from sampling phenomena.

Species-area curves can be used to evaluate the adequacy of sample size in a community data set (Fig. 4.5). This application is sometimes called the "species effort curve" or "species accumulation curve" (Angermeier & Smogor 1995, Hayek & Buzas 1997). In this case, whether the curve is produced by sampling phenomena alone or sampling plus ecological processes is immaterial. Normally, these species-area curves are constructed after preliminary sampling to help you determine sample size. Plant ecologists usually plot species number versus cumulative area of the sample. Animal ecologists often plot species number versus the number of individuals recorded. A relatively homogeneous area (e.g., a "stand" of vegetation) is sampled with a large number of sample units. This sample is subsampled to determine the average number of species as a function of size of the subsample. In theory, one selects a number of sample units beyond which additional sample units provide only small increases in the number of species encountered.

Using species-area curves to evaluate sample adequacy relies only on species presence, ignoring variation in abundance. You can incorporate information on abundance by calculating the average distance between the centroid of a subsample and the centroid of the whole sample (Fig. 4.5; output from PC-ORD v. 4). The more representative the subsample, the lower the distance between it and the whole sample. The interpretation is similar to the species-area curve, except that the distance curve has a fixed lower bound of zero. In the example in Figure 4.5, 40 subplots will yield over 100 species, with more subplots yielding relatively small increases in the number of species. Similarly, 40 subplots will yield a Sørensen distance of less than 0.1 (< 10%), measured between the centroid of the subsample and the centroid of the whole sample. Further increases in the size of the subsample render the subsample only slightly more similar to the whole sample.

Another use of species-area curves is for contrasting species richness between treatments or

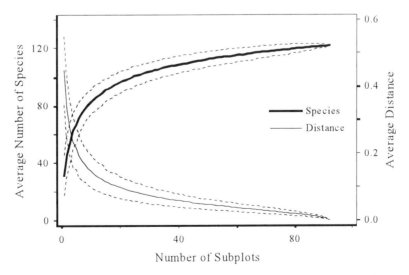

Figure 4.5. Species-area curve (heavy line) used to assess sample adequacy, based on repeated subsampling of a fixed sample (in this case containing 92 sample units and 122 species). Dotted lines represent ± 1 standard deviation. The distance curve (light line) describes the average Sørensen distance between the subsamples and the whole sample, as a function of subsample size.

other categories of sample units. For example, Clay and Holah (1999) plotted species-area curves with error bars, contrasting species diversity in two experimental treatments. Johansson and Gustafsson (2001) used species-area curves to compare richness of red-listed species between production forests and "woodland key habitats" in Sweden.

Another conceptual view of species-area curves is that they are statements of the relationship between species richness and spatial scale. This view is most directly applicable to samples with strict spatial nesting. Spatial scale can be broken into its components: grain, extent, and number of sample units (Palmer & White 1994). Grain, which refers to the size of the smallest sample unit, is one of the strongest influences on the species-area curve.

Much of the mid-20th century interest in species-area curves arose from an interest in island biogeography (MacArthur & Wilson 1967), with principal interest in species number vs. area of islands. As a result, the slope of the plot of log(S) vs. log($area$) is a parameter of interest in the literature of island biogeography.

Without going into any detail, a couple of basic equations are worth relating: Arrhenius (1921), popularized in a series of papers by F. W. Preston, proposed that:

$$S = cA^b$$

where S is the number of species, A is the area of the sample and c and b are fitted coefficients. In log form: $\log S = \log c + b \log A$

Gleason (1922) proposed a similar equation:

$$S = c + b \log A$$

Gleason's equation has found more acceptance in relatively homogeneous areas, while Arrhenius' equation is more widely used and works better in heterogeneous areas and in island biogeography. Rosenzweig (1995) pointed out a common misinterpretation of the intercept of the log form of the Arrhenius equation. See also papers by Buys et al. (1994), Connor and McCoy (1979), and Palmer and White (1994).

Software

Most diversity measures can be calculated fairly readily using spreadsheets or statistical software. These packages will not have built-in diversity measures, so you will have to construct your own.

PC-ORD reports some diversity measures. Species richness, the Shannon index, Pielou's J as a measure of evenness, and gamma diversity can be obtained by requesting row summaries. Beta diversity (β_w) is readily calculated from those values. Species-area curves can also be generated for data sets with PC-ORD using a built-in subsampling procedure.

What to report

Almost any community study should report basic diversity measures (alpha, beta, and gamma diversity) near the front of the results section (e.g., Table 4.3). This information helps the reader get a feel for the community and the properties of the data set. Your reader is probably aware that the choice of multivariate methods and their efficacy depends on the heterogeneity of your sample. These basic diversity measures are also useful to ecologists assembling statistics on diversity in various ecological settings.

If there are major subsets in the data, it is often desirable to report the diversity statistics for the subsets separately. Many readers will need a reminder of what you mean by alpha, beta, and gamma diversity. For example: "We applied using Whittaker's (1972) three kinds of diversity. Alpha diversity is calculated here as the average species richness per plot. Beta diversity is a measure of heterogeneity in the data, which we calculated with the ratio of the total number of species to the average number of species (gamma over alpha). Gamma diversity is the landscape-level diversity estimated as the total number of species across plots." (McCune et al. 1997).

Table 4.3. Species diversity of epiphytic macrolichens in the southeastern United States. Alpha, beta, and gamma diversity are defined in the text (table from McCune et al. 1997).

	N	Diversity measure		
		alpha	beta	gamma
On-frame				
Mountains	30	16.4	6.5	107
Piedmont	13	12.5	4.3	54
Coastal plain	19	11.9	5.5	66
Off-frame urban/industrial	17	9.2	4.6	64
QA plots	53	22.7	7.0	158

CHAPTER 5

Species on Environmental Gradients

The ideal and the real

Robert H. Whittaker's writings brought to the fore the concept of hump-shaped species responses to environmental gradients, through his ecological monographs (Whittaker 1956, 1960), review papers (1967, 1973), and his textbooks. Whittaker drew species responses as smooth, hump-shaped lines, some narrow, some broad, and varying in amplitude (Fig. 5.1).

These smooth, noiseless curves represent the **Gaussian ideal** response of species to environmental gradients. Under the Gaussian ideal, a species response is completely described by its mean position on the environmental gradient, its standard deviation along that gradient, and its peak abundance. Even if species followed the Gaussian ideal, community analysis would be difficult because two species following the Gaussian ideal will have a nonlinear relationship to each other, challenging our usual statistics based on linear models.

An even more idealistic model would be linear responses to environment (Fig. 5.2). The **linear ideal** has species rising and falling in straight lines in response to environmental gradients. Although the linear model is blatantly inappropriate for all but very short gradients, many of the most popular multivariate tools (e.g., principal components analysis and discriminant analysis) assume linear responses.

Even the Gaussian model has several critical shortcomings when compared with actual community data. Three major problems are common in community data:

1. Species response have the **zero truncation problem**.
2. **Curves are "solid"** due to the action of many other factors.
3. **Response curves can be complex**: polymodal, asymmetric, or discontinuous.

Each of these is explained below.

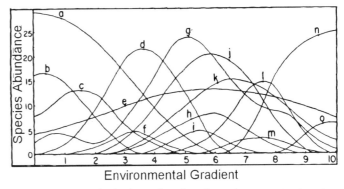

Figure 5.1. Hypothetical species abundance in response to an environmental gradient. Lettered curves represent different species. Figure adapted from Whittaker (1954).

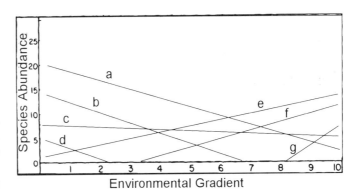

Figure 5.2. Hypothetical linear responses of species abundance to an environmental gradient. Lettered lines represent different species.

The zero truncation problem

Beals (1984, p. 6) introduced the term "zero truncation problem." Beyond the extremes of a species tolerance on an environmental gradient only zeros are possible (Fig. 5.3). Therefore, once a species is absent, we have no information on *how unfavorable* the environment is for that species. The dashed lines in Figure 5.3 indicate the ideal response curve, if the zero truncation problem did not exist.

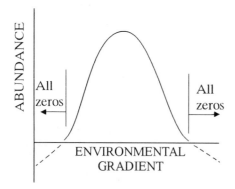

Figure 5.3. The zero truncation problem.

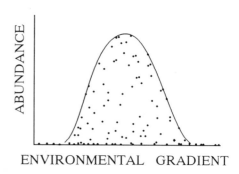

Figure 5.4. A "solid" response curve. Points represent a species abundances.

The zero truncation problem results from our choice of species abundance as a measure of performance. This precludes information on habitat quality beyond a species' range. In other words, no negative abundance values are possible; a species response curve is truncated at zero.

Solid curves

Species response curves are "solid" (Fig. 5.4). That is, a species is usually less abundant than its potential at a given point on an environmental gradient. Even values of zero abundance will be found near a species' optimum on the gradient. Species response curves are solid because species are limited by numerous factors in addition to the particular environmental gradient in question. These factors include other environmental variables, species interactions, chance, and dispersal limitations.

Complex curves

Species responses to environmental gradients can be complex: polymodal, asymmetric, or discontinuous. The distributions of black spruce (*Picea mariana*) and eastern redcedar (*Thuja occidentalis*) along a moisture gradient in parts of the American boreal forest are classic examples of bimodal species distributions (Curtis 1959, Loucks 1962). Black spruce occurs in soil pockets on dry rock outcrops as well as in wet *Sphagnum* bogs. In intermediate (mesic) sites with deeper, well-drained soils, black spruce is often a minor species or absent. These sites are dominated by white spruce (*Picea glauca*), sugar maple (*Acer saccharum*), or balsam fir (*Abies balsamea*).

More examples with real data

Most real data combine these three problems: zero-truncation, solid curves, and complex responses. One can see these problems either in species responses

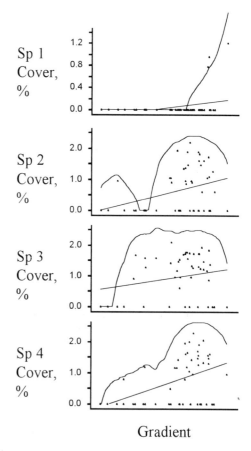

Figure 5.5. Scatterplot of abundance, measured as cover, of four species in relation to a gradient. Each point is a plot. Least-squares linear regression lines and fitted envelope lines are shown.

to directly measured environmental gradients (e.g., an elevation gradient) or on synthetic gradients extracted by ordination. Unfortunately, scatterplots of species abundance along environmental gradients are rarely published, but see Oksanen et al. (1988), Huisman et al. (1993), and Smith et al. (2001). In Figure 5.5, the gradient is a compositional gradient represented by an ordination axis. The envelope line was fit by PC-ORD as a smoothed line to include points falling within two standard deviations of a running mean along the gradient (PC-ORD flexibility parameter = 3). The straight lines are the least-squares regression lines. The correlation coefficients between Species 1, 2, 3, and 4 and the ordination axis are: 0.30, 0.34, 0.21, and 0.44, respectively. In this case, data are based on Derr (1994) and are average lichen species cover on twigs in shore pine (*Pinus contorta*) bogs in southeast Alaska.

Species 1 is absent except for a few SUs on one end of the gradient. Species 2 shows a hint of bimodality, but only one stray point occurs on the left side of the gradient. Species 3 is common throughout the gradient but is missing in a few SUs here and there. Species 4 shows a fairly consistent unimodal pattern, though it is absent in many SUs that fall in a favorable portion of the gradient.

Investigators rarely inspect, much less publish, scatterplots of species abundance along environmental or compositional gradients. Because the data are fraught with the problems described above, scatterplots of raw species abundance against environmental gradients are seldom an appealing summarization. Species responses to gradients are usually represented with smoothed curves or averaged along gradient segments (e.g., Beals 1969, Austin 1987; Fig. 5.1).

Describing species responses to environmental gradients is fundamental to developing and testing ecological theory, improving methods of community analysis, improving our use of indicator species in environmental assessments, predicting geographical and environmental distributions of species from sample surveys, and predicting the impacts of climate change on vegetation (Austin et al. 1994). How can we best represent these responses mathematically? Investigators have tried smoothing functions (Austin 1987), generalized linear modeling with third-order polynomials (Austin et al. 1990), beta functions (Austin et al. 1994, but see Oksanen 1997), maximum likelihood with a Gaussian response model (Oksanen et al. 1988), least squares with a Gaussian response model (op. cit.), weighted averaging (op. cit.), and logistic regression (Huisman et al. 1993). The complexity of the problem has defied a general satisfactory solution. Response curves are often skewed, sometimes polymodal, and the species optimum often lies outside the sampled range. In many cases, a Gaussian model is not appropriate.

Frequency distribution of species abundance

If you sample along one or more environmental gradients (similar to Figure 5.4), then plot a frequency distribution of abundance, you will often find a distribution resembling a negative exponential function (Fig. 5.6).

Note that the distribution is not even close to normal but instead has a large number of zero and small values tapering off to just a few large values. This distribution will be found for most species in most data sets. These strongly skewed curves are typical of species abundance data. Although Limpert et al. (2001) and others claim that abundance data typically follow a lognormal distribution, the large pile of zeros means that a log transformation cannot normalize the data.

Bivariate distributions

Community structure can be thought of as a set of associations between species. Examining the relationships among species, one pair of species at a time, illuminates a basic difficulty with community data.

Consider two species responding to an environmental gradient. If two species response curves lie near each other on the gradient, then we call them positively associated (Fig. 5.7A). Two species lying at a distance are negatively associated, in that sample units containing one of them will tend to not have the other (Fig. 5.7B). Causes of positive association between species include similar environmental requirements, similar responses to disturbance, and mechanistic dependence on each other, such as

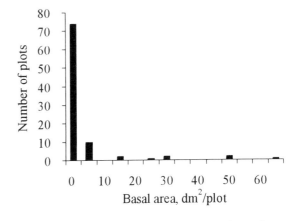

Figure 5.6. Frequency distribution of abundance for a typical tree species in a forest in southern Indiana. The abundance data are basal areas in about 90 plots. Note that most plots do not contain the species, but a few plots have large amounts of the species.

predation, parasitism, or symbiosis (e.g., a mycorrhizal fungus with its host). Causes of negative associations include differing environmental requirements, competition, spatial exclusion, and interspecific antagonisms.

Because many ecology textbooks have good descriptions of how to measure association among species (e.g., Greig-Smith 1983, Ludwig & Reynolds 1988), at least for presence-absence data, we skip that topic here. Instead, we introduce the concept of species association to show a fundamental difficulty with community data.

Even if two species closely follow each other in responses to an environmental gradient (Fig. 5.7A), the bivariate distribution is not linear (Fig. 5.7C). Similarly, two negatively associated species with noiseless responses to an environmental gradient (Fig. 5.7B) show a nonlinear pattern in 2-D species space (Fig. 5.7D). Most values lie on one of the two axes, except for the small hyperbolic part where the species overlap. If the realism is such that the species often show less abundance than their potential at a given point on the gradient, then "solid" curves result (Fig. 5.7E and 5.7F). The positive and negative relationships are not shown with a linear pattern in either case (Fig. 5.7G and 5.7H). Instead, we see two variants of the bivariate "dust bunny," as explained below.

The dust bunny distribution

With typical community data, a scatterplot of one species' abundance versus another usually has a "**dust bunny**" distribution (Fig. 5.8). This distribution resembles the fluffy accumulations of dirt and lint that are found in the corners of all but the tidiest spaces. The fact that the multivariate distribution of species data is such an extreme departure from multivariate normality (Fig. 5.9) has strong consequences for selecting methods for multivariate analysis. A typical example with real data (Fig. 5.10) is very similar to the hypothetical dust bunny distribution (Fig. 5.9, upper right).

Multivariate species space can be imagined as a p-dimensional hypercube. Extending the dust bunny to p dimensions, we find that most sample units lie along the corners of the p-dimensional space, just like the dust bunnies in the 3-space of a room.

A 3-D space has 2-D corners and 3-D corners. A p-dimensional space has 2-, 3-, 4-, ..., and p-dimensional corners. So a sample unit that contains 10 of 100 species in a sample lies in a 90-D corner of the 100-D space.

Dust bunnies do differ from distributions of natural communities in one respect. The origin is commonly vacant in community data sets, while the extreme corner of a physical space usually has the highest concentration of dust. Consider the consequences of emptying a sample unit (removing all species from it). This positions the sample unit at the origin of the p-dimensional space. Nature abhors a vacuum, so empty communities [0, 0, ... 0] are quickly colonized by one or more species (Fig. 5.11). Nevertheless, only a small subset of species colonizes the space, so that the trajectory of the sample unit in species space is along an edge in the space

The sparse matrix

The basic species × sample matrix used in community ecology usually has a large proportion of zeros. In other words, most samples contain a fairly small proportion of the species. Since nonzero values are few, the matrix can be called "sparse."

The community matrix becomes increasingly sparse as the range in environment encompassed by the sample increases. Note that of all the matrices commonly used in community ecology, only the species composition matrix is typically sparse. (A simple measure of this is beta diversity, measured as the total number of species divided by the average number of species per sample unit. If $\beta=1$ then the matrix is full. If $\beta=5$ then the average SU has one-fifth of all the species; see Chapter 4)

Ecologically, a joint absence is ambiguous at best, and in many cases it is misleading. Many commonly used methods for examining relationships among species treat joint absences as an indication of a positive relationship. Very dissimilar sample units can, therefore, appear similar by virtue of shared zeros. Two examples are the correlation coefficient and chi-square on a 2 × 2 matrix of presence-absence.

The more heterogeneous the data set, the greater the departure from bivariate normality and the worse the relationship between species is represented by a correlation coefficient (Fig. 5.12). As (0,0) values are added to a data set, a negative association between species can actually become positive. The reason for this becomes clear if you consider the consequence of adding (0,0)'s for the correlation between the two species in Figure 5.10.

Misinterpretation of shared zeros can induce gradient curvature in ordinations (Beals 1984). If an ordination method is based on a resemblance measure that is sensitive to shared absences, then opposing ends of a gradient are drawn toward each other in the ordination space, inducing a curvature (see Ch. 14 and Fig. 19.2).

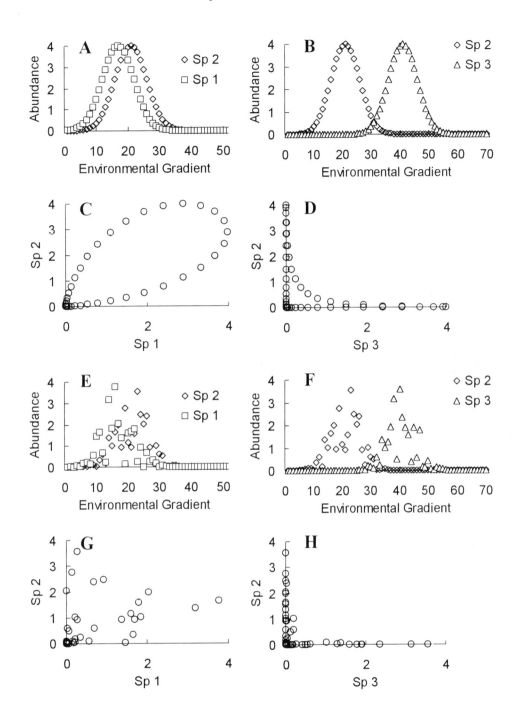

Figure 5.7. Bivariate species distributions from species responses to environmental gradients. Left column: two positively associated species. Right column: two negatively associated species. (A) — Positively associated species following Gaussian ideal curves. (B) — Negatively associated species following Gaussian ideal curves. (C) — Bivariate distribution of species abundances corresponding to A. (D) — Bivariate distribution of species abundances corresponding to B. (E) — Positively associated species with "solid" responses to the environmental gradient. (F) — Negatively associated species with "solid" responses to the environmental gradient. (G) — Bivariate distribution of species abundances corresponding to E. (H) — Bivariate distribution of species abundances corresponding to F.

Figure 5.8. The dust bunny distribution in ecological community data, with three levels of abstraction. **Background**: a dust bunny is the accumulation of fluff, lint, and dirt particles in the corner of a room. **Middle**: sample units in a 3-D species space, the three species forming a series of unimodal distributions along a single environmental gradient. Each axis represents abundance of one of the three species; each ball represents a sample unit. The vertical axis and the axis coming forward represent the two species peaking on the extremes of the gradient. The species peaking in the middle of the gradient is represented by the horizontal axis. **Foreground**: The environmental gradient forms a strongly nonlinear shape in species space. The species represented by the vertical axis dominates one end of the environmental gradient, the species shown by the horizontal axis dominates the middle, and the species represented by the axis coming forward dominates the other end of the environmental gradient. Successful representation of the environmental gradient requires a technique that can recover the underlying 1-D gradient from its contorted path through species space.

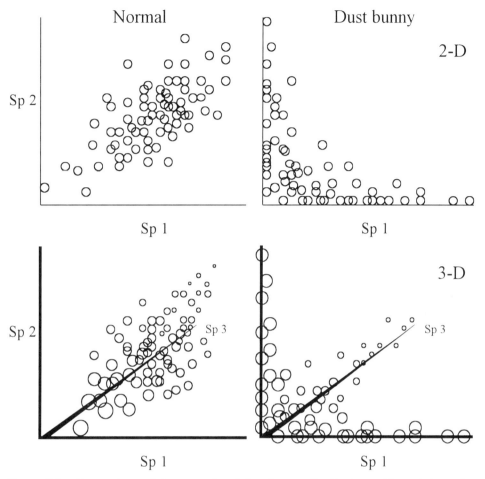

Figure 5.9. Comparison of the normal and dust bunny distributions. Upper left.— the bivariate normal distribution forms an elliptical cloud most dense near the center and tapering toward the edges. The more strongly correlated the variables, the more elongate the cloud. Upper right.— the bivariate dust bunny distribution has most points lying near one of the two axes. A distribution like this results from two overlapping "solid" Gaussian curves (Fig. 5.4). Lower left.— the multivariate normal distribution (in this case 3-D) forms a hyperellipsoid most dense in the center. Elongation of the cloud is described by correlation between the variables. Lower right.— the multivariate dust bunny has most points lying along the inner corners of the space. Two positively associated species would have many points lying along the wall of the hypercube defined by their two axes.

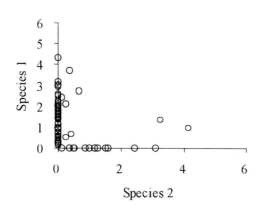

Figure 5.10. Plotting abundance of one species against another reveals the bivariate dust bunny distribution. Note the dense array of points near the origin and along the two axes. This bivariate distribution is typical of community data. Note the extreme departure from bivariate normality.

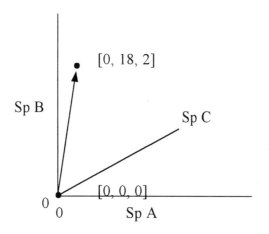

Figure 5.11. Nature abhors a vacuum. A sample unit with all species removed is usually soon colonized. The vector shows a trajectory through species space. The sample unit moves away from the origin (an empty sample unit) as it is colonized. In this case, species B and a bit of species C colonized the sample unit. As in this example, successional trajectories tend to follow the corners of species space.

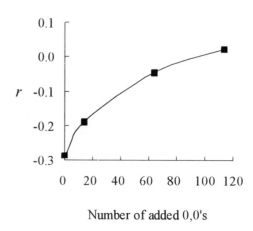

Figure 5.12. The consequence for the correlation coefficient of adding (0,0) values between species.

As the heterogeneity of our sample increases, our distance measures lose sensitivity. This is discussed further under "Distance Measures" (Chapter 6).

Summary

Box 5.1 summarizes the basic properties of ecological community data. These properties influence the choice of data transformations, analytical methods, and interpretation of results.

Partial solutions to the zero-truncation problem can be found at every level in the analysis (data adjustments and transformations, distance measures, methods of data reduction, and hypothesis testing; Table 5.1), yet most of these solutions are not available in the major statistical packages. Multivariate data sets in most other fields do not usually have this problem; hence their lack of consideration by the major statistical packages. This is changing, however, Recent versions of SPSS and SAS include some tools that are useful with this kind of matrix.

Box 5.1. Basic properties of ecological community data.

1. Presence or abundance (cover, density, frequency, biomass, etc.) is used as a measure of species performance in a sample unit.

2. Key questions depend on how abundances of species relate (a) to each other and (b) to environmental or habitat characteristics.

3. Species performance over long environmental gradients tends to be hump-shaped or more complex, sometimes with more than one hump.

4. The zero-truncation problem limits species abundance as a measure of favorability of a habitat. When a species is absent we have no information on how unfavorable the environment is for that species.

5. Species performance data along environmental gradients form "solid" curves because species fail for many reasons other than the measured environmental factors.

6. Abundance data usually follow the "dust bunny" distribution, whether univariate or multivariate; the data rarely follow normal or lognormal distributions.

7. Relationships among species are typically nonlinear.

Table 5.1. Solutions to multivariate analytical challenges posed by the basic properties of ecological community data (Box 5.1). The various classes of problems and their solutions are explained in the remainder of this book.

Class of problem	Example solution, appropriate for community data	Solutions based on a linear model, usually inappropriate for community data
Measure distances in multidimensional space	Sørensen distance (proportionate city-block distance)	Euclidean distance or correlation-based distance
Test hypothesis of no multivariate difference between two or more groups (one-way classification)	MRPP or Mantel test, using Sørensen distance	one-factor MANOVA
Single factor repeated measures, randomized blocks, or paired sample	blocked MRPP	randomized complete block MANOVA, repeated measures MANOVA
Partition variation among levels in nested sampling	nonparametric MANOVA (=NPMANOVA)	univariate nested ANOVA
Two-factor or multi-factor design with interactions	NPMANOVA	MANOVA
Evaluate species discrimination in one-way classification	Indicator Species Analysis	Discriminant analysis
Extract synthetic gradient (ordination)	Nonmetric multidimensional scaling (NMS) using Sørensen (Bray-Curtis) distance	Principal components analysis (PCA)
Assign scores on environmental gradients to new sample units, on basis of species composition	NMS scores	linear equations from PCA

CHAPTER 6

Distance Measures

Background

The first step of most multivariate analyses is to calculate a matrix of distances or similarities among a set of items in a multidimensional space. This is analogous to constructing the triangular "mileage chart" provided with many road maps. But in our case, we need to build a matrix of distances in hyperspace, rather than the two-dimensional map space. Fortunately, it is just as easy to calculate distances in a multidimensional space as it is in a two-dimensional space.

This first step is extremely important. If information is ignored in this step, then it cannot be expressed in the results. Likewise, if noise or outliers are exaggerated by the distance measure, then these unwanted features of our data will have undue influence on the results, perhaps obscuring meaningful patterns.

Distance concepts

Distance measures are flexible:

- Resemblance can be measured either as a distance (dissimilarity) or a similarity.
- Most distance measures can readily be converted into similarities and vice-versa.
- All of the distance measures described below can be applied to either binary (presence-absence) or quantitative data.
- One can calculate distances among either the rows of your data matrix or the columns of your data matrix. With community data this means you can calculate distances among your sample units (SUs) in species space or among your species in sample space.

Figure 6.1 shows two species as points in sample space, corresponding to the tiny data set below (Table 6.1). We can also represent sample units as points in species space, as on the right side of Figure 6.1, using the same data set.

There are many distance measures. A selection of the most commonly used and most effective measures are described below. It is important to know the domain of acceptable data values for each distance measure (Table 6.2). Many distance measures are not compatible with negative numbers. Other distance measures assume that the data are proportions ranging between zero and one, inclusive.

Table 6.1. Example data set. Abundance of two species in two sample units.

Sample unit	Species	
	1	2
A	1	4
B	5	2

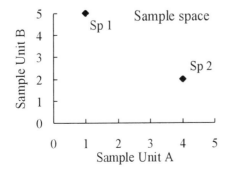
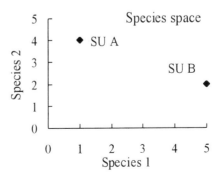

Figure 6.1. Graphical representation of the data set in Table 6.1. The left-hand graph shows species as points in sample space. The right-hand graph shows sample units as points in species space.

Table 6.2. Reasonable and acceptable domains of input data, x, and ranges of distance measures, $d = f(x)$.

Name (synonyms)	Domain of x	Range of $d = f(x)$	Comments
Sørensen (Bray & Curtis; Czekanowski)	$x \geq 0$	$0 \leq d \leq 1$ (or $0 \leq x \leq 100\%$)	proportion coefficient in city-block space; semimetric
Relative Sørensen (Kulczynski; Quantitative Symmetric)	$x \geq 0$	$0 \leq d \leq 1$ (or $0 \leq x \leq 100\%$)	proportion coefficient in city-block space; same as Sørensen but data points relativized by sample unit totals; semimetric
Jaccard	$x \geq 0$	$0 \leq d \leq 1$ (or $0 \leq d \leq 100\%$)	proportion coefficient in city-block space; metric
Euclidean (Pythagorean)	all	non-negative	metric
Relative Euclidean (Chord distance; standardized Euclidean)	all	$0 \leq d \leq \sqrt{2}$ for quarter hypersphere; $0 \leq d \leq 2$ for full hypersphere	Euclidean distance between points on unit hypersphere; metric
Correlation distance	all	$0 \leq d \leq 1$	converted from correlation to distance; proportional to arc distance between points on unit hypersphere; cosine of angle from centroid to points; metric
Chi-square	$x \geq 0$	$d \geq 0$	Euclidean but doubly weighted by variable and sample unit totals; metric
Squared Euclidean	all	$d \geq 0$	metric
Mahalanobis	all	$d \geq 0$	distance between groups weighted by within-group dispersion; metric

Distance measures can be categorized as metric, semimetric, or nonmetric. A **metric** distance measure must satisfy the following rules:

1. The minimum value is zero when two items are identical.
2. When two items differ, the distance is positive (negative distances are not allowed).
3. Symmetry: the distance from objects A to object B is the same as the distance from B to A.
4. Triangle inequality axiom: With three objects, the distance between two of these objects cannot be larger than the sum of the two other distances.

Semimetrics can violate the triangle inequality axiom. Examples include the Sørensen (Bray-Curtis) and Kulczynski distances. Semimetrics are extremely useful in community ecology but obey a non-Euclidean geometry. **Nonmetrics** violate one or more of the other rules and are seldom used in ecology.

Distance measures

The equations use the following conventions: Our data matrix **A** has q rows, which are sample units and p columns, which are species. Each element of the matrix, $a_{i,j}$, is the abundance of species j in sample unit i. Most of the following distance measures can also be used on binary data (1 or 0 for presence or absence). In each of the following equations, we are calculating the distance between sample units i and h.

Euclidean distance

$$ED_{i,h} = \sqrt{\sum_{j=1}^{p}(a_{i,j} - a_{h,j})^2}$$

This formula is simply the Pythagorean theorem applied to p dimensions rather than the usual two dimensions (Fig. 6.2).

City-block distance (= Manhattan distance)

$$CB_{i,h} = \sum_{j=1}^{p}|a_{i,j} - a_{h,j}|$$

In city-block space you can only move along one dimension of the space at a time (Fig. 6.2). By analogy, in a city of rectangular blocks, you cannot cut diagonally through a block, but must walk along either of the two dimensions of the block. In the mathematical space, size of the blocks does not affect distances in the space. Note also that many equal-length paths exist between two points in city-block space.

Euclidean distance and city-block distance are special cases (different values of k) of the Minkowski metric. In two dimensions:

$$Distance = \sqrt[k]{x^k + y^k}$$

where x and y are distances in each of two dimensions. Generalizing this to p dimensions, and using the form of the equation for ED:

$$Distance_{i,h} = \sqrt[k]{\sum_{j=1}^{p}(a_{i,j} - a_{h,j})^k}$$

Note that $k = 1$ gives city-block distance, $k = 2$ gives Euclidean distance. As k increases, increasing emphasis is given to large differences in individual dimensions.

Correlation

The correlation coefficient (r) is cosine α (third panel in Fig. 6.2) where the origin of the coordinate system is the mean species composition of a sample unit in species space (the "centroid"; see Fig. 6.2). If the data have not been transformed and the origin is at (0,0), then this is a noncentered correlation coefficient.

For example, if two sample units lie at $180°$ from each other relative to the centroid, then $r = -1 =$

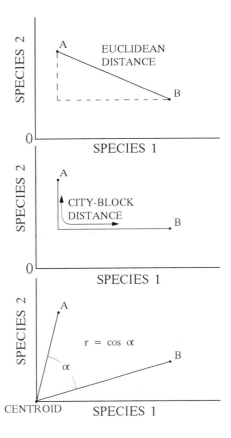

Figure 6.2. Geometric representations of basic distance measures between two sample units (A and B) in species space. In the upper two graphs the axes meet at the origin; in the lowest graph, at the centroid.

$\cos(180°)$. Two sample units lying on the same radius from the centroid have $r = 1 = \cos(0°)$. If two sample units form a right angle from the centroid then $r = 0 = \cos(90°)$.

The correlation coefficient can be rescaled to a distance measure of range 0-1 by

$$r_{\text{distance}} = (1 - r)/2$$

Proportion coefficients

Proportion coefficients are city-block distance measures expressed as proportions of the maximum distance possible. The Sørensen, Jaccard, and QSK coefficients described below are all proportion coefficients. One can represent proportion coefficients as the overlap between the area under curves. This is easiest to visualize with two curves of species abundance on an environmental gradient (Fig. 6.3). If A is the area under one curve, B is the area under the other, and w is the overlap (intersection) of the two areas, then the Sørensen coefficient is $2w/(A+B)$. The Jaccard coefficient is $w/(A+B-w)$.

Written in set notation:

$$Sorensen\ similarity = \frac{2(A \cap B)}{(A \cup B) + (A \cap B)}$$

$$Jaccard\ similarity = \frac{A \cap B}{A \cup B}$$

Proportion coefficients as distance measures are foreign to classical statistics, which are based on squared Euclidean distances. Ecologists latched onto proportion coefficients for their simple, intuitive appeal despite their falling outside of mainstream statistics. Nevertheless, Roberts (1986) showed how proportion coefficients can be derived from the mathematics of fuzzy sets, an increasingly important branch of mathematics. For example, when applied to quantitative data, Sørensen similarity is the intersection between two fuzzy sets.

Sørensen similarity (also known as "BC" for Bray-Curtis coefficient) is thus shared abundance

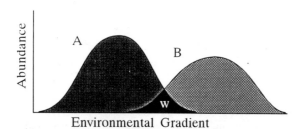

Figure 6.3. Overlap between two species abundances along an environmental gradient. The abundance shared between species A and B is shown by w.

divided by total abundance. It is frequently known as "$2w/(A+B)$" for short. This logical-seeming measure was also proposed by Czekanowski (1913). Originally used for binary (0/1) data, it works equally well for quantitative data. It is often called the "**Bray-Curtis coefficient**" (Faith et al. 1987) when applied to quantitative data, as in Bray and Curtis (1957). Rewriting the equation as a dissimilarity (or distance) measure, dissimilarity between items i and h is:

$$D_{i,h} = \frac{\sum_{j=1}^{p} |a_{ij} - a_{hj}|}{\sum_{j=1}^{p} a_{ij} + \sum_{j=1}^{p} a_{hj}}$$

Another way of writing this, where MIN is the smaller of two values is:

$$D_{ih} = 1 - \frac{2 \sum_{j=1}^{p} \text{MIN}(a_{ij}, a_{hj})}{\sum_{j=1}^{p} a_{ij} + \sum_{j=1}^{p} a_{h_j}}$$

One can convert this dissimilarity (or any of the following proportion coefficients) to a percentage dissimilarity (PD):

$$Sorensen\ distance = BC_{ih} = PD_{ih} = 100\ D_{ih}$$

Jaccard dissimilarity is the proportion of the combined abundance that is not shared, or $w / (A + B - w)$ (Jaccard 1901):

$$JD_{ih} = \frac{2 \sum_{j=1}^{p} |a_{ij} - a_{hj}|}{\sum_{j=1}^{p} a_{ij} + \sum_{j=1}^{p} a_{hj} + \sum_{j=1}^{p} |a_{ij} - a_{hj}|}$$

Quantitative symmetric dissimilarity (also known as the **Kulczynski** or **QSK** coefficient; see Faith et al. 1987):

$$QSK_{ih} = 1 - \frac{1}{2} \left[\frac{\sum_{j=1}^{p} \text{MIN}(a_{ij}, a_{hj})}{\sum_{j=1}^{p} a_{ij}} + \frac{\sum_{j=1}^{p} \text{MIN}(a_{ij}, a_{hj})}{\sum_{j=1}^{p} a_{hj}} \right]$$

Although Faith et al. (1987) stated that this measure has a "built-in standardization," it is not a standardized city-block distance in the same way that relative Euclidean distance is a standardized measure of Euclidean distance. In contrast with the "relative Sørensen distance" (below), the QSK coefficient gives different results with raw data versus data standardized by SU totals (Ch. 9). After such relativization, however, QSK gives the same results as Sørensen, relative Sørensen, and city-block distance.

Relative Sørensen (also known as relativized Manhattan coefficient in Faith et al. 1987) is mathematically equivalent to the Bray-Curtis coefficient on data relativized by SU total. This distance measure builds in a standardization by sample unit totals, each sample unit contributing equally to the distance measure. Using this relativization shifts the emphasis of the analysis to proportions of species in a sample unit, rather than absolute abundances.

$$D_{ih} = 1 - \left[\sum_{j=1}^{p} MIN\left(\frac{a_{ij}}{\sum_{j=1}^{p} a_{ij}}, \frac{a_{hj}}{\sum_{j=1}^{p} a_{hj}} \right) \right]$$

An alternate version, using an absolute value instead of the MIN function, is also mathematically equivalent to Bray-Curtis coefficient on data relativized by SU total:

$$D_{ih} = \frac{1}{2} \sum_{j=1}^{p} \left| \frac{a_{ij}}{\sum_{j=1}^{p} a_{ij}} - \frac{a_{hj}}{\sum_{j=1}^{p} a_{hj}} \right|$$

Note that with standardization or relativization of the data before application of the distance measure, many of the city-block distance measures become mathematically equivalent to each other. After relativization by sample unit totals, PD (Bray-Curtis) = CB = QSK = Relative Sørensen.

Relative Euclidean distance (RED)

RED is a variant of ED that eliminates differences in total abundance (actually totals of squared abundances) among sample units. RED ranges from 0 to the square root of 2 when all abundance values are nonnegative (i.e., a quarter hypersphere). Visualize the SUs as being placed on the surface of a quarter hypersphere with a radius of one (Fig. 6.4). RED is the chord distance between two points on this surface.

$$RED_{ih} = \sqrt{\sum_{j=1}^{p} \left[\left(\frac{a_{ij}}{\sqrt{\sum_{j=1}^{p} a_{ij}^2}} \right) - \left(\frac{a_{hj}}{\sqrt{\sum_{j=1}^{p} a_{hj}^2}} \right) \right]^2}$$

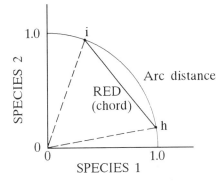

Figure 6.4. Relative Euclidean distance is the chord distance between two points on the surface of a unit hypersphere.

RED builds in a standardization. It puts differently scaled variables on the same footing, eliminating any signal other than relative abundance. Note that the correlation coefficient also accomplishes this standardization, but arccos(r) gives the arc distance on the quarter hypersphere, not the chord distance. Also, with the correlation coefficient the surface is a full hypersphere, not a quarter hypersphere, and the center of the hypersphere is the centroid of the cloud of points, not the origin.

Chi-square distance

The chi-square distance measure is used in correspondence analysis and related ordination techniques (Chardy et al. 1976, Minchin 1987a). Although Faith et al. (1987) found that this distance measure performed poorly, it is important to be aware of it, since it is the implicit distance metric in some of the more popular ordination techniques (detrended correspondence analysis and canonical correspondence analysis).

Let

a_{h+} = total for sample unit h (i.e., $\sum_{j=1}^{p} a_{hj}$)

a_{i+} = total for sample unit i (i.e., $\sum_{j=1}^{p} a_{ij}$)

a_{+j} = total for species j (i.e., $\sum_{i=1}^{n} a_{ij}$)

then the chi-square distance (Chardy et al. 1976) is

$$\chi^2_{ih} = \sqrt{\sum_{j=1}^{p} \frac{1}{a_{+j}} \left[\frac{a_{hj}}{a_{h+}} - \frac{a_{ij}}{a_{i+}} \right]^2}$$

Note that this distance measure is similar to Euclidean distance, but it is weighted by the inverse of the species totals. If the data are prerelativized by sample unit totals (i.e., $b_{ij} = a_{ij}/a_{i+}$), then the equation simplifies to:

$$\chi^2_{ih} = \sqrt{\sum_{j=1}^{p} \frac{(b_{hj} - b_{ij})^2}{a_{+j}}}$$

The numerator is the squared difference in relative abundance. It is expressed as a proportion of the species total (the denominator) and summed over all species.

Minchin (1987a) offered the following critique of this distance measure:

> The appropriateness of Chi-squared distance as a measure of compositional dissimilarity in ecology may be questioned (Faith et al. 1987). The measure accords high weight to species whose total abundance in the data is low. [Conversely, it de-emphasizes abundant species.] It thus tends to exaggerate the distinctiveness of samples containing several rare species. Unlike the Bray-Curtis coefficient and related measures, Chi-squared distance does not reach a constant, maximal value for sample pairs with no species in common, but fluctuates according to variations in the representation of species with high or low total abundances. These properties of Chi-squared distance may account for some of the distortions observed in DCA ordinations.

Mahalanobis distance (D^2)

Mahalanobis distance D_{fh}^2 is used as a distance measure between two groups (f and h). It is commonly used in discriminant analysis and in testing for outliers. If $\bar{a}_{if}$ is the mean for the ith variable in group f, and w_{ij} is an element from the inverse of the pooled within-groups covariance matrix, representing variables i and j, then

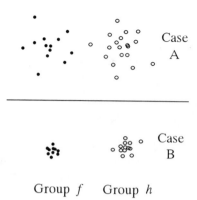

Figure 6.5. Illustration of the influence of within-group variance on Mahalanobis distance.

$$D^2_{fh} = (n-g) \sum_{i=1}^{p} \sum_{j=1}^{p} w_{ij} \cdot (\bar{a}_{if} - \bar{a}_{ih})(\bar{a}_{jf} - \bar{a}_{jh})$$

where n is the number of sample units, g is the number of groups, and $i \neq j$. Note that differences are weighted more heavily by w_{ij} when variables i and j are uncorrelated. Thus, Mahalanobis distance corrects for the correlation structure of the original variables (the dimensions of the space). The built-in standardization means that it is independent of the measurement units of the original variables.

In which case in Figure 6.5 are groups f and h more distant? Because the Mahalanobis distance inversely weights the distance between centroids by the variance, the two groups are more distant in Case B, even though the centroids are equidistant in the two cases.

Note the conceptual similarity to an F-ratio of between- to within-group variance. Indeed, the Mahalanobis distance can be used to calculate an F-test for multivariate differences between groups. Similarly, it can be used to test for outliers by calculating the distance between each point and the cloud of remaining points.

Performance of distance measures

Loss of sensitivity with heterogeneity

Performance of distance measures can be evaluated by comparing the relationship between environmental distance (distance along an environmental gradient, such as elevation) vs. sociological distance (the difference in communities as reflected by the distance in species space). This method of evaluating distance measures was used by Beals (1984), Faith et al. (1987), De'ath (1999a), and Boyce and Ellison (2001). If species respond noiselessly to environmental gradients and the environmental gradients are known, then we seek a perfect linear relationship between distances in species space and distances in environmental space. Any departure from that relationship represents a partial failure of our distance measures.

Two examples help clarify the variability in the relationship between distance in species space and environmental space. These examples are based on synthetic data sets with a known underlying structure and noiseless responses of species to two environmental gradients.

The first example is an "easy" data set, consisting of 25 sample units and 16 species. It is easy because the beta diversity is fairly low (average Sørensen distance among SUs = 0.59; 1.3 half changes), the sample unit totals are fairly even (coefficient of variation (CV) of SU totals = 17%), and the species are all similarly abundant (CV of species totals = 37%).

Despite this being an easy data set, all of the distance measures show a curvilinear relationship with environmental distance (Fig. 6.6). Specifically, we see the loss in sensitivity of our distance measures at large environmental distances. The problem is least apparent in the Sørensen, chi-square, and Jaccard distances. The problem is worst with the correlation distance, where the curve not only flattens at high environmental distances, but starts to decline at the highest distances. The drop in the curve for the correlation coefficient (actually $(1 - r)/2$ which converts r into a distance rather than a similarity measure) is due to interpreting shared zeros (0,0) as positive association.

The second example is a more difficult data set, consisting of 100 sample units and 25 species. The difficulty has nothing to do with the size of the data set. Rather its beta diversity is higher (average Sørensen distance among SUs = 0.79; 2.3 half changes), the sample unit totals vary more widely than in the previous example (CV of SU totals = 40%), and the species vary realistically in abundance (CV of species totals = 183%).

Again, all of the distance measures lose sensitivity with increasing environmental distance (Fig. 6.7). This loss is greatest for distance based on the correlation coefficient. Euclidean distance not only loses sensitivity at high distances, but introduces considerable error, even at moderate distances. Note also that Euclidean distance shows no fixed upper bound for sample units that have nothing in common.

Sørensen distance loses sensitivity over a distance about half the length of the environmental gradients. The flat top on the Sørensen scatterplot results because it has a fixed maximum for SUs having no species in common. Many ecologists consider this a desirable, intuitive property for species data. Chi-square distance performs reasonably well at small environmental distances but misinterprets many distant SUs as being close in species space.

Transformation of the data to binary (presence-absence) in both examples results in a more linear relationship with most distance measures. This was shown for Euclidean distance and Sørensen distance with real data from an elevation gradient (Beals 1984).

Given the apparently poor performance of all of the distance measures, it is remarkable that multivariate analysis is able to extract clear, sensible patterns (Fig. 6.8). We are rescued by the redundancy in the data — all ordination and classification techniques benefit from this redundancy in the data. Where two species fail to be informative of a difference, another two species *are* informative. Some ordination techniques, nonmetric multidimensional scaling (NMS) in particular, are able to linearize the relationship between distance in species space and distance in a reduced ordination space. NMS has an advantage over other ordination techniques: it is based on *ranked* distances, which improves its ability to extract information from the nonlinear relationships illustrated in the two examples.

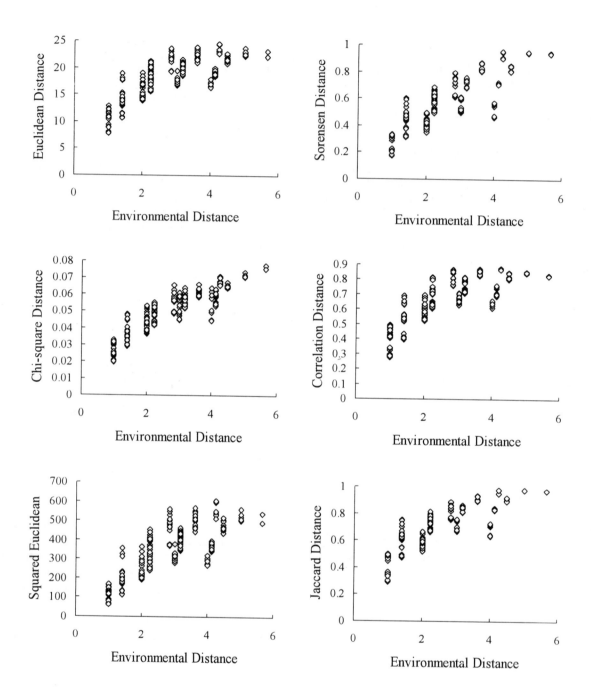

Figure 6.6. Relationship between distance in species space for an "easy" data set, using various distance measures and environmental distance. The graphs above are based on a synthetic data set with noiseless species responses to two known underlying environmental gradients. The gradients were sampled with a 5 × 5 grid. This is an "easy" data set because the average distance is reasonably small (Sørensen distance = 0.59; 1.3 half changes), all species are similar in abundance (CV of species totals = 37%), and sample units have similar totals (CV of SU totals = 17%).

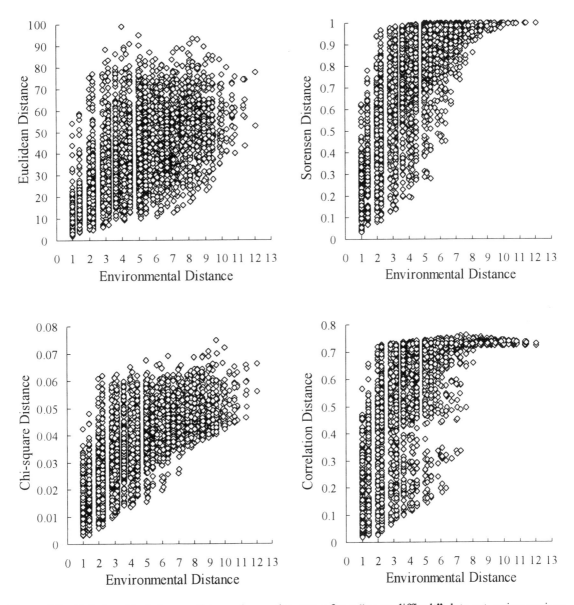

Figure 6.7. Relationship between distance in species space for a "more difficult" data set, using various distance measures, and environmental distance. The graphs above are based on a synthetic data set with noiseless species responses to two known underlying environmental gradients. The gradients were sampled with a 10 x 10 grid. This is a "more difficult" data set because the average distance is rather large (Sørensen distance = 0.79; 2.3 half changes), species vary in abundance (CV of species totals = 183%), and sample units have moderately variable totals (CV of SU totals = 40%).

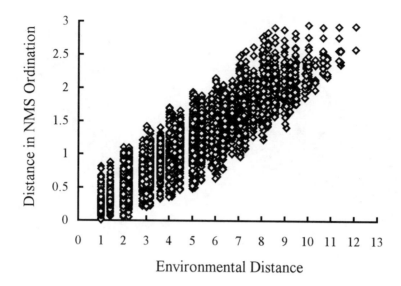

Figure 6.8. Distance in a 2-D nonmetric multidimensional scaling ordination (NMS) in relation to environmental distances, using the same data set as in Figure 6.7. Note how the ordination overcame the limitation of the Sørensen coefficient at expressing large distances.

Availability

The Sørensen (Bray-Curtis) index has repeatedly been shown to be one of the most effective measures of sample or species similarity, yet it is not widely available in general statistical software.

Compatibility

The primary disadvantage of city-block distance measures is that they are not compatible with many standard multivariate analyses (e.g., discriminant analysis, canonical correlation, and canonical correspondence analysis). This becomes less and less important, however, as non-Euclidean alternatives become increasingly available. Certain methods of classification and ordination are effective with ecological data (e.g., Bray-Curtis ordination and nonmetric multidimensional scaling), in part because they are amenable to distance measures that perform well with ecological data.

Theoretical basis

Beyond the choice of a proportion coefficient or not (Box 6.1) and the choice of a relativized distance measure or not, there is little basis in ecological theory for selecting one distance measure over another. The rationale for our choice is primarily empirical: we should select measures that have shown superior performance, based on the other criteria listed here. One important theoretical difference between Euclidean and city-block distance is, however, apparent. Long Euclidean distances in species space are measured through an uninhabitable portion of species space — in other words the straight-line segments tend to pass through areas of impossibly species-rich and overly full communities. In contrast, city-block distances are measured along the edges of species space — exactly where sample units lie in the dust bunny distribution.

Intuitive criteria

Does Euclidean or city-block distance better match our intuition on how community distances should be measured? Consider the following examples.

In city-block space, the importance of a gradient is proportional to the number of species responding to it. For example, assume that 20 species respond only to gradient X and 4 species only to gradient Y. In city-block space, gradient X is 5 times as important as gradient Y. In Euclidean space, gradient X is squareroot-of-5 times as important as gradient Y. Which space matches your intuition?

With Euclidean distance, large differences are weighted more heavily than several small differences (Box 6.2). This results in greater sensitivity to outliers with Euclidean distance than with city-block distance measures. For example, assume we have four species and three sample units, A, B, and C. The data and differences in abundance of each species for each pair of sample units are listed in Box 6.2.

Geodesic distance

Performance of all of the traditional distance measures declines as distances in species space increase (Figs. 6.7 and 6.8). An innovative solution to the problem of measuring long distances in nonlinear structures is the geodesic distance (Tenenbaum et al. 2000). This concept is similar to the "shortest path" adjustments to a distance matrix (Williamson 1978, 1983; Clymo 1980; Bradfield & Kenkel 1987; De'ath 1999a). Williamson summed distances between sample unit pairs representing the shortest path between two distant SUs, but only applied this to SUs with no species in common. Bradfield and Kenkel (1987) added flexibility by varying the threshold for the number of species in common. Bradfield and Kenkel found better results with a lower threshold; i.e., adjusting a larger proportion of the distance matrix. De'ath (1999a) further extended the method by using city-block distance measures, changing the threshold to a quantitative dissimilarity value, and allowing multiple-step paths between very distant SUs.

A geodesic distance between two points is measured by accumulating distances between nearby points. Tenenbaum et al. (2000) used Euclidean distances, but geodesic distances can be built from other distance measures. "Nearby" can be defined as a fixed radius or as the n nearest neighbors. Geodesic distances should be able to find effectively the curvature of compositional gradients in species space. Geodesic distances are one of the most promising new methodological developments. A key issue will be objectively defining "nearby" to optimize the recovery of patterns in ecological communities.

The difference between Tenenbaum's geodesic distance and the ecologists' shortest path (SP) methods can be visualized with an analogy to crossing a stream dotted with stepping stones. We want to find the shortest route from a particular point on one bank to a particular point on the opposite bank. The SP method must find a *single* stepping stone that gets us across the stream in the two shortest possible leaps (one to the stepping stone and one to the far bank). The geodesic method, however, defines a comfortably small step, then seeks the shortest series of steps without ever exceeding that small step length. The geodesic method thus considers the whole array of stepping stones, while the SP method can consider only one stone at a time.

A problem with the SP method is that if the stream is broader than two leaps, then no single stepping stone will work. This corresponds to two SUs so different that there is no third SU that shares species with both of them. De'ath (1999a) solved this problem by allowing multiple passes of the SP method, in essence allowing multiple stepping stones.

Despite the excellent ordinations in Bradfield and Kenkel (1987), Boyce and Ellison (2001), and De'ath (1999a), the geodesic distances and related methods have not been widely adopted, probably because they have not been included in popular software packages. Whether Tenenbaum et al.'s (2000) geodesic distances offer further improvements over the SP methods used by ecologists remains to be seen.

Box 6.1. Comparison of Euclidean distance with a proportion coefficient (Sørensen distance). Relative proportions of species 1 and 2 are the same between Plots 1 and 2 and Plots 3 and 4.

Data matrix containing abundances of two species in four plots.

	Sp1	Sp2
Plot 1	1	0
Plot 2	1	1
Plot 3	10	0
Plot 4	10	10

Example calculations of distance measures for Plots 3 and 4. ED = Euclidean distance, PD = Sørensen distance as percentage

$$ED_{3,4} = \sqrt{(10-10)^2 + (10-0)^2} = 10$$

$$PD_{3,4} = \frac{100\left(|10-10| + |10-0|\right)}{10+20} = 33.3\,\%$$

Sørensen Distance matrix, expressed as percentages.

	Plot 1	Plot 2	Plot 3	Plot 4
Plot 1	0			
Plot 2	33.3	0		
Plot 3	81.8	83.3	0	
Plot 4	90.5	83.3	33.3	0

Euclidean distance matrix

	Plot 1	Plot 2	Plot 3	Plot 4
Plot 1	0			
Plot 2	1.0	0		
Plot 3	9.0	9.1	0	
Plot 4	13.4	12.7	10.0	0

The Sørensen distance between Plots 1 and 2 is 0.333 (33.3%), as is the Sørensen distance between Plots 3 and 4, as illustrated below. In both cases the shared abundance is one third of the total abundance. In contrast, the Euclidean distance between Plots 1 and 2 is 1, while the Euclidean distance between Plots 3 and 4 is 10. Thus the Sørensen coefficient expresses the shared abundance as a proportion of the total abundance, while Euclidean distance is unconcerned with proportions.

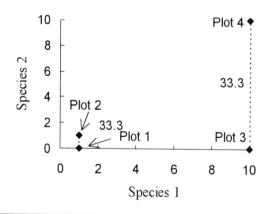

Box 6.2. Example data set comparing Euclidean and city-block distances, contrasting the effect of squaring differences versus not.

Hypothetical data: abundance of four species in three sample units (SU).

	Sp			
SU	1	2	3	4
A	4	2	0	1
B	5	1	1	10
C	7	5	3	4

Sample units A,B: species differences d = 1, 1, 1, 9 for each of the four species.
Sample units A,C: species differences d = 3, 3, 3, 3

Pair of SUs	Distance Measure	
	Euclidean	City-block
AB	9.165	12
AC	6	12

The simple sum of differences (city-block distance) is the same for AB and AC. Euclidean distance sums the squared differences, so that difference of 9 is given more emphasis with Euclidean distance than with city-block distance. Thus, the Euclidean distance between A and B is larger than the distance between A and C. The city-block distance between A and B is the same as that between A and C. Which distance measure matches your intuition?

Part 2. Data Adjustments

CHAPTER 7

Data Screening

Error checking

It often happens that a data entry error is discovered well into the analysis process, forcing you to repeat the analysis. Thus, it is very important to carefully check your data.

The best way to check hand-entered data is to read your data sheet to somebody who is checking a printout of the entered data. Another method is to enter your data twice, then compare the files. Various tools are available for comparing files, depending on the file type. For example, most word processors have a way of searching for differences between two versions of a file.

Data produced by instrumentation also should be checked carefully. Methods for doing so depend on the nature of the data and instrument. Common errors to look for are out-of-range values (e.g., latitudes or longitudes outside of the study area) and missing value indicators. If there is partial redundancy between variables, they can be cross-checked for consistency.

After the data are entered and checked, generate summary statistics to further screen for errors:

- Are the row and column means reasonable?
- Are the sample sizes correct?
- Are the ranges for the variables reasonable?

These checks are easy. A small investment of time has a potentially big return in not having to redo analyses and writing if a mistake is discovered late in the process of analysis or review.

Missing data

How you should handle missing data depends on the **amount** and **pattern** of missing data,

Amount of missing Data

If you have large amounts of missing data the most basic question is, "Should the existing data be analyzed?" If only small amounts of data are missing, researchers will usually salvage the study by selective deletion of variables or cases, or by an objective substitution of values for the missing values (discussed below).

Pattern of missing data

If missing values are concentrated in particular rows or columns, the most reasonable solution is to delete the offending rows or columns. In community ecology, it is very common to have missing values in certain environmental variables.

Species data sets often have at least a few "species" that are ambiguous in one way or another (e.g., seedlings too small to identify). Sometimes these are treated as species in themselves (perhaps "unidentified seedlings" or "unknown sparrows" are distributed nonrandomly with respect to other species or environment). But sometimes they are discarded from the data with the viewpoint that they are likely to be ecologically meaningless amalgams of species that differ in ecological responses to their environment.

If missing values are few, well distributed, and essentially random with respect to groups of data or underlying gradients, then it is often reasonable to take corrective steps.

Corrective steps

The most straightforward method for eliminating missing values is to delete specific rows or columns (cases or variables). But sometimes, these rows or columns cannot be sacrificed because they are critical to the point of the study. In these cases, many researchers will estimate missing values. If you do this, it is important to report what you did.

Three ways to estimate missing values are listed below:

- **Use prior knowledge.** You may have a very good basis for making a well-educated guess. Of course you need to be extremely cautious in doing so. Most scientists would be uncomfortable with this.

- **Insert means (or medians)**. This is usually preferable because no guessing is involved. One might use the overall mean for a given variable or the mean for a given variable within a group. Introduction of a few means usually has little effect on the outcome of an analysis. It will, however, reduce the variance of the variable and will often weaken the correlation structure slightly.
- **Use regression**. It is possible to use one variable or a set of variables to estimate the missing values. For example, soil pH and conductivity are often strongly correlated. It may be fairly harmless to estimate one missing conductivity value from pH, as long as that relationship is not in itself the focus of the analysis. But if the predictors are also part of the matrix being analyzed, then this introduces a circularity: you may be partially creating the correlation structure that you later detect.

It is best to repeat your analyses with and without the corrective steps. Did the corrective steps alter your conclusions? If so, you should try to understand why the results were altered and whether your corrective steps were justified.

Outliers

Outliers are sample units with extreme values for individual variables (univariate outliers) or sample units that have an unusual combination of values for more than one variable (multivariate outliers). Outliers are a matter of concern because they can have a large effect on the outcome of an analysis. Outliers are often so strong as to greatly affect the conclusions of a study. Analysts must, therefore, take steps to detect and understand their outliers.

Univariate outliers are easy to detect and describe (Fig. 7.1), but multivariate outliers are more tricky. Multivariate outliers need not also be univariate outliers (see example below).

Your approach to handling outliers should depend on the purposes of your study. If hypothesis testing is required, particularly careful description and handling

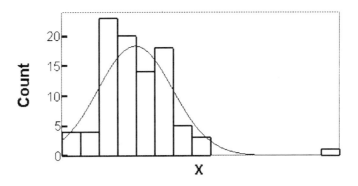

Figure 7.1. Frequency distribution showing a univariate outlier falling 5.5 standard deviations above the mean.

of outliers should be used, because they can have an unduly large effect on the resultant probabilities. Outliers may also strongly influence descriptive uses of multivariate analysis. In many cases, the outliers will show up as such in graphical output of the analysis, for example, isolated points in an ordination space or isolated sample units in a cluster analysis.

Causes of outliers

Some common causes of outliers in community ecology are listed below:

- One or a few very large values. This is much more frequent in ecology than undue influence from very small values. Look for variables (columns) with exceptionally strong univariate skewness. Obtain descriptive statistics that include skewness. (Note, however, that species as variables in most community data sets are often strongly skewed regardless of transformation; see Chapter 9).
- High SU total. Obtain descriptive statistics including SU totals.
- Empty or nearly empty SU. Look for SUs that have exceptionally low totals.
- Coincidence of rare species. This kind of outlier often shows up with chi-square distance but less often with the other distance measures. The disparity between an outlier analysis with chi-square distances and that with other distance measures can, therefore, be used to point to this kind of outlier.

Table 7.1. Steps to detect, describe, and reduce the influence of outliers.

Type	Detect →	Describe →	Reduce Influence
Univariate	Histograms, boxplots, normal probability plots, etc.	Note which variables and sample units are involved.	Check for accurate data entry. If ok, consider whether cases are part of the target population you intended to sample. If not, you can drop the outliers. If they are part of your target population, consider transformations. Also consider analysis with and without the outlier(s) so that you better understand their effect on the outcome.
Multivariate	**A**. Compute Mahalanobis distance (distance from a sample unit to the group of remaining sample units). Use a very conservative probability, e.g., $p < 0.001$. Use a chi-square table with degrees of freedom equal to the number of variables. If data are grouped, seek outliers in each group. OR **B**. Calculate average distance, using your choice of distance measure, from each sample unit to every other sample unit. Examine SUs that fall greater than some number of standard deviations (say 2 or 3) above the mean for average distance. In PC-ORD you can do this by selecting *Summary	Outlier Analysis*. OR **C**. Look for outliers as isolated entities in cluster analysis or ordinations.	If there are only a few outlying sample units (cases), examine each case. If there are many outliers, create a dummy grouping variable where outlier cases are given a 1 and others get dummy=0. Use the dummy variable as the grouping variable in discriminant analysis or as a dependent variable in multiple regression.	

- Combination of variables (i.e., a multivariate outlier not determined by univariate outliers). This can be inferred only if the other possibilities are excluded. Without knowledge of the species (or other variables) involved, the contributing variables can be difficult to pin down. If you are familiar with the variables, look for counter-intuitive combinations of values.

Detecting, describing, and reducing the influence of outliers

Table 7.1 summarizes an approach described by Tabachnik and Fidell (1989). In general, you should take steps to detect outliers, then describe them, then take steps to reduce their influence. Much more detail is in Tabachnik and Fidell (2001).

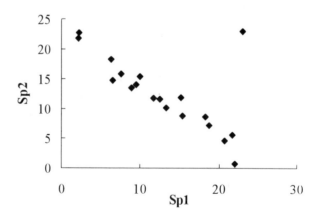

Figure 7.2. A bivariate outlier that is not a univariate outlier for either of the two variables Sp1 and Sp2.

Bivariate example

Assume you have a data set with abundances recorded for two species in 20 sample units (Fig. 7.2). We obtain the Mahalanobis distance from each point to the remaining points. It may not be obvious how to do this, so we give an example using SPSS. First we created a dummy variable from the case number.

```
COMPUTE Dummy = $casenum .
```

This dummy variable is the dependent variable in a multiple regression against both abundance variables. It is simply a convenience, because multivariate outliers with respect to the independent variables are unaffected by the dependent variable. The regression is executed in SPSS with the following commands.

```
REGRESSION
  /DEPENDENT dummy
  /NOORIGIN
  /METHOD=ENTER sp1 sp2
  /RESIDUALS=outliers(mahal)/.
```

The last part of the command requests summary statistics for residuals and a list of Mahalanobis distances.

The point in the upper right of Figure 7.2 has a Mahalanobis distance of 16.5, exceeding chi-square for two degrees of freedom at alpha = 0.001. On the other hand, the point falls only 1.55 and 1.75 standard deviations above the mean for the variables Sp1 and Sp2. We conclude, therefore, that the point in the upper right is a bivariate outlier, but it is not a univariate outlier for either variable.

Multivariate example

An outlier was introduced into an artificial data set containing 16 species with smooth unimodal responses to two independent gradients. The gradients were "sampled" with a 5 × 5 grid of 25 points. For one of those points, the smooth pattern was disrupted by substituting values that were anomalous relative to the pattern, but with a magnitude within the range of the other values. Distance measures were calculated as listed in Table 7.2.

One can plot a frequency distribution of average distances to each sample unit (Fig. 7.3). In this example the average distance is based on relative Euclidean distance. The bar at the extreme right represents the outlier sample unit with an average distance 3.2 standard deviations greater than the overall mean distance (see Table 7.2).

Table 7.2. Deviations of the mean distance to an outlying sample unit from the average distance among sample units, using five distance measures.

Distance Measure	Standard Deviations
Euclidean	3.42
Relative Euclidean	3.20
Chi-square	4.47
Sørensen	2.92
Mahalanobis	2.69

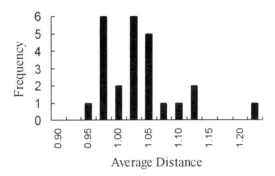

Figure 7.3. Frequency distribution of average relative Euclidean distances to particular sample units with a total of 25 sample units. The sample unit represented by the rightmost bar is 3.2 standard deviations above the mean of the average distances.

Using a procedure in SPSS similar to the bivariate example, the outlier has a Mahalanobis distance of 23.04. Looking up this value in a table of chi-square distributions, with 16 degrees of freedom, the associated p-value is between 0.05 and 0.10. This sample unit is not, therefore, considered an outlier against the criterion of alpha = 0.001. While the other distance measures do not provide a p-value, their large size suggests that this point will have considerable influence if methods using those distance measures are selected.

CHAPTER 8

Documenting the Flow of Analyses

Documenting what you did and why is an important activity, but most people slight it. While analyzing data, most people are anxious to go on to the next step. It seems painfully slow to stop and record each step as it is taken, much less pause to record observations as they emerge and how those observations led to the next step.

But face the facts: (1) Community analysis never proceeds in a simple 3-step linear process: enter the data, run the analysis, write the report. Rather, it is a process of probing and exploring the properties and peculiarities of an individual data set until one comes to understand what techniques best reveal the underlying structure. (2). Nine out of ten times you will have to repeat some or all of the analysis, having discovered some problem, either with your data or with what you did at an early stage. Given these facts, it is important to leave a clear paper trail.

A record of our path allows us to easily retrace our steps without having to rethink the many steps of logic and file manipulations by which we arrived at point B from point A. Also, if you've done a good job of documenting the flow of analysis, you can easily pinpoint the file to which you need to return.

There are two basic ways of documenting the flow of data analysis, **flow charts** and **analysis logs**. Examples of both are given below. You should use one or the other or both, or some variant comfortable for you.

Flow charts

Although there is flowcharting software, it is usually inconvenient to use during data analysis. The process is easily represented by boxes and arrows on paper (Figs. 8.1 & 8.2).

Follow a set of graphical conventions to make the charts easy to follow. For example consider the following simple rules:

- Files are contained in rectangles.
- Operations and the programs that did the operations are written alongside arrows to or from files.
- File types in many operating systems are identified by their extensions, which we consistently apply to all projects. Some of these file types are standard – for example, the *.wk1 extension in Windows implies that the file is a spreadsheet matching the Lotus 1-2-3 version 2 software. This extension is widely recognized by spreadsheet, database, and statistical software. Other file types are personal; for example, McCune uses the following extensions:

*.raw	PC-ORD compact format
*.spp	species coding files
*.txt	output files
*.gph	PC-ORD graph files
(etc.)	

Windows users can formally define these or other file types so that when their icon is double clicked, they are opened by a preselected program.

Analysis logs

An analysis log is a written-text record of the sequence of events. They can be kept on paper, in spreadsheets, or in word processors. The key elements of each entry are input file names, output file names, the action taken, and a brief statement of the result and how it leads to the next step in the analysis.

After the flowchart pages, you can find an example of an analysis log (Fig. 8.3) followed by a blank form that you are welcome to photocopy and use.

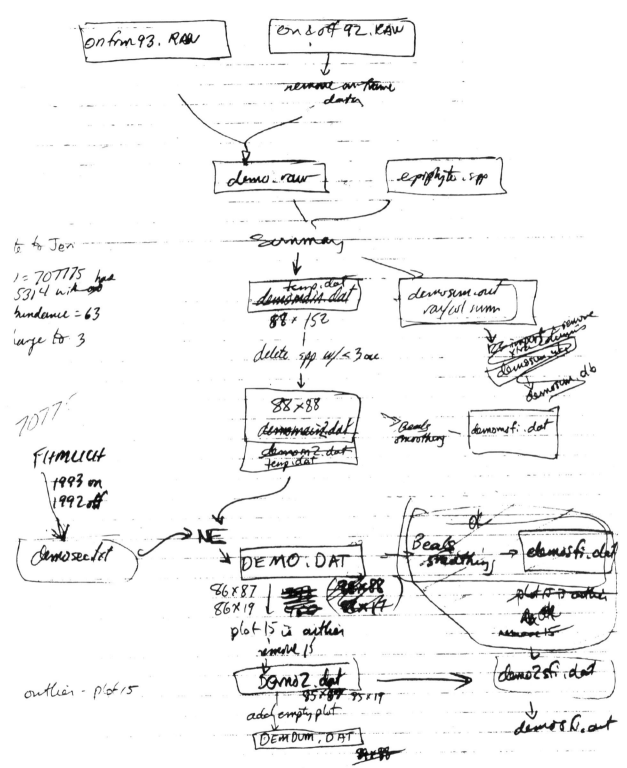

Figure 8.1. Example flowchart, as it happens... kind of messy.

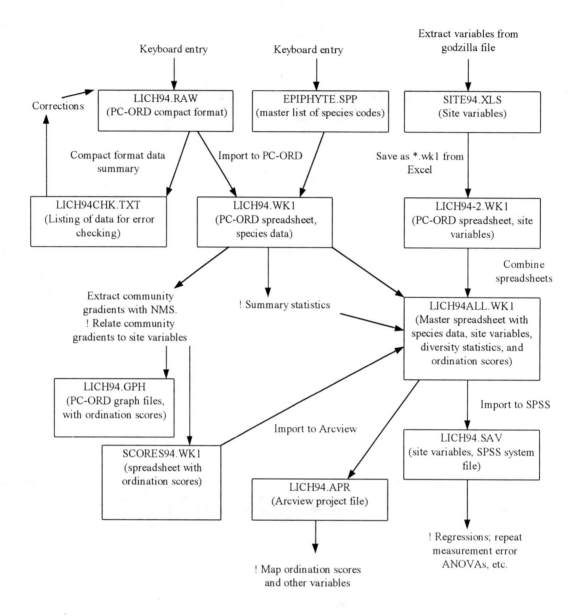

Figure 8.2. Example flowchart that is more formalized. Data files are indicated by boxes. Products or goals are indicated with exclamation points (!).

Documenting the Flow

Project __FHM SE demo__   Analysis Log Date __Sept. 1994__

Input file(s)	Action	Result or conclusion	Output file(s)
demo.wk1t	NMS, city-block dist., 2-D	Outlier was obvious (plot 15)	-
demo.wk1	Beals smoothing	-	demosfi.wk1
demosfi.wk1	NMS	outlier still obvious	-
demo.wk1	deleted outlier after checking why	-	demo2.wk1
demo2.wk1	NMS, 6D to 1D to check dimensionality	Use 2D. See table.	-
demo2sfi.wk1	ditto	Use 2D. See table.	-
demo2.wk1	NMS 2D	climatic gradient + air quality gradient, but not entirely orthogonal	demo2.gph
demo2sfi.wk1	NMS 2D	same as above, but strengthened relationships with climate and poll.	demo2sfi.gph
Demo.wk1	Added dummy plot containing no spp to better represent the most polluted extreme.	-	demodum.wk1
Demodum.wk1	Beals smoothing	failed — math error	-
demodum.wk1	NMS 2D	Junk! Empty plot is in one spot, rest of plots in a pile in another spot.	-
Demo2.wk1	Changed second matrix — reassigned Plot18 to poll=0. Is swamp plot near Charleston causing dependence of poll / climate?	-	demo3.wk1
Demo3.wk1	Beals smoothing	-	demo3sfi.wk1
demo3sfi.wk1	NMS 2D	Did not change orthogonality, but spp scores are likely improved. Use as final.	Demo3sfi.gph

Figure 8.3. Example of an analysis log.

Project _____ Analysis Log Date _____

Input file(s)	Action	Result or conclusion	Output file(s)

CHAPTER 9

Data Transformations

Most data sets benefit by one or more data transformations. The reasons for transforming data can be grouped into statistical and ecological reasons:

Statistical
- improve assumptions of normality, linearity, homogeneity of variance, etc.
- make units of attributes comparable when measured on different scales (for example, if you have elevation ranging from 100 to 2000 meters and slope from 0 to 30 degrees)

Ecological
- make distance measures work better
- reduce the effect of total quantity (sample unit totals) to put the focus on relative quantities
- equalize (or otherwise alter) the relative importance of common and rare species
- emphasize informative species at the expense of uninformative species.

Monotonic transformations are applied to each element of the data matrix, independent of the other elements. They are "monotonic" because they change the values of the data points without changing their rank. **Relativizations** adjust matrix elements by a row or column standard (e.g., maximum, sum, mean, etc.). One transformation described below, Beals smoothing, is unique in being a **probabilistic transformation** based on both row and column relationships. In this chapter, we also describe other adjustments to the data matrix, including **deleting rare species, combining entities,** and calculating **first differences** for time series data.

It is difficult to overemphasize the potential importance of transformations. They can make the difference between illusion and insight, fog and clarity. To use transformations effectively requires a good understanding of their effects, and a clear vision of your goals.

Notation.— In all of the transformations described below,

x_{ij} = the original value in row i and column j of the data matrix

b_{ij} = the adjusted value that replaces x_{ij}.

Domains and ranges

Bear in mind that some transformations are unreasonable or even impossible for certain types of data. Table 9.1 lists the kinds of data that are potentially usable for each transformation.

Monotonic transformations

Power transformation

$$b_{ij} = x_{ij}^p$$

Different parameters (exponents) for the transformation change the effect of the transformation; $p = 0$ gives presence/absence, $p = 0.5$ gives square root, etc. The smaller the parameter, the more compression applied to high values (Fig. 9.1).

The square root transformation is similar in effect to, but less drastic than, the log transform. Unlike the log transform, special treatment of zeros is not needed. The square root transformation is commonly used. Less frequent is a higher root, such as a cube root or fourth root (Fig. 9.1). For example, Smith et al. (2001)

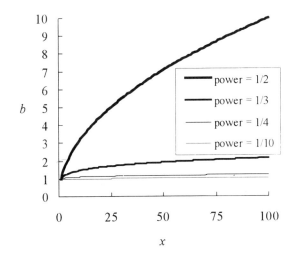

Figure 9.1. Effect of square root and higher root transformations, $b = f(x)$. Note that roots higher than three are essentially presence-absence transformations, yielding values close to 1 for all nonzero values.

Table 9.1. Domain of input and range of output from transformations.

	Reasonable and acceptable domain of x	Range of $f(x)$
MONOTONIC TRANSFORMATIONS		
x^0 (power)	all	0 or 1 only
$x^{1/2}$ (power)	nonnegative	nonnegative
$\log(x)$	positive	all
$(2/\pi)\cdot\arcsin(x)$	$0 \leq x \leq 1$	0 to 1 inclusive
$(2/\pi)\cdot\arcsin(x^{1/2})$	$0 \leq x \leq 1$	0 to 1 inclusive
SMOOTHING		
Beals smoothing	0 or 1 only	0 to 1 inclusive
ROW/COLUMN RELATIVIZATIONS		
general	nonnegative	0 to 1 inclusive
by maximum	nonnegative	0 to 1 inclusive
by mean	all	all
by standard deviates	all	generally between -10 and 10
binary by mean	all	0 or 1 only
rank	all	positive integers
binary by median	all	0 or 1 only
ubiquity	nonnegative	nonnegative
information function of ubiquity	nonnegative	nonnegative

applied a cube root to count data, a choice supported by an optimization procedure. Roots at a higher power than three nearly transform to presence-absence: nonzero values become close to one, while zeros remain at zero.

Logarithmic transformation

$$b_{ij} = \log(x_{ij})$$

Log transformation compresses high values and spreads low values by expressing the values as orders of magnitude. Log transformation is often useful when there is a high degree of variation within variables or when there is a high degree of variation among attributes within a sample. These are commonly true with count data and biomass data.

Log transformations are extremely useful for many kinds of environmental and habitat variables, the lognormal distribution being one of the most common in nature. See Limpert et al. (2001) for a general introduction to lognormal distributions and applications in various sciences. They claim that the abundance of species follows a truncated lognormal distribution, citing Sugihara (1980) and Magurran (1988). While the nonzero values of community data sets often resemble a lognormal distribution, excluding zeros often amounts to ignoring half of a data set. The lognormal distribution is fundamentally flawed when applied to community data because a zero value is, more often than not, the most frequent abundance value for a species. Nevertheless, the log transformation is extremely useful in community analysis, providing that one carefully handles the problem of log(0) being undefined.

To log-transform data containing zeros, a small number must be added to all data points. If the lowest nonzero value in the data is one (as in count data), then it is best to add one before applying the transformations:

$$b_{ij} = \log(x_{ij} + 1)$$

If, however, the lowest nonzero value of x differs from one by more than an order of magnitude, then adding one will distort the relationship between zeros and other values in the data set. For example, biomass data often contain many small decimal fractions (values such as 0.00345 and 0.00332) ranging up to fairly large values (in the hundreds). Adding a one to the whole data set will tend to compress the resulting distribution at the low end of the scale. The order-of-magnitude difference between 0.003 and 0.03 is lost if you add a one to both values before log transformation: log(1.003) is about the same as log(1.03).

The following transformation is a generalized procedure that (a) tends to preserve the original order of magnitudes in the data and (b) results in values of zero when the initial value was zero. Given:

Min(x) is the smallest nonzero value in the data

Int(x) is a function that truncates x to an integer by dropping digits after the decimal point

c = order of magnitude constant = Int(log(Min(x)))

d = decimal constant = $\log^{-1}(c)$

then the transformation is

$$b_{ij} = \log(x_{ij} + d) - c$$

Subtracting the constant c from each element of the data set after the log transformation shifts the values such that the lowest value in the data set will be a zero.

For example, if the smallest nonzero value in the data set is 0.00345, then

log(min(x)) = -2.46

c = int(log(min(x))) = -2

$\log^{-1}(c) = 0.01$.

Applying the transformation to some example values:

If $x = 0$,
then $b = \log(0+0.01)-(-2)$,
therefore $b = 0$.

If $x = 0.00345$,
then $b = \log(0.00345+0.01)-(-2)$,
therefore $b = 0.128$.

Arcsine transformation

$$b_{ij} = 2/\pi * \arcsin(x_{ij})$$

The constant $2/\pi$ scales the result of arcsin(x) [in radians] to range from 0 to 1, assuming that 0 $\leq x \leq 1$. The function "arcsin" is the same as $\sin^{-1}$ or inverse sine. Data **must** range between zero and one, inclusive. If they do not, you should relativize before selecting this transformation.

Unlike the arcsine-squareroot transformation, an arcsine transformation is usually counterproductive in community ecology, because it tends to spread the high values and compress the low values (Fig. 9.2). This might be useful for distributions with negative skew, but community data almost always have positive skew.

Arcsine squareroot transformation

$$b_{ij} = 2/\pi * \arcsin\left(\sqrt{x_{ij}}\right)$$

The arcsine-squareroot transformation spreads the ends of the scale for proportion data, while compressing the middle (Fig. 9.2). This transformation is recommended by many statisticians for proportion data, often improving normality (Sokal and Rohlf 1995). The data must range between zero and one, inclusive. The arcsine-squareroot is multiplied by $2/\pi$ to rescale the result so that it ranges from 0 to 1.

The **logit** transformation, $b = \ln(x/(1-x))$, is also sometimes used for proportion data (Sokal and Rohlf 1995). However, if $x = 0$ or $x = 1$, then the logit is undefined. Often a small constant is added to prevent

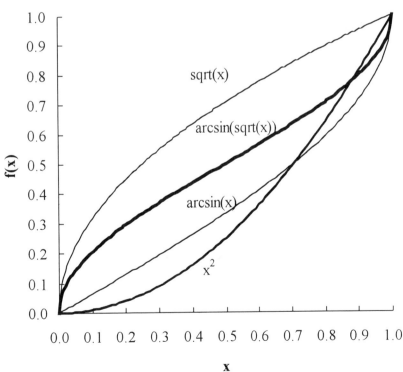

Figure 9.2. Effect of several transformations on proportion data

ln(0) and division by zero. Alternatively, empirical logits may be used (see Sokal and Rohlf 1995:762). Because zeros are so common in community data, it seems reasonable to use the arcsine squareroot or squareroot transformations to avoid this problem.

Beals smoothing

Beals smoothing replaces each cell in the community matrix with a probability of the target species occurring in that particular sample unit, based on the joint occurrences of the target species with the species that are actually in the sample unit. The purpose of this transformation (also known as the sociological favorability index, Beals 1984) is to relieve the "zero-truncation problem" (Beals 1984). This problem is nearly universal in community data sets and most severe in heterogeneous community data sets that contain a large number of zeros (i.e., most samples contain a fairly small proportion of the species). Beals smoothing replaces presence/absence or other binary data with quantitative values that represent the "favorability" of each sample for each species, regardless of whether the species was present in the sample. The index evaluates the favorability of a given sample for species i, based on the whole data set, using the proportions of joint occurrences between the species that do occur in the sample and species i.

$$b_{ij} = \frac{1}{S_i} \sum_k \left(\frac{M_{jk}}{N_k} \right)$$

where S_i is the number of species in sample unit i, M_{jk} is the number of sample units with both species j and k, and N_k is the number of sample units with species k. This transformation is illustrated in Box 9.1.

This transformation is essentially a smoothing operation designed for community data (McCune 1994). As with any numerical smoothing, it tends to reduce the noise in the data by enhancing the strongest patterns. In this case the signal that is smoothed is the pattern of joint occurrences in the data. This is an extremely powerful transformation that is particularly effective on heterogeneous or noisy data. Caution is warranted, however, because, as for any smoothing function, this transformation can produce the appearance of reliable, consistent trends even from a series of random numbers.

This transformation should **not** be used on data sets with few zeros. It also should not be used if the data are quantitative and you do not want to lose this information.

Beals smoothing can be slow to compute. If you have a large data set and a slow computer, be sure to allocate plenty of time. This transformation is available in PC-ORD but apparently not in other packages for statistical analysis.

Relativizations

"To relativize or not to relativize, that focuses the question." (Shakespeare, ????)

Relativizations rescale individual rows (or columns) in relationship to some criterion based on the other rows (or columns). Any relativization can be applied to either rows or columns.

Relativization is an extremely important tool that all users of multivariate statistics in community ecology MUST understand. There is no right or wrong answer to the question of whether or not to relativize UNTIL one specifies the question and examines the properties of the data.

If the row totals are approximately equal, then relativization by rows will have little effect. Consistency of row totals can be evaluated by the coefficient of variation (CV) of the row totals (Table 9.2). The CV% is calculated as 100*(standard deviation / mean). In this case, it is the standard deviation of the row totals divided by the mean of the row totals.

Table 9.2. Evaluation of degree of variability in row or column totals as measured with the coefficient of variation of row or column totals.

CV, %	Variability among rows (or columns)
< 50	Small. Relativization usually has small effect on qualitative outcome of the analysis.
50-100	Moderate (with a correspondingly moderate effect on the outcome of further analysis).
100-300	Large. Large effect on results.
> 300	Very large.

Box 9.1. Example of Beals smoothing

Data matrix **X** before transformation (3 sample units × 5 species):

	sp1	sp2	sp3	sp4	sp5	S_i
SU1	1	0	1	1	1	4
SU2	0	0	0	1	0	1
SU3	1	1	0	0	0	2
N_j	2	1	1	2	1	

S_i = number of species in sample unit i.
N_j = number of sample units with species j.

Construct matrix **M**, where M_{jk} = number of sample units with both species j and k.
(Note that where $j = k$, then $M_{jk} = N_j$).

Species k

	1	2	3	4	5
Species j 1	2				
2	1	1			
3	1	0	1		
4	1	0	1	2	
5	1	0	1	1	1

Construct new matrix **B** containing values transformed with Beals smoothing function:

$$b_{ij} = \frac{1}{S_i} \sum_k \left(\frac{M_{jk}}{N_k} \right) \text{ for all } k \text{ with } x_{ik} \neq 0$$

Data after transformation (**B**):

	sp1	sp2	sp3	sp4	sp5
SU1	0.88	0.13	0.75	0.88	0.75
SU2	0.50	0.00	0.50	1.00	0.50
SU3	1.00	0.75	0.25	0.25	0.25

Example for sample unit 1 and species 2:
 $b_{1,2} = 1/4 \ (1/2 + 0/1 + 0/2 + 0/1)$
 $b_{1,2} = 0.25 \ (0.5)$
 $b_{1,2} = 0.125$ (rounded to 0.13 in matrix above)

Example for sample unit 3 and species 2:
 $b_{3,2} = 1/2 \ (1/2 + 1/1)$
 $b_{3,2} = 0.5 \ (1.5)$
 $b_{3,2} = 0.75$

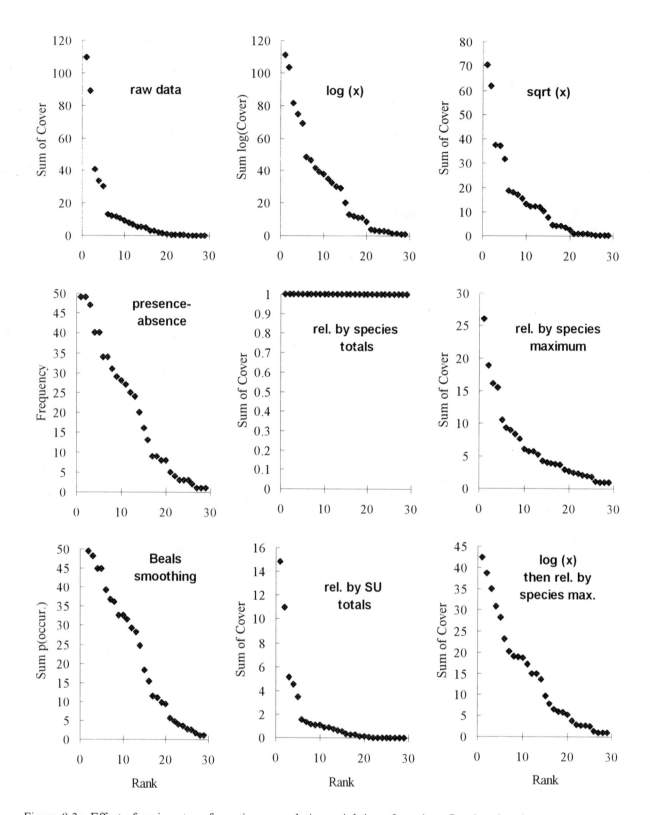

Figure 9.3. Effect of various transformations on relative weighting of species. Species abundance was measured on a continuous, quantitative scale. "Rank" is the order of species ranked by their abundance.

If the row or column totals are unequal, one must decide whether to retain this information as part of the analysis or whether to remove it by relativizing. One must justify this decision on biological grounds, not on its effect on the CV of row or column totals. For example, consider two quadrats with identical proportions of three species, but one quadrat has a total cover of 1% and the other has a total cover of 95%. If the data are relativized, then the quadrats appear similar or identical. If they are not relativized, then distance measures will consider them to be very different. Which choice is correct? The answer depends on the question. Does the question refer to proportions of different species or is the total amount also important? If the latter is true, the data should not be relativized.

An example demonstrates how relativization can change the focus of the analysis. Menges et al. (1993) reported rates of vegetation change based on both relativized and nonrelativized tree species data, beginning with a matrix of basal area of each species in remeasured permanent plots. They used absolute rates to emphasize structural changes (e.g., increase in basal area of existing species) and relative rates to emphasize shifts in species composition (changes in the relative proportions of species).

Relativization is often used to put variables that were measured in different units on an equal footing. For example, a data set may contain counts for some species and cover for other species. In forest ecology, one may wish to combine basal area data for trees with cover data for herbs. If the species measured in different units are to be analyzed together, then one must relativize the data such that the quantity for each species is expressed as a *proportion* of some total or maximum abundance.

Relativizations can have a huge effect on the relative weighting of rare and abundant species. Raw quantitative data on a continuous scale tends to have a few abundant species and many rare species (Fig. 9.3). A multivariate analysis of these raw data might emphasize only a few species, ignoring most of the species. Log or square-root transformation of the data usually moderates the imbalance, while relativization by species totals can eliminate it completely (Fig. 9.3). This is, however, a drastic transformation. Rare species often occur haphazardly, so that giving them a lot of weight greatly increases the noise in the analysis.

General relativization

By rows:

$$b_{ij} = \frac{x_{ij}}{\left(\sum_{j=1}^{q} x_{ij}^{p}\right)^{1/p}}$$

By columns:

$$b_{ij} = \frac{x_{ij}}{\left(\sum_{i=1}^{n} x_{ij}^{p}\right)^{1/p}}$$

for a matrix of n rows and q columns.

The parameter, p, can be set to achieve different objectives. If $p = 1$, relativization is by row or column totals. This is appropriate when using analytical tools based on city-block distance measures, such as Bray-Curtis or Sørensen distance. If $p = 2$, you are "standardizing by the norm" (Greig-Smith 1983, p. 248). Using $p = 2$ is the Euclidean equivalent of relativization by row or column totals. It is appropriate when the analysis is based on a Euclidean distance measure. The same effect can be achieved by using "relative Euclidean distance" (see Chapter 6).

Relativization by maximum

$$b_{ij} = x_{ij}/\text{xmax}_j$$

where rows (i) are samples and columns (j) are species, xmax_j is the largest value in the matrix for species j. As for relativization by species totals, this adjustment tends to equalize common and uncommon species. Relativization by species maxima equalizes the heights of peaks along environmental gradients, while relativization by species totals equalizes the areas under the curves of species responses.

Many people have found this to be an effective transformation for community data. A couple of cautions should be heeded, however: (1) very rare species can cause considerable noise in subsequent analyses if not omitted; (2) this and any other statistic based on extreme values can accentuate sampling error.

Adjustment to mean

$$b_{ij} = x_{ij} - \bar{x}_i$$

The row or column mean is subtracted from each value, producing positive and negative numbers. If relativized by rows, the means are row means; if by columns, the means are column means. The negative numbers obviate proportion-based distance measures, such as Sørensen and Jaccard. This unstandardized centering procedure can have detrimental effects on analysis of community data. It tends to emphasize to values of zero more than does the raw data. Also, more variable species are reduced in importance relative to more constant species.

Adjustment to standard deviate

$$b_{ij} = (x_{ij} - \bar{x}_j) / s_j$$

where s_j is the standard deviation within column j.

Each transformed value represents the number of standard deviations that it differs from the mean, often known as "z scores." As for all of the relativizations, this transformation can be applied to either rows or columns. It is, however, usually applied to variables (columns). This transformation results in all variables having mean = 0 and variance = 1.

Because this transformation produces both positive and negative numbers, it is NOT compatible with proportion-based distance measures, such as Sørensen's. While this transformation is of limited utility for species data, it can be a very useful relativization for environmental variables, placing them on equal footing for a variety of purposes.

Binary with respect to mean

$$b_{ij} = 1 \text{ if } x_{ij} > \bar{x}, \quad b_{ij} = 0 \text{ if } x_{ij} \leq \bar{x}$$

An element is assigned a zero if its value is less than or equal to the row or column mean, $\bar{x}$. The element is assigned a one if its value is above the mean. Applied to species (columns), this transformation can be used to contrast above-average conditions with below-average conditions. The transformation therefore emphasizes the optimal parts of a species distribution. It also tends to equalize the influence of common and rare species. Applied to sample units, it emphasizes dominant species and is likely to eliminate many species, particularly those that rarely, if ever, occur in high abundances.

Rank adjustment

Matrix elements are assigned ranks within rows or columns such that the row or column totals are constant. Ties are assigned the average rank of the tied elements. For example, the values 1, 3, 3, 9, 10 would receive ranks 1, 2.5, 2.5, 4, 5.

This transformation should be applied with caution. For example, most community data have many zeros. These zeros are counted as ties. Because the number of zeros in each row or column will vary, zeros will be transformed to different values, depending on the number of zeros in each row or column. For example, the values 0, 0, 0, 0, 6, 9 would receive the ranks 2.5, 2.5, 2.5, 2.5, 5, 6, while the values 0, 0, 6, 9 would receive the ranks 1.5, 1.5, 3, 4.

Binary with respect to median:

$$b_{ij} = 1 \text{ if } x_{ij} > \text{median}, \quad b_{ij} = 0 \text{ if } x_{ij} \leq \text{median}$$

The transformed values are zeros or ones. An element is assigned a zero if its value is less than or equal to the row or column median. The element is assigned a one if its value is greater than the row or column median. This transformation can be used to emphasize the optimal parts of a species range, at the same time equalizing to some extent the weight given to rare and dominant species. The *Rank adjustment* caution also applies to this relativization because it too is based on ranks.

Weighting by ubiquity

$$b_{ij} = U_j x_{ij} \quad \text{where} \quad U_j = N_j / N$$

If rows are samples, columns are species, and relativization is by columns, more ubiquitous species are given more weight. Under these conditions, N_j is the number of samples in which species j occurs and N is the total number of samples.

Information function of ubiquity

$$b_{ij} = I_j x_{ij}$$

where

$$I_j = -p_j \cdot \log(p_j) - (1 - p_j) \log(1 - p_j)$$

and $p_j = N_j / N$ with N_j and N as defined above.

To illustrate the effect of this relativization, assume that rows are samples, columns are species, and relativization is by columns. Maximum weight is applied to species occurring in half of the samples because those species have the maximum information content, according to information theory. Very common and rare species receive little weight. Note that if there are empty columns, the transformation will fail because the log of zero is undefined.

Double relativizations

The relativizations described above can be applied in various combinations to rows then columns or vice-versa. When applied in series, the last relativization necessarily mutes the effect of the preceding relativization.

The most common double relativization was first used by Bray and Curtis (1957). They first relativized by species maximum, equalizing the rare and abundant species, then they relativized by SU total. This and other double relativizations tend to equalize emphasis among SUs and among species. This comes at a cost of diminishing the intuitive meaning for individual data values.

Austin and Greig-Smith (1968) proposed a "contingency deviate" relativization. This measures the deviation from an expected abundance. The expected abundance is based on the assumption of independence of the species and the samples. Expected abundance is calculated from the marginal totals of the $n \times p$ data set, just as if it were a large contingency table:

$$b_{ij} = x_{ij} - \frac{\sum_{j=1}^{p} x_{ij} \sum_{i=1}^{n} x_{ij}}{\sum_{j=1}^{p} \sum_{i=1}^{n} x_{ij}}$$

The resulting values include both negative and positive values and are centered on zero. The row and column totals become zero. Because this transformation produces negative numbers, it is incompatible with proportion-based distance measures.

One curious feature of this transformation is that zeros take on various values, depending on the marginal totals. The meaning of a zero is taken differently depending on whether the other elements of that row and column create large or small marginal totals. With sample unit × species data, a zero for an otherwise common species will be given more weight (i.e., a more negative value). This may be ecologically meaningful, but applied to rows the logic seems counter-intuitive: a species that is absent from an otherwise densely packed sample unit will also be given high weight.

Deleting rare species

Deleting rare species is a useful way of reducing the bulk and noise in your data set without losing much information. In fact, it often enhances the detection of relationships between community composition and environmental factors. In PC-ORD, you select deletion of columns "with fewer than N nonzero numbers." For example, if $N = 3$, then all species with less than 3 occurrences are deleted. If $N = 1$, all empty species (columns) are deleted.

Deleting rare species is clearly inappropriate if you wish to examine patterns in species diversity. Cao et al. (1999) correctly pointed this out but confused the issue by citing proponents of deletion of rare species who were concerned with extracting patterns with multivariate analysis, not with comparison of species diversity. None of the authors they criticized suggested deleting rare species prior to analysis of species richness.

For multivariate analysis of correlation structure (in the broad sense), it is often helpful to delete rare species. As an approximate rule of thumb, consider deleting species that occur in fewer than 5% of the sample units. Depending on your purpose, however, you may wish to retain all species or eliminate an even higher percentage.

Some analysts object to removal of rare species on the grounds that we are discarding good information. Empirically this can be shown true or false by using an external criterion of what is "good" information. You can try this yourself. Use a familiar data set that has at least a moderately strong relationship between communities and a measured environmental factor. Ordinate (Part 4) the full data set, rotate the solution to align it with that environmental variable (Ch. 15), and record the correlation coefficient between the environmental variable and the axis scores. Now delete all species occurring in just one sample unit. Repeat the ordination→rotation→correlation procedure. Progressively delete more species (those only in two sample units, etc.), until only the few most common species remain. Now plot the correlation coefficients against the number of species retained (Fig. 9.4).

In our experience the correlation coefficient usually peaks at some intermediate level of retention of species (Fig. 9.4). When including all species, the noise from the rare ones weakens the structure slightly. On the other hand, when including only a few dominant species, too little redundancy remains in the data for the environmental gradient to be clearly expressed.

A second example compared the effect of stand structures on small mammals using a blocked design (D. Waldien 2002, unpublished). Fourteen species were enumerated in 24 stands, based on trapping data, then relativized by species maxima. The treatment effect size was measured with blocked MRPP (Ch. 24), using the A statistic (chance-corrected within-group agreement). Rare species were successively deleted, beginning with the rarest one, until only half of the species remained. In this case, removal of the four rarest species increased slightly the apparent effect size

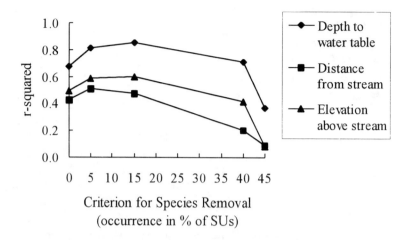

Figure 9.4. Correlation between ordination axis scores and environmental variables can often be improved by removal of rare species. In this case, the strength of relationship between hydrologic variables and vegetation, as measured by r^2, is maximized with removal of species occurring in fewer than 5-15% of the sample units, depending on the hydrologic variable. The original data set contained 88 species; 59, 35, 16, and 9 species remained after removal of species occurring in fewer than 5, 15, 40, and 45% of the sample units, respectively. Data are courtesy of Nick Otting (1996, unpublished).

(Fig. 9.5). The fifth and sixth rarest species, however, were distinctly patterned with respect to the treatment, so their removal sharply diminished the apparent effect size.

Another objection to removal of rare species is that you cannot test hypotheses about whole-community structure, if you exclude rare species. Certainly this is true for hypotheses about diversity. But it also applies to other measures of community structure. Statistical hypothesis tests are always, in some form or another, based on evaluating the relative strength of signal and noise. Because removal of rare species tends to reduce noise, the signal is more likely to be detected. This can be taken as an argument against removal of rare species because it introduces a bias toward rejecting a null hypothesis. Alternatively, one can define beforehand the community of interest as excluding the rare species and proceed without bias.

Mark Fulton (1998, unpublished) summarized the noise vs. signal problem well:

> Noise and information can only be defined in the context of a question of interest. An analogy: we are sitting in a noisy restaurant trying to have a conversation. From the point of view of our attempting to communicate, the ambient sound around us is 'noise.' Yet that noise carries all kinds of information — that clatter over to the left is the bus person clearing dishes at the next table; the laughter across the room is in response to the punchline of a fairly good joke; with a little attention you can hear what the two men in business suits two tables over are arguing about; and that rumble you just heard is a truck full of furniture turning the corner outside the restaurant. But none of this information is relevant to the conversation, and so we filter it out without thinking about the process much.
>
> Vegetation analysis is a process of noise filtering right from the very start. Data collection

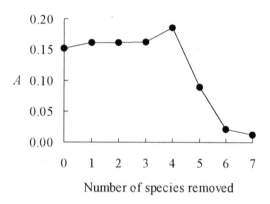

Figure 9.5. Response of A statistic (blocked MRPP) to removal of rare species from small mammal trapping data. A measures the effect size of the treatments, in this case different stand structures.

itself is a tremendous filtering process. We decide what NOT to measure. Any transformations we do on the data — whether weighting, rescaling, or deletion of rare species — is also a filtering process. Ordination itself is a further filter. The patterns in the whole *n*-dimensional mess are of less interest than a carefully selected reduction of those patterns. The point is, as scientists, we need to do this process of information selection and noise reduction carefully and with full knowledge of what we are doing. There is no single procedure which will always bring out the information of interest. Data selection, transformation, and analysis can only be judged on how well they work in relation to the questions at hand.

Combining entities

Aggregate sample units (SUs) can be created by averaging existing SUs. Each new entity is the "centroid" of the entities that you average. See Greig-Smith (1983, p. 286) for comments on ordinating groups of SUs. In general, community SUs should not be averaged unless they are very similar. If SUs are heterogeneous, then the average species composition tends to fall outside the variation of the SUs, the averages being unnaturally species-rich.

If your rows are SUs, and you also have an environmental matrix, you should also calculate centroids for environmental data. Be careful if you have categorical environmental variables. Depending on how the categories are structured, averaging the categories can be meaningless.

Difference between two dates

Before-and-after data on species abundance obtained by revisiting the same SUs can be analyzed as differences, rather than the original quantities. If a_{ij1} and a_{ij2} are the abundances of species j in sample unit i at times 1 and 2, then the difference between dates is

$$b_{ij} = a_{ij2} - a_{ij1}$$

The transformed data represents changes through time. Even with species abundance data, this transformation yields variables that are more or less normally distributed with means near zero and with both positive and negative numbers. After this transformation, be sure not to use methods that demand nonnegative numbers: proportion coefficients (such as Sørensen) as distance measures and techniques based on Correspondence Analysis (CA, RA, CCA, DCA, Twinspan). On the other hand, PCA and other techniques calling for multivariate normal data and linear relationships among variables will work far better on such a matrix than they would with either matrix alone.

First difference of time series

If your data form a time series (sample units are repeatedly evaluated at fixed locations), you may want to ordinate the **differences** in abundance between successive dates rather than the raw abundances:

$$b_{ij} = a_{ij,t+1} - a_{ij,t}$$

for a community sampled at times t and $t+1$. This is simply the extension through time of the idea described in the preceding section. This transformation can be called a "first difference" (Allen et al. 1977) because it is analogous to the first derivative of a time series curve. With community data, a matrix of first differences represents changes in species composition. If we visualize changes in species composition as vectors in species space, the matrix of differences represents the lengths and directions of those vectors. A matrix of "second" differences would represent the rates of acceleration (or deceleration) of sample units moving through species space.

The matrix of first differences takes into account the direction of compositional change. For example, assume that the plankton in a lake go through two particular compositional states in the fall, then go through the same compositional states in the spring, but in the opposite direction. The difference between the two fall samples is not, therefore, the same as the difference between the spring samples, even though the absolute values of the differences are equal. Analyzing the signed difference is logical, but other possibilities exist. Allen et al. (1977) analyzed the absolute differences, creating a matrix of species' contributions to community change, without regard to the direction of the change:

$$b_{ij} = |a_{ij,t+1} - a_{ij1,t}|$$

If environmental variables are recorded at each date, you might analyze species change from time t to $t+1$ in relationship to the state of the environment at time t. Alternatively, you could apply the first difference transformation to the environmental variables as well, to analyze the question of how community change is related to environmental change. On the other hand, variables that are constant through time for a given sample unit through time (e.g., location or treatment variables) could be retained without transformation.

Note that the statistical properties of these differences are radically different from the original data. For more information, see the preceding section on differences between two dates.

A general procedure for data adjustments

Species data

While one can easily grasp the logic of a particular data adjustment, the number of combinations and sequences can be bewildering. Although it is impossible to write a step-by-step cookbook that covers all possible data sets and goals, we suggest a general procedure for data adjustments that will be applicable to many community data sets (Table 9.3). For more details on steps 2, 3, and 4, consult the preceding pages. For more detail on step 5, consult the section on outliers in Chapter 7.

The sequence of actions is important. For example, we check for outliers last, because many apparent outliers will disappear, depending on the monotonic transformations or relativizations that are used.

Table 9.3. Suggested procedure for data adjustments of species data matrices.

Action to be considered	Criteria	

1. Calculate descriptive statistics. <u>Repeat this</u> after each step below. (In PC-ORD run *Row & column summary*)
    Beta diversity (community data sets)
    Average skewness of columns
    Coefficient of variation (CV, %)
        CV of row totals
        CV of column totals
	Always	

2. Delete rare species (< 5% of sample units)	Usually applied to community data sets, unless contrary to study goals	
3. Monotonic transformation (if applied to species, then usually applied uniformly to all of them, so that all are scaled the same)	A. Average skewness of columns (species) B. Data range over how many orders of magnitude? (Count and biomass data often are extreme.) C. Beta diversity. (Consider presence/absence transformation for community data when β is high.)	
4. Row or column relativizations	What is the question? Are units for all variables the same? Is relativization built into the subsequent analysis? CV of row totals CV of column totals What distance measure do you intend to use? Note: regardless of your decision to relativize or not, you should state your decision and justify it briefly on biological grounds.	
5. Check for outliers based on the average distance of each point from all other points. Calculate standard deviation of these average distances. Describe outliers and take steps to reduce influence, if necessary	standard deviation ---------- < 2 2 - 2.3 2.3 - 3 > 3	degree of problem ---------------------- no problem weak outlier moderate outlier strong outlier

Environmental data

Adjustments of environmental data depend greatly on their intended use, as indicated in Table 9.4. Categorical and binary variables in general need no adjustment, but one should always examine quantitative environmental variables.

Table 9.4. Suggested procedure for data adjustments of quantitative variables in environmental data matrices.

Action to be considered	Criteria
1. Calculate descriptive statistics for quantitative variables. <u>Repeat this</u> after each step below. (In PC-ORD run *Row & column summary*) Skewness and range for each variable (column)	Always
2. Monotonic transformation (applied to individual variables, depending on need)	Consider log or square root transformation for variables with skewness > 1 or ranging over several orders of magnitude. Consider arcsine squareroot transformation for proportion data.
3. Column relativizations	Consider column relativization (by norm or standard deviates) if environmental variables are to be used in a distance-based analysis that does not automatically relativize the variables (for example, using MRPP to answer the question: do groups of sample units defined by species differ in environmental space?). Column relativization is not necessary for analyses that use the variables one at a time (e.g., ordination overlays) or for analyses with built-in standardization (e.g., PCA of a correlation matrix).
4. Check for univariate outliers and take corrective steps if necessary.	Examine scatterplots or frequency distributions or relativize by standard deviates ("z scores") and check for high absolute values.

Part 3. Defining Groups with Multivariate Data

CHAPTER 10

Overview of Methods for Finding Groups

A natural starting point for learning multivariate analysis is to study ways to group objects using quantitative criteria. Humans naturally organize the world around them by putting things into groups.

The problem is fairly complex with community data, because each object has so many characteristics (the species). The problem can be visualized geometrically as sample units scattered in species space. Our goal is to place the objects (entities) into groups. The problem is complicated because the objects are in multidimensional space, yet we can see their relationships in only two or three dimensions at a time.

Goodall (1973b) summarized the geometry of the problem:

> Considering [sample units] to be represented by points in a multidimensional system of which the axes are the variables by which the [sample units] may be described, classification amounts to the division of this space into discrete cells by hypersurfaces. If the distribution of points is interrupted by discontinuities or regions of low density, the dividing hypersurfaces should follow these discontinuities. Even if there are no such discontinuities, division of vegetation space into arbitrary classes may still serve practical purposes.

To find and define groups is to classify, whether or not this classification is formalized into community types. Thousands of pages written by ecologists debate whether discrete community types exist in nature. How does the conclusion of this debate affect classification of communities? Miles (1979, p. 65) aptly summarized:

> The study of vegetation presents a dilemma which is common to many branches of biological science. On the one hand, vegetation shows endless variation in composition in time and space. Hence any classification of it has to use arbitrary criteria, and the different units thus identified inevitably intergrade. On the other hand, in order to study vegetation, or any other biological phenomenon, it is necessary to create order, to identify small units which it is possible to study. It is important to recognize that any classification is only a working hypothesis, an ad hoc fiction necessary to advance scientific understanding, but whose usefulness is limited to the particular situation for which it was formulated. Unfortunately, the essential purpose of classification and its intrinsic limitations seem often to have been overlooked.

General strategies

Hierarchical methods find groups that are composed of subgroups. In other words, groups are nested within groups, and this is represented in a **dendrogram** (tree diagram; Fig. 10.1). **Nonhierarchical** methods seek an optimum structure for a specific number of groups not necessarily related to any other level of grouping. The optimum is defined statistically, for example minimizing the within-group variance. Hierarchical methods also seek an optimum structure, but these methods are constrained by groups necessarily being formed of subgroups.

The choice of hierarchical vs. nonhierarchical methods depends partly on the goals of a study and partly on the conceptual preference of the investigator. Requiring that groups are nested in one another imposes a constraint on the analysis that may result in a suboptimal solution for a given number of groups. On the other hand, a hierarchical structure appeals conceptually because the consequences of choosing, for example, 4, 5, or 6 groups are more clearly seen and stated.

Polythetic methods use multiple species (or other attributes or variables) as a basis for deciding on each fusion or division of items. **Monothetic** methods base each division or fusion on a single species or variable. But all monothetic methods select the criterion variable after examining all of the variables. Monothetic methods are typically divisive (defined below).

Goodall (1973b) described the geometrical difference between polythetic and monothetic methods. In monothetic methods, a hyperplane divides the species space perpendicular to the axis defined by the chosen species. Each dividing plane is perpendicular to the others. In contrast, polythetic methods can partition the space along oblique planes or even curved surfaces.

Agglomerative methods build groups hierarchically from the bottom up. Groups form by fusion rather than division, according to the similarity or dissimilarity of the items. **Divisive** methods start with all items in a single group. This group is divided into two groups, which in turn may be repeatedly subdivided.

The choice of agglomerative vs. divisive methods in practice depends as much on availability of software and familiarity of the method as quality of the result. Agglomerative methods are fast and readily available. Most divisive methods are monothetic, which is usually perceived as a drawback because less information is used at each step than with polythetic methods.

Hierarchical agglomerative cluster analysis

Hierarchical, polythetic, agglomerative cluster analysis is described in detail in the next chapter and briefly summarized here:

1. Calculate distance matrix. This is a matrix of distances between each pair of entities. From a data matrix of n objects $\times p$ attributes we calculate a $n \times n$ distance matrix. These are distances in a p-dimensional hyperspace. You can use any method for calculating distances that you like, as long as it is compatible with subsequent steps (see Hierarchical Clustering section).

2. Merge two groups, selecting the two groups by some criterion of minimum distance.
3. Combine the attributes of the entities in the two groups that were fused.
4. Merge the next two groups, etc.
5. Usually the results are shown as a dendrogram (Fig. 10.1) scaled by distance between groups or the "objective function".

Scaling dendrograms

Very often, dendrograms are scaled by the simple distance function. At each fusion point, there is a distance between the groups being fused. These distances can be used to scale the fusion points in the tree.

A more informative scaling of the dendrogram is Wishart's (1969) objective function. This function measures the information lost at each step in hierarchical cluster analysis. As groups are fused, the amount of information decreases until all groups are fused and no information remains.

The objective function (E) is the sum of the error sum of squares from each centroid to the items in that group:

$$E = \sum_{t=1}^{T} E_t$$

where t indexes the T clusters and E_t is the error sum of squares for cluster t. Each E_t is found by:

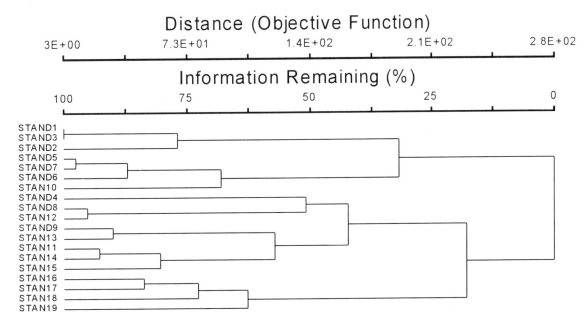

Figure 10.1. Example dendrogram scaled by Wishart's objective function and percent of information remaining.

$$E_t = \sum_{i=1}^{k_t} \sum_{j=1}^{p} (x_{ijt} - \bar{x}_{jt})^2$$

In this equation, x_{ijt} is the value of the jth variable for the ith point of cluster t containing k_t points, and $\bar{x}$ is the mean of the jth variable for cluster t. Note that the deviations are zero at the beginning. The sum of squared deviations is maximal when all groups are fused. At this point, the objective function is equal to the sum of the squared deviations from the overall centroid in p-dimensional space.

The objective function can be rescaled from 0% to 100% of information:

% information remaining = 100(SST - E)/SST

where SST is the total sum of squares. SST is the same as E when all groups are fused into one. Conceptually the information remaining is similar to an r^2 value (coefficient of determination), but in this application, maximizing r^2 is not an objective. Rather, think of the objective as seeking a compromise between minimizing the number of groups and maximizing the information retained.

You can apply the objective function to any clustering method, but most statistical packages plot the dendrogram on a distance scale. Depending on the linkage method, this can result in "reversals" where fusion of two groups occurs at a lower distance than a previous fusion (Fig. 10.2). Reversals never happen when the dendrogram is scaled by the objective function.

Dendrograms can be compared to each other with a variety of methods (Podani & Dickinson 1984). The comparison is usually based on correlations between new matrices describing the topological relationship between items in the dendrogram, one matrix for each dendrogram.

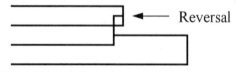

Figure 10.2. Reversal in a dendrogram.

Where to cut the dendrogram

Using a dendrogram to define groups requires cutting (pruning) the dendrogram, the position of the cut defining the number of resulting groups. There are a number of objective procedures for deciding where to cut, but these usually depend on rather arbitrary criteria. Ecologists often cut off the dendrogram at some "natural" break point that will give a level of aggregation appropriate to the goals of the study. "Natural" groups have long stems in the dendrogram.

Cutting the dendrogram demands a compromise between homogeneity of the groups and the number of groups. Cutting the dendrogram into a large number of groups yields better within-group homogeneity but defeats the goal of producing a small number of groups. A very small number of groups may result in such diffuse or heterogeneous groups that they do not seem biologically meaningful.

One promising method for community data was designed by Dufrene and Legendre (1997). They calculated the indicator value for species according to their method at each step in the procedure (Ch. 25). They then selected the level that maximized the collective indicator values of the species.

Pros and cons of hierarchical agglomerative methods

Entities can be misclassified because later fusions depend on earlier fusions. The hierarchical constraint means that the solution is seldom optimal for a given number of groups. For example, if sample A is similar to B, but diverges from B in the dendrogram because of the samples with which they merge, they may end up in different main groups. You can correct misclassifications by following up with discriminant analysis to identify misclassified entities.

A hierarchical structure is appealing for multilevel classifications. A prior decision on number of groups is not necessary. You can let the data structure guide you in this, considering discreteness of the groups and the amount of information retained.

Note that cluster analysis is not an explicitly dimensional solution — there is no inherent dimensionality in a dendrogram. Imagine a dendrogram as a child's mobile (Fig. 10.3). In your mind, pick up the mobile by the basal stem so that the parts are free to rotate on their stems. This lack of inherent dimensionality in a dendrogram is an advantage for dealing with very heterogeneous data where there are more than two strong dimensions.

For very heterogeneous data you might, therefore, use cluster analysis to partition your data into more manageable groups. The relationships within each group could be explored with ordination techniques.

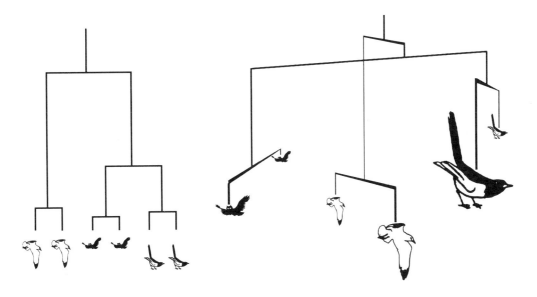

Figure 10.3. A dendrogram is an inherently nondimensional representation. Imagine the branches as free to pivot, like a child's mobile.

Hierarchical divisive: Association analysis

Association analysis in community ecology is based on 2 × 2 contingency tables of presence-absence data, one table for each pair of species. Goodall (1953a) used species associations to classify samples, the first attempt at numerical classification of community data. This paper inspired improvements by Williams and Lambert (1959), resulting in Association Analysis. See Ludwig & Reynolds (1988) for a summary more complete than the brief description below.

1. Calculate chi-square for each 2 × 2 contingency table, a table for each pair of species.
2. Sum the chi-squares for each species.
3. Choose the species with the highest sum of chi-squares as the basis for dividing the data set.
4. Split data into two groups, those with and those without the indicator species.
5. Repeat steps 1-4, subdividing each group as many times as desired.

This method is monothetic because each division is based on a single species rather than all species at once. This may be appealing if your world view has a single indicator species as a criterion. Before computers were common, the method was appealing because the calculations were relatively easy. This allowed association analysis by hand, even with a moderate-sized data set.

Multivariate regression trees

The preceding methods seek groups by analyzing data in a single matrix. You can also define a hierarchical group structure by analyzing the relationship *between* two matrices. De'ath (2002) used multivariate regression trees (MRT) as a hierarchical divisive method of clustering, dividing sample units based on species-environment relationships. (See Chapter 29 for more information on classification and regression trees.) At each division of sample units, a specific environmental variable and a division point along that variable are selected to minimize "impurity," the compositional variation within the resulting groups. Impurity of the groups can be measured as the sum of the squared Euclidean distances from individual sample units to the centroids in species space. The method is made useful for community data by using distance-based MRT (db-MRT; De'ath 2002). Rather than measuring impurity as the sum of squared Euclidean distances from a centroid, db-MRT measures impurity as the sum of squared within-group interpoint distances without reference to a centroid. This approach can be applied to *any* distance measure, allowing us to avoid the poor performance of Euclidean distance with community data. The db-MRT measure of impurity is closely allied to Anderson's (2001) method of calculating sums of squares in non-parametric multivariate analysis of variance (NPMANOVA; Chapter 24). The two differ mainly in that we define groups a priori in NPMANOVA, while in db-MRT we search the environmental data for a grouping principle.

Nonhierarchical methods

With nonhierarchical methods, you specify the number of groups, then items are placed into that number of groups, attempting to optimize some statistical characteristic of the groups. You can request a series of group sizes, but the groups will not usually nest.

"*K*-means" method is the most widely known method of the many forms of non-hierarchical clustering. Its chief advantage is that memory requirements are small so that very large data sets can be analyzed efficiently; we need not store the whole distance matrix as in hierarchical methods. This is an "N routine" rather than an "N-squared routine," such as agglomerative cluster analysis. N is the number of objects being clustered. Computation time for an "N routine" increases linearly with N, rather than with N^2. For example, if an analysis with 1000 objects takes 1 hour on your computer, then multiplying the number of objects by 10 (to 10,000) would result in a run taking 100 hours for an N-squared routine, but only 10 hours for an N routine.

Clustering by *k*-means is, therefore, useful for very large data sets. One of the most common applications is clustering pixels in images from remote sensing, based on reflectance in multiple bands of wavelengths of light. Clusters of points are sought in a hyperspace defined by reflectance of different wavelengths.

The basic steps for a *k*-means analysis are:

1. Skewer the *p*-dimensional space with a diagonal line from the origin to the outermost corner. (A thought problem: what defines the outermost corner of species space? Answer: a hypothetical sample unit with all species present in their maximum amount.)
2. Divide the diagonal into *k* segments of equal length.
3. For each SU, find the closest segment (i.e., the segment with shortest distance from the midpoint of that segment to the SU) and assign the SU to the group defined by that segment.
4. Calculate the centroid for each group of SUs.
5. For each SU find the closest group centroid and assign the SU to that group.
6. Go back to step 4, iterating a fixed number of times or until a criterion level of stability is reached (usually based on the number of points changing position from one group to another).

TWINSPAN

TWINSPAN is short for "two-way indicator species analysis." This polythetic divisive method partitions an ordination space (Ch. 12).

Evaluating quality of groups

Chaining. Chaining is the addition of single items to existing groups. A completely chained dendrogram has not served the primary purpose of dividing the data into a smaller number of groups.

Many people assess chaining just qualitatively. One can, however, calculate a value for percent chaining (McCune & Mefford 1999). The average path length in the dendrogram is compared with the minimum possible average path length (no chaining) and the maximum possible average path length (complete chaining; Fig. 10.4). An alternative method (Williams et al. 1966) accumulates the differences in subgroup size — the more equal the size of the subgroups, the less the chaining.

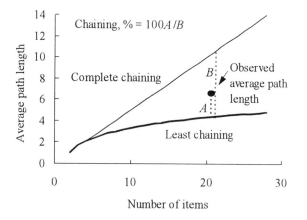

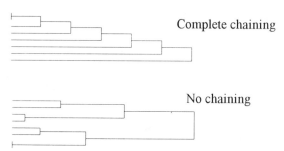

Figure 10.4. Use of average path length to measure percent chaining in cluster analysis. Path length is the number of nodes between the tip of a branch and the trunk.

Although percent chaining is an easy and objective criterion for evaluating dendrograms, you should not make fine distinctions based solely on degree of chaining. More important are criteria based on interpretability, comparison to an independent attribute, and whether or not the method is "space conserving" (Ch. 11). For example, you should not conclude that method A was better than B just because chaining was 10% vs. 15%. You should, however, avoid extreme chaining (greater than 25%).

Interpretability. Does the dendrogram make sense? Does it speak to you? Study the structure of the dendrogram and compare it to your raw data. Describe the differences between groups (see below).

Comparison to an independent attribute. If you have data on one or more variables that you are confident should be related to your groups, this may provide a way of evaluating the quality of the groups. Do the groups separate in accordance with this variable? Of course this cannot be a variable that was used in constructing the groups.

Describing group differences

Usually we want more information than simply an assignment of items to groups. Typically, we also wish to characterize the groups. The following descriptive techniques and diagnostics are informative and easy to use:

- Means and variances for each variable in each group
- Discriminant analysis (Ch. 26) enables you to determine the variables that best separate the groups. This is particularly useful for non-species (i.e., environmental) data.
- Dufrêne and Legendre's (1997) indicator species analysis (applicable only to species data; Ch. 25) evaluates the faithfulness and concentration of abundance of species in each group.
- Overlaying groups onto ordinations (Ch. 13) often illuminates the relationships among groups more clearly than does the dendrogram.

Testing for group differences

Beware of the circularity of testing for differences among groups using variables that were used to define the groups. It is fine to test your groups against a different variable or set of variables. Assuming you are not making a circular test, consider the following choices:

1. **Univariate** (for example, do groups defined by species composition differ with respect to a particular environmental variable?).
   A. **Parametric**: one-way analysis of variance
   B. **Nonparametric**: Kruskal-Wallis
2. **Multivariate** (for example, do groups defined by tree species composition differ with respect to composition of insect species?).
   A. **Parametric**: Discriminant analysis (discussed later) or MANOVA.
   B. **Nonparametric** (well suited to community data): MRPP and related techniques (Ch. 24). MRPP gives you a p-value for the hypothesis that the differences between groups are no tighter than expected for randomly chosen groups of the same data.

Analyze the variation within groups

Cluster analysis can be used to partition a data set into more homogenous subsets, each of which could be analyzed on its own. For example, say you had data on stream biota from all kinds of streams throughout a large, heterogeneous region. To study the relationship between water quality variables and biota, you would probably want to divide your data into a few big classes of streams, then analyze the water quality question for each class of stream. You could divide your data arbitrarily (say by stream order), or you could let a cluster analysis guide your decision.

CHAPTER 11

Hierarchical Clustering

Background

One of the simplest approaches to constructing hierarchical classifications of objects (entities) is to sequentially merge objects or groups of objects with other objects or groups. This is called agglomeration.

Agglomerative cluster analysis has a long history of use in ecology (Goodall 1973b). Some early applications were by Field and McFarlane (1968), Mountford (1962), and Williams et al. (1966).

Agglomerative strategies can be divided into two groups: (1) those that optimize some property of a group of entities and (2) those that optimize the route by which the groups are attained. Lance and Williams (1967, 1968) named these *clustering strategies* and *hierarchical strategies* respectively.

When to use it

Agglomerative cluster analysis is a useful tool almost anytime groups are sought from multivariate ecological data, provided they can be represented by a distance matrix. See the preceding chapter for pros and cons of all hierarchical agglomerative methods. One can cluster any of the metaobjects of community analysis in a space defined by any other metaobject (see Chapter 2). Most common is clustering of sample units based on species abundance or presence-absence.

How it works

The general algorithm for clustering was described by Williams (1967, 1968), Wishart (1969), and Post and Sheperd (1974). These were the basis of hierarchical cluster analysis in PC-ORD.

A dissimilarity matrix of order $n \times n$ (n = number of entities) is calculated and each of the elements is squared. The algorithm then performs n-1 loops (clustering cycles) in which the following steps are done:

1. The smallest element (d_{pq}^2) in the dissimilarity matrix is sought (the groups associated with this element are S_p and S_q).

2. The objective function E_n (the amount of information lost by linking up to cycle n; see preceding chapter) is incremented according to the rule

$$E_n = E_{n-1} + \tfrac{1}{2} d_{pq}^2 \qquad [E_0 = 0]$$

3. Group S_p is replaced by $S_p \cup S_q$ by recalculating the dissimilarity between the new group and all the other groups (practically this means replacing the pth row and column by new dissimilarities).

4. Group S_q is rendered inactive and its elements assigned to group S_p.

After joining all items, the procedure is complete. The pattern of fusions can be shown in a dendrogram (tree diagram; Fig. 11.1). See the preceding chapter for information on scaling the dendrogram.

Properties of the linkage methods

The properties of the linkage methods ("sorting strategies") depend on the type of dissimilarity measure used. Two classes of distance measures (Ch. 6) are Euclidean metrics (absolute and relative distance) and proportion coefficients (Sørensen and Jaccard). There are three general properties of hierarchical strategies: combinatorial or not, compatible or not, and space conserving or not.

Combinatorial or noncombinatorial

Lance and Williams (1967) distinguished between combinatorial and noncombinatorial strategies depending on how the dissimilarity matrix is reduced as clustering proceeds. Consider two groups S_p and S_q, with n_p and n_q elements, respectively, whose squared dissimilarity (d_{pq}^2) is the smallest in the dissimilarity matrix; that is, they are the next to be fused into a group. The union of the two groups is defined as $S_p \cup S_q = S_r$. We now calculate the dissimilarity (squared) between S_r and all other groups, for example S_i. Before fusion, the values of d_{pr}^2, d_{qr}^2, d_{pq}^2, n_p, n_q and n_r are known, and if d_{ir}^2 can be calculated from these, then the original data need not be stored after the first set of measures have been calculated. Such a strategy is **combinatorial**.

Table 11.1. Summary of combinatorial coefficients used in the basic combinatorial equation.

Linkage method	α_p	α_q	β	γ
Nearest neighbor	0.5	0.5	0	-0.5
Farthest neighbor	0.5	0.5	0	0.5
Median	0.5	0.5	-0.25	0
Group average	n_p/n_r	n_q/n_r	0	0
Centroid	n_p/n_r	n_q/n_r	$-\alpha_p \alpha_p$	0
Ward's method	$\dfrac{n_i + n_p}{n_i + n_r}$	$\dfrac{n_i + n_q}{n_i + n_r}$	$\dfrac{-n_i}{n_i + n_r}$	0
Flexible beta	$(1 - \beta)/2$	$(1 - \beta)/2$	β	0
McQuitty's method	0.5	0.5	0	0

n_p = number of elements in S_p
n_q = number of elements in S_q
n_r = number of elements in $S_r = S_p \cup S_q$
n_i = number of elements in S_i $i = 1, n$ except $i \neq p$ and $i \neq q$

Table 11.2. Summary of properties of linkage methods and distance measures.

Linkage method	Euclidean distance (absolute and relative)		Sørensen distance $(1 - 2w/a+b)$	
	Combinatorial compatible?	Space contracting, expanding, or conserving?	Combinatorial compatible?	Space contracting, expanding, or conserving?
Nearest neighbor (single linkage)	yes	contracting	yes	contracting
Farthest neighbor (complete linkage)	yes	expanding	yes	expanding
Median (Gower's method)	yes	contracting	no	unknown
Group average (average linkage)	yes	conserving	yes	conserving
Centroid (weighted group)	yes	contracting	no	contracting
Ward's method (Orloci's method)	yes	conserving	no	unknown
Flexible beta	yes	flexible	yes	flexible
McQuitty's method	yes	contracting	no	unknown

The basic combinatorial equation is

$$d_{ir}^2 = \alpha_p d_{ip}^2 + \alpha_q d_{iq}^2 + \beta d_{pq}^2 + \gamma |d_{ip}^2 - d_{iq}^2|$$

where values of α_p, α_q, β, and γ determine the type of sorting strategy (Table 11.1). Think of these parameters as weights that define how distances from two groups are fused into a set of new distances for the new group.

A **noncombinatorial** strategy is one in which the new dissimilarities cannot be calculated from the previous ones; they must be recalculated from the data. After each step the whole dissimilarity matrix must be recalculated. The advantages of a combinatorial strategy are, therefore, computational: noncombinatorial strategies are slower and require more memory than combinatorial strategies.

Compatible or incompatible

A **compatible** strategy is one in which the dissimilarities calculated later in the analysis are calculated in the same fashion as the initial dissimilarity matrix. An example of an incompatible strategy would be to choose Sørensen (Bray-Curtis) dissimilarity along with a hierarchical method that calculates the new intergroup dissimilarities as Euclidean distances. Incompatible strategies should be considered experimental at present.

The mathematical incompatibility of city-block distance measures with popular linkage methods, such as Ward's method, has not stopped people from applying them together on ecological community data. Often these people feel that the results are as good as or better than those resulting from the same linkage method with Euclidean distance measures. To better understand the tradeoffs involved, we need a careful comparison of the performance of various distance measures, including incompatible distance measures, under controlled conditions; in other words, based on simulated but realistic community data sets having known underlying structure.

On theoretical grounds, one would expect choice of distance measure to be less important in cluster analysis than in ordination. With cluster analysis we are usually defining groups that occupy relatively localized areas of species space (short distances), as opposed to trying to do an ordination that spans a big chunk of species space (long distances). Because all distance measures tend to lose sensitivity as environmental and community differences increase, the emphasis on joining similar items first in cluster analysis postpones this distance measure problem to the final fusions.

Space-conserving or space-distorting

The initial dissimilarity matrix can be thought of as defining distances in a space with certain properties conferred by the choice of dissimilarity measure. As groups form, measures of intergroup distances may alter the original properties of the space. If the properties of the original space are preserved, then the strategy is **space-conserving**. With certain strategies the space in the vicinity of a group may become expanded or contracted. Such strategies are **space-distorting**. Chaining is the result of a **space-contracting** strategy. A group will appear to move nearer to some or all of the remaining elements and the chance that an element will be added to an existing group rather than act as the nucleus of a new group is increased. **Space-expanding** strategies produce "nonconformist" groups of peripheral elements; that is, groups appear to move farther apart and individual elements not yet in a group are more likely to form nuclei of new groups. A summary of the properties of the most common hierarchical methods is in Table 11.2.

What to report

- ☐ Distance measure.
- ☐ Linkage method.
- ☐ If and how misclassified items were sought and corrected (see preceding chapter).
- ☐ Dendrogram (in some cases).
- ☐ How the dendrogram was scaled, if a dendrogram is reported.
- ☐ If the dendrogram is used to define groups, describe the criteria for "pruning" the dendrogram. Specify the amount of information retained at that level (see "Scaling dendrograms" in previous chapter).

Examples

This simple example of Ward's method is based on a data matrix (Table 11.3) containing abundances of two species in four plots. This is the same data matrix used in Box 6.1 (Distance Measures). The example uses Euclidean distances (Table 11.4), followed by a comparison of dendrograms using Euclidean and Sørensen distances (Figs. 11.1 & 11.2).

Table 11.3. Data matrix

	Sp1	Sp2
Plot 1	1	0
Plot 2	1	1
Plot 3	10	0
Plot 4	10	10

Table 11.4. Squared Euclidean distance matrix

	Plot 1	Plot 2	Plot 3	Plot 4
Plot 1	0	1	81	181
Plot 2	1	0	82	162
Plot 3	81	82	0	100
Plot 4	181	162	100	0

Cluster step 1:

Combine group 2 (plot 2) into group 1 (plot 1) at level $E = 0.5$. This fusion produces the least possible increase in Wishart's objective function (below).

Now group 1 contains plots 1 and 2. Three groups remain: 1, 3, and 4 (Note that groups are identified by the sequence number of the first item in that group.)

Calculate Wishart's objective function (E), which sums the error sum of squares for each group t:

$$E_t = \sum_{i=1}^{k_t}\sum_{j=1}^{p}(x_{ijt} - \bar{x}_{jt})^2$$

Most software uses a different method than given below, calculating changes in E from the distance matrix. In this case, however, we used the longer calculation for conceptual clarity. After this first step, only one group has more than one item, so $E = E_1$. Group $t = 1$ has $k = 2$ items and we sum across each of $p = 2$ species.

$$E_1 = \sum_{i=1}^{2}\sum_{j=1}^{2}(x_{ij1} - \bar{x}_{j1})^2$$
$$= (1-1)^2 + (1-1)^2 + (0-0.5)^2 + (1-0.5)^2$$
$$= 0.5$$

The distance matrix shrinks from 4 × 4 to 3 × 3. Only the terms involving group 1 are recalculated. We use the combinatorial equation for this. For example, the distance between the new group $r = 1$ and group $i = 3$ is formed from the distance between the old groups $p = 1$ and $q = 2$ and group 3.

$$d_{3,1}^2 = \alpha_1 d_{3,1}^2 + \alpha_2 d_{3,2}^2 + \beta d_{1,2}^2 + \gamma |d_{3,1}^2 - d_{3,2}^2|$$

Obtain the coefficients for this equation by applying the formulas for Ward's method from Table 11.1:

$$\alpha_1 = \frac{1+1}{1+2} = \frac{2}{3} \quad \alpha_2 = \frac{1+1}{1+2} = \frac{2}{3}$$
$$\beta = -\frac{1}{3} \quad \gamma = 0$$

So

$$d_{3,1}^2 = \frac{2}{3}(81) + \frac{2}{3}(82) - \frac{1}{3}(1) = \frac{325}{3} = 108.3$$

The calculation is repeated for the distance between the first and fourth groups, and the revised distance matrix (Table 11.5) is complete.

Table 11.5. Revised distance matrix after the first fusion.

	Plots 1+2	Plot 3	Plot 4
Plots 1+2	0	108.3	228.3
Plot 3	108.3	0	100
Plot 4	228.3	100	0

Cluster step 2:

Combine group 4 into group 3 at level 50.5

Now group 3 contains plots 3 and 4

Two groups remain: 1 and 3

Cluster step 3:

Combine group 3 into group 1 at level 151.7

Now group 1 contains all plots, so clustering is finished.

The resulting dendrogram is in Figure 11.1.

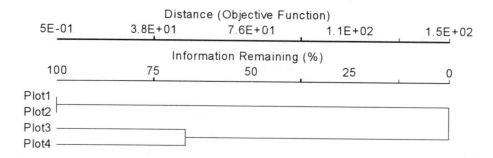

Figure 11.1. Agglomerative cluster analysis of four plots using Ward's method and Euclidean distance. The data matrix is given in Table 11.3.

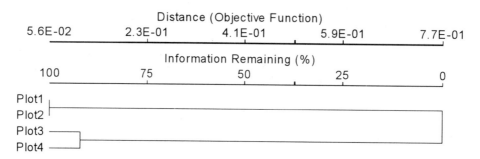

Figure 11.2. Agglomerative cluster analysis of four plots using Ward's method and Sørensen distance. The data are the same as for Figure 11.1.

Compare this dendrogram (Fig. 11.1) with that for the same method but with Sørensen (Bray-Curtis) distances (Fig. 11.2). Note that plots 3 and 4 appear more similar using this proportion coefficient for distance than with Euclidean distance. Plots 3 and 4 have the same relative proportions of species 1 and 2 (see Box 6.1 under "Distance Measures").

Species groups can be defined by cluster analysis of the transposed community matrix, but the approach differs somewhat from the normal clustering of sample units. Species tend to group according to abundance rather than by similar ecological tolerances (Gittens 1965, Webb et al. 1967, Stephenson et al. 1970, Goldsmith 1973). Austin & Belbin (1982) proposed a solution to this problem based on calculating a special two-stage distance matrix before clustering. The problem can also be partly relieved by relativization, as described below.

McCune et al. (2000) sampled epiphytic species in various locations in an old-growth conifer forest. For most of their analyses they deleted species occurring in fewer than three quadrats, yielding a matrix of 72 quadrats × 61 species. For cluster analysis, however, they applied a stricter criterion for removing rare species, retaining only those species with five or more occurrences, because species with only a few occurrences provide little reliability in assigning them to groups. The transposed 44 species × 72 quadrat matrix was relativized by species sums of squares to de-emphasize clustering based on total abundance alone. Without this step, cluster analysis of species often results in abundant species tending to group together. Instead they sought groupings by habitat preferences. They used Ward's method of clustering based on a Euclidean distance matrix. Although Sørensen distance is generally preferred for community analysis, the use of cluster analysis to seek local group structure renders many of the differences between distance measures unimportant. Sørensen distance is, in theory, incompatible with Ward's method, but the latter is a reliable, effective method of clustering. The resulting dendrogram (Figure 11.3) was scaled by Wishart's objective function converted to a percentage of information remaining.

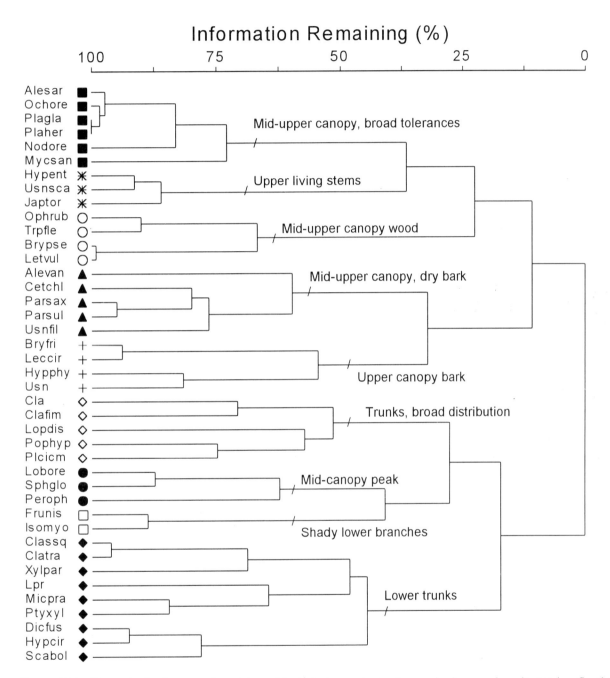

Figure 11.3. Example dendrogram from hierarchical cluster analysis of a species by sample unit matrix. Symbols indicate species groups formed by pruning the dendrogram (see "/" marks). The species are epiphytic lichens and bryophytes. Each species group is accompanied by an interpretation of the habitat associated with those species groups (McCune et al. 2000).

McCune et al. concluded: "The dendrogram from cluster analysis of species was trimmed at 9 groups. This level of grouping provided an good compromise between loss of information (about 45% retained) and providing a simple, interpretable summary of ecological affinities among species (Fig. 11.3). The species groups are described below."

Variations (linkage methods)

Nearest neighbor

Nearest neighbor (= single linkage) was the first hierarchical method. The distance between two groups is defined as the smallest distance between two elements, one from each group. Nearest-neighbor linkage results in "straggling" clusters (Fig. 11.4) and usually fails with large populations due to chaining (Sokal & Sneath 1963, Proctor 1966, Williams et al. 1966). The method is space-contracting.

Farthest neighbor

Farthest neighbor (= complete linkage) is exactly the opposite of the nearest neighbor method in that the distance between two groups is defined as the largest distance between two elements, one from each group. This method is space-expanding (Sokal & Sneath 1963, Mac-Naughton-Smith et al. 1964, Proctor 1966). It can be used when an intense grouping strategy is needed. Unfortunately, the space-expanding nature of this method imposes the appearance of distinct groups, even when there are none.

Nearest-neighbor or farthest-neighbor **may produce irregular results** because the dissimilarity is determined by only two elements, and group structure is not taken into account. The remaining linkage methods consider group structure.

Median

Median clustering (= Gower's method = WPGMC) defines the distance between any cluster S_i and the cluster that results from the fusion of clusters S_p and S_q as the distance from the centroid of S_i to the midpoint of the line joining the centroids of S_p and S_q. Median tends to chain for large populations (Gower 1967; Fig. 11.4). Each cluster is represented by a single point at the midpoint of the line between the two groups that are being fused. This gives equal weight to groups of different sizes, unlike the centroid, which is weighted by the number of items in each group.

Group average

Group average is equivalent to the unweighted pair-group method (UPGMA) of Sokal and Michener (1958) and was the first strategy to account for group structure. This distance between groups is the average of all distances for all pairs of individuals, one from each group (the average between-group distance). This method finds spherical groups and is reasonably space-conserving (Sokal & Michener 1958, Sokal & Sneath 1963, Proctor 1966, Gower 1967). It is, however, more prone to chaining (Fig. 11.4) than Ward's method, another space-conserving technique.

Centroid

With centroid linkage (= unweighted centroid), the intergroup distance is calculated as the distance between the centroids (mean coordinate vectors) of the two groups. Centroid clustering tends not to chain as much as nearest-neighbor (Fig 11.4; Proctor 1966, Williams et al. 1966). It is, nevertheless, space contracting.

Ward's method

Ward's method (= hierarchical grouping = minimum variance method = Orloci's method) is based on minimizing increases in the error sum of squares. The error sum of squares is defined as the sum of the squares of distances from each individual to the centroid of its group. The fusion of groups S_p and S_q yields the least increase in the error sum of squares. The same method was proposed independently by Orloci (1967a). This method finds minimum-variance spherical clusters (Ward 1963, Orloci 1967a, Wishart 1969).

Ward's method is an effective, useful tool and is one of the few space-conserving linkage methods (Table 11.2). The main drawback for application in community ecology is its incompatibility with Sørensen (Bray-Curtis) distance. Nevertheless, many analysts report satisfactory results with this combination. The results do not, however, guarantee adherence to the underlying principle of minimizing the within-group error sum of squares. To achieve a result similar to Ward's method, but applying Sørensen distance without offending any principles, consider using flexible-beta linkage with $\beta = -0.25$.

Clustering with Ward's method and Euclidean distance is often improved by relativizing the data. Relativizing the data can focus the analysis in various directions (see Chapter 9).

Flexible beta

Flexible beta linkage (= flexible clustering) takes advantage of the inherent flexibility in the combinatorial equation (Fig. 11.4). This method is combinatorial by definition. It is derived from the basic combinatorial equation by the four constraints:

$$(\alpha_p + \alpha_q + \beta = 1; \alpha_p = \alpha_q; \beta < 1; \gamma = 0)$$

where β is supplied by the user. It is compatible with Euclidean distance although the strictly Euclidean nature is lost. Lance and Williams (1967) felt that its incompatibility with non-metric measures is also of no importance; in other words, it is conceptually compatible with Sørensen and other semi-metrics or non metrics.

Flexible-beta linkage is flexible because the user can control its space-distorting properties. As β approaches +1, it is increasingly space-contracting. Chaining approaches 100% as β approaches +1. As β approaches zero and then becomes negative, the method ceases to be space-contracting and becomes increasingly space-expanding and the elements more intensely grouped (Fig. 11.4). A value of $\beta = -0.25$ gives results similar to Ward's method (Lance & Williams 1967).

McQuitty's method

McQuitty's method is defined combinatorially by the constraints $\alpha_p = \alpha_q = 0.5$, $\beta = \gamma = 0$. This is the same as flexible beta with $\beta = 0$ and tends to chain with large populations (McQuitty 1966).

Recommendations

Avoid distortion by selecting one of the few space-conserving methods: Ward's method, flexible beta with $\beta = -0.25$, or group average. The first two are preferred, having less propensity to chain. To maximize defensibility, select a compatible distance measure (e.g., relative Euclidean distance for Ward's method or Sørensen for flexible beta.). For experimental purposes, consider using Ward's method with Sørensen distance.

Chapter 11

Flexible beta, beta = -1

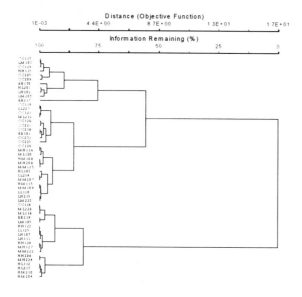

Flexible beta, beta = 0

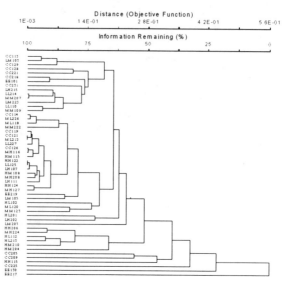

Flexible beta, beta = -0.25

Flexible beta, beta = 0.99

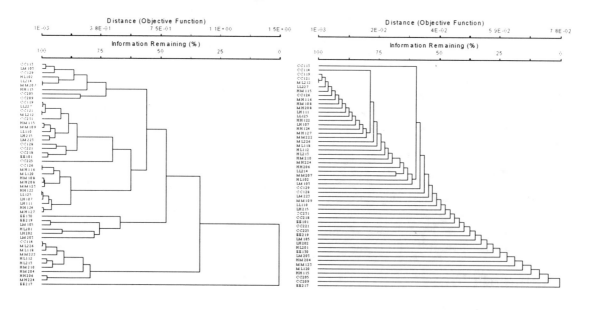

Figure 11.4. Example of effect of linkage method on dendrogram structure. Note how strongly the degree of chaining depends on the linkage method.

Hierarchical Clustering

Ward's Method

Nearest Neighbor

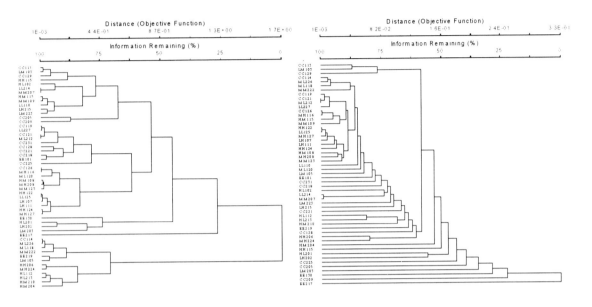

Farthest Neighbor

Group Average

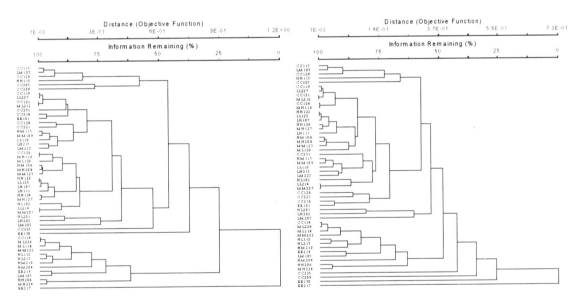

Figure 11.4, cont.

Median Centroid

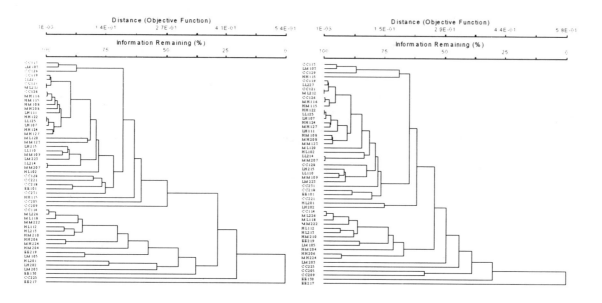

McQuitty's

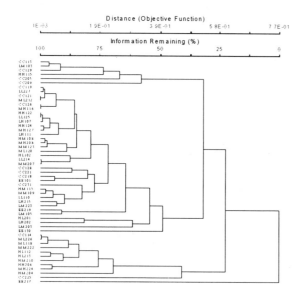

Figure 11.4, cont.

CHAPTER 12

Two-Way Indicator Species Analysis

Background

Two-way indicator species analysis (TWINSPAN; Hill 1979b, Gauch and Whittaker 1981) simultaneously classifies species and sample units. The method was designed specifically for sample unit × species data. TWINSPAN repeatedly divides a correspondence analysis (Ch. 19; CA or RA; Hill 1973b) ordination space. TWINSPAN was originally distributed as a program in the Cornell Ecology Program Series (Hill 1979b). Gauch (1982, pp. 201-203) described the procedure in accessible terms.

The main attraction of TWINSPAN is the resulting two-way ordered table that summarizes variation in species composition. Species and sample units are simultaneously ranked along a dominant gradient, sample units across the top and species down the side (Fig. 12.1). Major divisions in the data are indicated by a pattern of digits in the margins. The interior of the table contains the abundance class of each species in each sample unit. Abundance classes are defined by "pseudospecies cut levels." The structure of the table is similar to phytosociological tables that are part of the Braun-Blanquet tradition (Braun-Blanquet 1965, Mueller-Dombois & Ellenberg 1974).

Note that TWINSPAN seeks groups in species data, and reports indicator species for those groups. It cannot, however, be used to find indicator species for predefined groups.

When to use it

Ecologists should not use TWINSPAN, except in the very special case of when a two-way ordered table (Fig. 12.1) is needed for a data set with a simple, one-dimensional underlying structure. Presence of a single strong underlying gradient can be manifest from prior knowledge or it can be detected with non-metric multidimensional scaling (Ch. 16).

TWINSPAN has inherited a number of faults from its parent method, correspondence analysis (CA). Like CA, TWINSPAN performs poorly with more than one important gradient. This limitation is particularly awkward when trying to interpret the ordered table. Unlike cluster analysis (Ch. 11), which has no inherent dimensionality in the solution, TWINSPAN cannot effectively represent complex data sets in its one-dimensional framework (van Groenewoud 1992, Belbin & McDonald 1993). Groups of sample units in typical community data set will often be better represented with cluster analysis than TWINSPAN.

Another problem is that the algorithm of TWINSPAN is very complex. The most complete published description of the method (Hill 1979b) lacks sufficient detail to reproduce the method. Many arbitrary but potentially important decisions in designing TWINSPAN, evident only in the computer code, have not been explained. By contrast, the transparency of other methods for cluster analysis is quite appealing.

Some of the complexity of TWINSPAN derives from the use of "pseudospecies." Pseudospecies were invented as a work-around to convert a nonquantitative method into a quantitative method. If we want a quantitative method, why not use a method that deals with quantitative data directly, rather than through the back door?

Despite these problems, TWINSPAN is still frequently used (e.g., FAUNMAP Working Group 1996). In most cases, no attempt is made to justify the use of TWINSPAN as opposed to other clustering methods.

Table 12.1. Pros and cons of TWINSPAN.

Pros	Cons
Conceptual appeal of two-way ordered table	Two-way table effectively displays only 1-D pattern
	Performs poorly with large heterogeneous data sets
	Underlying CA method implies chi-square distance
	"Pseudospecies" needed to make method semi-quantitative
	Algorithm complex and difficult to communicate

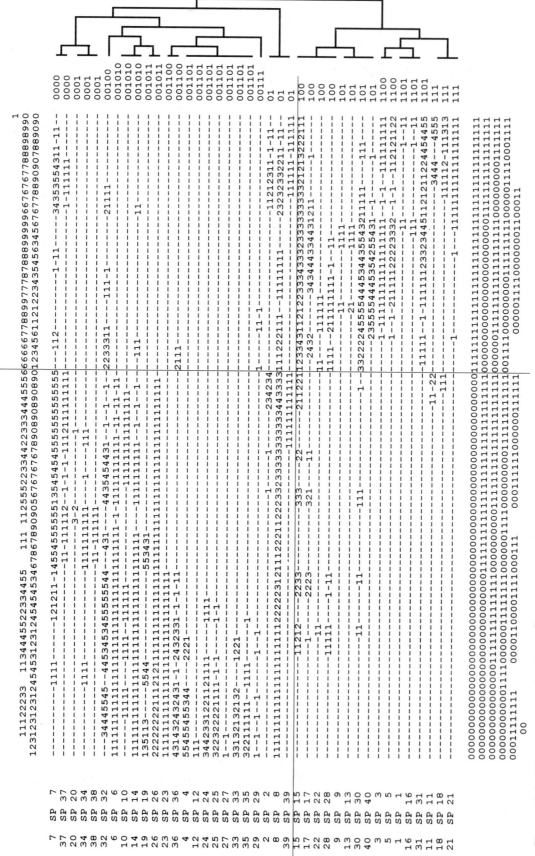

Figure 12.1. The two-way ordered table produced by TWINSPAN. Sample unit numbers are arrayed across the top by stacking the digits, species numbers and names are arrayed along the left side. The patterns of zeros and ones below and on the right side indicate successive divisions of sample units and species, respectively. These can be converted into dendrograms, as illustrated on the right side. The superimposed lines that cross the table indicate the first division of sample units and species.

How it works

Method for division

The algorithm for making divisions is quite complex. A single dichotomy requires three successive ordinations:

1. The primary ordination is by CA (Ch. 19). The first CA axis is divided near the middle. "Differential species" are defined by differential frequencies on the two sides of the axis.

2. This difference is then amplified by constructing a refined ordination of sites. This is essentially a weighted averaging ordination based on the differential species. Weights are assigned to species which are preferential to one side or the other of the division. The position of the division is then refined by summing the weights of the most preferential species.

3. The "indicator ordination" is then calculated based on the preferential species for the new dichotomy.

Indicator species are then defined by selecting a minimal set of species that will reproduce the ordination as closely as possible. The purpose of defining the indicator species is to aid interpretation of the dichotomies.

Species are classified using a similar strategy, taking advantage of the initial simultaneous ordination of both sites and species.

Pseudospecies

Strictly speaking, TWINSPAN operates on presence-absence (0/1) data. The quantitative nature of the data, however, can be preserved somewhat by the careful definition of "pseudospecies." TWINSPAN approximates quantitative abundance data by creating a variable number of "pseudospecies" representing abundance classes. The "pseudospecies cut levels" are used to define the ranges of the abundance classes.

Table 12.2. Examples of pseudospecies cut levels.

Cut levels	Purpose
0, 2, 5, 10, 20	for percentage data
0, 0.02, 0.05, 0.10, 0.20	for proportion data
0	presence-absence

Suppose you have data ranging from 0 to 100% cover. You need to define cover classes from the data. The default cut levels are 0, 2, 5, 10, and 20, and are reasonable for your data in this case (Table 12.2). These are cutoff points for pseudospecies 1, 2, 3, 4, and 5. A species (say its code name is Testsp) that had an abundance of 8% in a given sample unit would be interpreted as "present" for three pseudospecies: Testsp 1, Testsp 2, and Testsp 3 (the digit after the species name indicates the pseudospecies level). An input value of 1.5% would result in only one pseudospecies, Testsp 1, being "present" at these cut levels.

Selection of appropriate cut levels for pseudospecies is important for retaining the quantitative nature of your data. For example, if you had relativized your data such that all of the values fell between 0 and 1, then use of the default cut levels would result in your data being treated as presence/absence. This would occur because only the first pseudospecies level would be used. If you want to retain the quantitative information in your data, cut levels such as 0.0, 0.02, 0.05, 0.10, and 0.20 would be appropriate (Table 12.2).

What to report

- ☐ Justification for its use.
- ☐ Software version used (note computational problem described below under "Variations").
- ☐ Pseudospecies cut levels.
- ☐ Any departures from the default options.
- ☐ The two-way ordered table (optional).
- ☐ Selected indicator species (optional).

Examples

An example of the two-way ordered table is explained below. See McCune & Mefford (1999) for an extended example.

In the final two-way ordered table (Fig. 12.1), species names are on the left side of the table, while sample unit numbers are along the top (the digits stacked vertically). Sample units are numbered by sequence in the data. The patterns of zeros and ones on the right and bottom sides define the dendrogram of the classifications of species and sample units, respectively. For example, the first division separated the 40 sample units on the right (61, 62, 63, 64, etc.) from the remaining 60 sample units. Similarly, the first division separated the 16 lower species (SP 15, 17, 22, etc.) from the remaining 24. The interior of the table contains the abundance class of each species in each sample. Abundance classes are defined by pseudospecies cut levels. In this example, "SP 32" in sample unit 11 received a 3 in the table because it met the third cut level.

A dendrogram can be drawn by hand using the following rules. Consider the digits on the right side of the table. These digits specify the dendrogram of species (rows). The first column of zeros and ones shows the first division; rows receiving zeros were separated from rows receiving ones. Rows receiving the same digit are more similar to each other than rows in different groups. Moving to the right, the next column shows the next two divisions. Each of the preceding groups is again divided, as shown by the pattern of zeros and ones. You can continue these sequential divisions until you reach the rightmost columns. These groups were too small to divide further. You can sketch in a traditional dendrogram to the right of the TWINSPAN table. But it differs a little from the dendrogram you would get by cluster analysis, because in TWINSPAN, the finest branches of the tree usually have more than one item on them.

The examples in Figures 12.1 and 12.2 illustrate TWINSPAN's difficulty in portraying more than one dimension in the ordered table. This synthetic data set with known underlying structure has been used to illustrate distortions in various ordination methods (McCune 1994, 1997). Species respond noiselessly to two environmental gradients of approximately equal importance.

TWINSPAN must represent the data structure in a 1-D array of sample units and species (Fig. 12.1). How does it deal with a 2-D data structure (Fig. 12.2A)? Follow the 1-D sequence of sample units in the ordered table through the plane of the 2-D environment (Fig 12.2C). Although a good 1-D portrayal of the 2-D pattern is impossible, the more crisscrossed the lines, the more jumbled is the representation in the ordered table. Despite the difficulty of the 1-D representation, in this case TWINSPAN divided the data into four reasonable groups (Fig. 12.2D), comparable in quality to groups produced by agglomerative cluster analysis using Ward's method and relative Euclidean distance (Fig. 12.2B). In other cases in the literature, TWINSPAN has performed relatively poorly (Belbin & McDonald 1993).

Variations

Tausch et al. (1995) discovered a computational problem in TWINSPAN, resulting in an instability OF the results. They found that results from TWINSPAN using quantitative data depended on the order of sample units in the data set. Differences in the divisions between reorderings of the data were fewest in the first divisions but increased in later divisions. This problem was found for both quantitative and presence-absence data. The order-dependent instability diminished if rare species were removed. In contrast, Tausch et al. found the results of hierarchical agglomerative clustering to be independent of the order of sample units in the data.

Oksanen and Minchin (1997) showed that the cause of instability in TWINSPAN was the lax criterion for stability of the CA solutions. Because TWINSPAN in the PC-ORD package was based on the original Cornell Ecology Programs, it too contained these problems. The problems were fixed in PC-ORD, beginning with version 2.11. PC-ORD now applies Oksanen and Minchin's (1997) "super strict" criteria of tolerance = 0.0000001 and maximum number of iterations = 999, replacing the old criteria of tolerance = 0.003 and maximum iterations = 5.

A variant called COINSPAN is similar to TWINSPAN, but divides axes created by canonical correspondence analysis (Carleton et al. 1996).

Figure 12.2. Representation of a synthetic data set with a 2-D underlying environment by TWINSPAN. A.— A data set of 100 sample units arrayed on a 10 × 10 grid along two independent environmental gradients. Species have noiseless hump-shaped responses (not shown) to the gradients. B. — Four groups of sample units found by agglomerative cluster analysis, using Ward's method and relative Euclidean distance. C. — TWINSPAN forces the 2-D gradient into a single gradient of sample units in the ordered table (Fig. 12.1), as shown by the path of arrows through the sampling grid. Dotted lines indicate breaks between groups at the four-group level. D.— Division of the sample units into four groups by TWINSPAN.

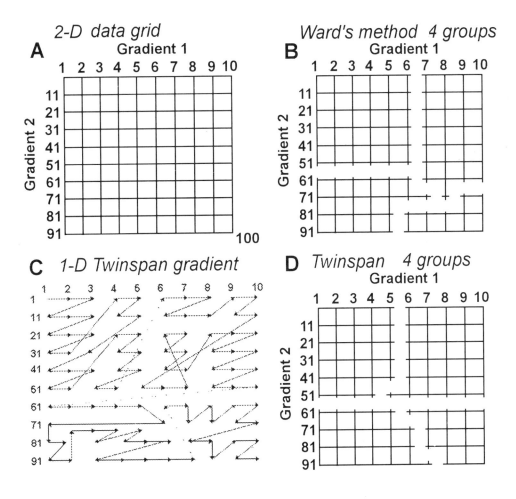

Part 4. Ordination

CHAPTER 13

Introduction to Ordination

Ordination in a literal sense is simply arranging items along a scale (axis) or multiple axes. There are many purposes for doing so, but normally the items are arranged as a way of graphically summarizing complex relationships, extracting one or a few dominant patterns from an infinite number of possible patterns.

The process of extracting those axes is called "ordination" because it results in a placement of objects along an axis or dimension (as in the "ordinate" in the Cartesian coordinate system). What makes this possible is that the variables are correlated (in the broad sense). Ordination thrives on the complex network of intercorrelations that can make multiple regression a nightmare.

Ordination is most often used in ecology to seek and describe pattern. Although we commonly think of ordination as a means of generating hypotheses about underlying mechanisms, ordination can also be used to test hypotheses (see Chapter 23). Ordination is used in one form or another in most sciences. For example, Arkin et al. (1997) used multidimensional scaling to elucidate complex chemical kinetics.

In community ecology, the most common use of ordination is to describe the strongest patterns in species composition. When the underlying factors are thought to vary continuously, this activity is often termed "gradient analysis." The underlying concept is that species abundance varies along environmental or historical gradients. Species composition changes as you go up the mountain and species composition changes with time. Whether these changes are continuous or in "fits and starts" is not important at this point. Gradient analysis is, in general, capable of detecting either kind of pattern.

One occasionally hears the criticism that users of ordination tend to manipulate the results to fit their preconceptions. A related criticism is that ordination reveals only the obvious patterns that we already knew. While both of those criticism surely apply to ordination (and any other tools of science) at times, more often ordination helps people to see their data more clearly. Ordination often helps to:

- select the most important factors from multiple factors imagined or hypothesized,
- separate strong, important patterns from weak ones, and
- reveal unforeseen patterns and suggest unforeseen processes.

Direct vs. indirect gradient analysis

With **direct gradient analysis**, items (most often sample units) are positioned (ordinated) according to measurements of environmental factors in those sample units. We use direct gradient analysis to learn how species are distributed along specific gradients of interest (e.g., Austin et al. 1984, 1990, 1994; Beals 1969; Carleton 1990; Minchin 1989; Oksanen et al. 1988; Oksanen 1997). In contrast, **indirect gradient analysis** positions sample units according to covariation and association among the species. Hybrid methods ordinate sample units in species space, but the ordination is constrained or determined by correlations with environmental variables. Two tools for this kind of "constrained ordination" are canonical correspondence analysis (Ch. 21) and canonical correlation.

Beals (1984) clearly explained the appeal of indirect gradient analysis (= sociological ordination) over direct gradient analysis (= environmental ordination):

> Species differences between two samples do reflect their environmental differences, but in a highly integrated fashion, which includes differences in biotic interactions and in historical events. The environmental differences are automatically scaled according to overall species response. Therefore the ordination with the clearest species patterns reflects the environmental space the way the biotic community interprets it.

Beals (1984) contrasts this with environmental ordination:

> The disadvantage of environmental ordination is that one must prejudge which are the important environmental factors to the vegetation or the fauna. An environmental ordination may omit

Introduction to Ordination

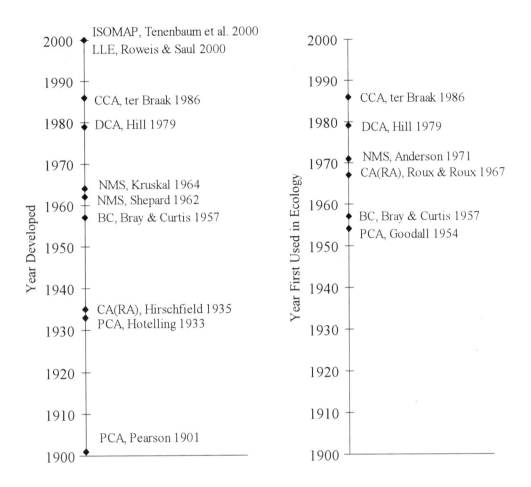

Figure 13.1. Timeline of development of some ordination methods and their application in ecology. BC = Bray-Curtis ordination, CA(RA) = correspondence analysis (reciprocal averaging), CCA = canonical correspondence analysis, DCA = detrended correspondence analysis, ISOMAP = isometric feature mapping, LLE = locally linear embedding, NMS = nonmetric multidimensional scaling, and PCA = principal components analysis.

important variables; it is often biased toward those factors most easily measured; measured variables may be scaled wrong; and biotic patterns imposed by competition, predation, and other interactions are ignored.

Direct gradient analysis can be univariate if species abundances or sample units are plotted against a single factor. Direct gradient analysis can be multivariate, when species or sample units are plotted with respect to two or more factors at once. An example of univariate direct gradient analysis would be plotting species abundance versus elevation, based on a transect up a mountain. Another example would be graphing species abundance against water pollution levels. Some of the most famous gradient analyses are Whittaker's bivariate summary diagrams of community variation in mountainous topography (Whittaker 1956, 1960). He plotted communities as zones on axes of elevation vs. topographic position.

Multivariate gradients are more difficult to visualize than univariate gradients. Environmental gradients that are combinations of measured environmental variables, rather than single environmental variables, can be constructed. While this may seem to increase the power of the analysis, we sacrifice the appeal of direct gradient analysis for simplicity and ease of communication.

Indirect gradient analysis is typically performed with an ordination method. Many of these have been used in community ecology. A timeline for the advent of some of the most popular techniques in ecology is in Figure 13.1.

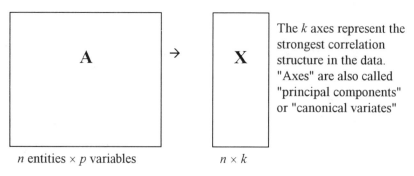

Figure 13.2. Reduction of a p-dimensional matrix to a k-dimensional ordination.

Ordination as data reduction

"Data reduction" means to summarize a data set containing many variables into a smaller number of composite (synthetic) variables. Ordination is a method of data reduction that results in continuous synthetic variables (axes). In contrast, classification reduces a multivariate data set into discrete classes.

We begin with a data set of n entities by p variables (dimensions), then try to effectively represent the information in the data set in a smaller number of dimensions, k (Fig. 13.2). Each of the new dimensions is a synthetic variable representing as much of the original information as possible. The only way this can be effective is if the variables covary. Information that is omitted in the reduced-dimensional space is the "residual" variation. This residual represents, one hopes, primarily noise and the influence of minor factors.

Gnanadesikan and Wilk (1969) summarized nicely the goals of data reduction:

> The issue in reduction of dimensionality in analyzing multiresponse situations is between attainment of simplicity for understanding, visualization and interpretation, on the one hand, and retention of sufficient detail for adequate representation, on the other hand.
>
> Reduction of dimensionality can lead to parsimony of description or of measurement or of both. It may also encourage consideration of meaningful physical relationships between the variables, as for example, summarizing (mass, volume)-data in terms of density = mass/volume.

After extracting the synthetic variables or axes, one usually attempts to relate them to other variables. In community ecology the most common procedure is to first summarize a matrix of sample units by species into a few axes representing the primary gradients in species composition. Those axes are then related to measured environmental variables.

Many other approaches are possible. A matrix of resource use by different species may be summarized into a few dominant patterns of resource use, which can then be related to characteristics of the species (longevity, size, mode of reproduction, etc.).

Ordination methods can also be used to provide an operational definition for a process, concept, or phenomenon that is difficult or impossible to measure directly. For example, an ordination of species compositions or symptomology along an air pollution gradient may be used to define biological impacts of air pollution. Positions on the ordination axes can then be used to describe changes in future air pollution impacts.

Ordination diagrams

Typically an ordination diagram is a 2-dimensional plot of one synthetic axis against another (Fig. 13.3). Ideally, the distance between points in the ordination space is proportional to the underlying distance measure. Do not look for regression-like patterns in the array of points. In most methods, the axes are uncorrelated *by definition*.

There are always some people in every audience who are confused by ordination diagrams. Alleviate this problem by telling your audience:

- What the points represent (sample units? species? quadrats? stand-level averages?)
- That the distances between points in the ordination are approximately proportional to the dissimilarities between the entities.

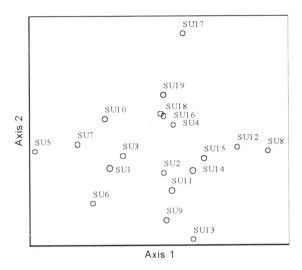

Figure 13.3. A two-dimensional ordination of sample units in species space. Distances between sample units approximate dissimilarity in species composition.

Ordination of sample units vs. species

In community ecology, we typically plot sample units in species space (normal analysis), species in sample space (transpose analysis), or both sample units and species in a combined space. Sample units are accurately represented as points because they define a single invariant combination of species and their abundances. Species, on the other hand, occupy a volume of sample unit space. By representing species as points in sample unit space, we are representing an average or typical position in sample space but ignoring the breadth of a species' distribution in sample space. Consider two species, one ubiquitous but the other occurring in just a few sample units. Despite the large difference in the distribution of the species in the sample, they might be represented as nearby points in the ordination diagram.

The situation is more accurately portrayed if sample units are represented as points and species are represented as overlays on the normal analysis. Species distributions can be clearly shown by scaling symbol sizes to reflect the presence or abundance of species (see example later in this chapter). The tradeoff is that such overlays are possible for only one species at a time. In contrast, by plotting species either as points or in a joint plot, many species can be shown at once.

How many axes?

The question of "how many axes?" can be rephrased as "how many discrete signals can be detected against a background of noise?" As the complexity of a data set grows, the number of discrete signals should increase, but the noise also increases. Occasionally, one encounters data sets with a single overriding pattern, but in these cases, the investigator is usually well aware of that pattern. More often one is faced with several possible underlying factors, but their relative importance is unknown. In these cases, two- or three-dimensional ordinations are most common, regardless of the ordination technique. Very few ecologists have dared to venture into the uncertain waters of four or more dimensions. The risk of fabricating interpretations of spurious or weak patterns increases and the results are much more difficult to display and discuss. Some guidance in answering "how many axes" is offered by considering the statistical significance of ordination axes.

Statistical significance of ordination axes

When viewing an ordination, it is common for the analyst to wonder if the displayed structure is stronger than expected by chance. In most cases, the structure is so strong that there is little question of this. But in questionable cases, we do have tools to help assess how many axes, if any, are worth interpreting. Some methods are just suggestive, rather than being bound to statistical theory (for example, rules-of-thumb for eigenvalue size in PCA; Ch. 14). Other methods estimate the probability of a type I error: rejecting the hypothesis of "no structure" when in fact there is no structure stronger than expected by chance alone.

The way we test significance depends in part on the analytical method, so this topic is addressed in this book under particular methods. Some common solutions for assessing accuracy, consistency, and randomness of ordination results are in Chapter 22.

Relating variables to ordination results

Relating variables to your ordinations is the principle means for interpreting them. You can relate variables from the matrix used to construct the ordination or any other variables measured for the same entities as are in the ordination. Two main ways to relate a variable to an ordination are overlays and correlations.

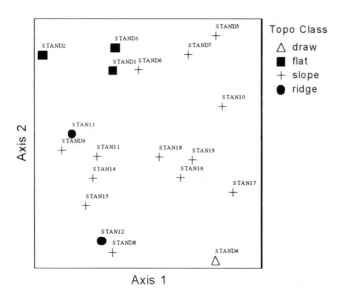

Fig. 13.4. Overlay with different symbols for different topographic classes of sample units (stands).

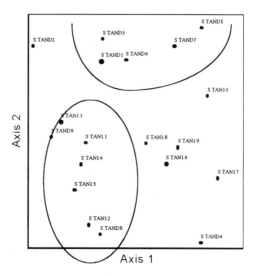

Fig. 13.5. Overlay with related points enclosed by loops or arcs.

Overlays

Overlays are a flexible way of seeing whether a variable is patterned on an ordination because you are not limited to detecting only linear relationships. For example, consider a species with a hump-shaped response to an environmental gradient expressed by Axis 1. The correlation coefficient between that species abundance and Axis 1 may be near zero, even though the species shows a clear unimodal response. This kind of pattern can easily be seen with an overlay, although it is obscured by a coefficient that expresses only a linear relationship.

Typically only one variable is overlaid at a time. Each point on the ordination is substituted by a symbol or value representing the size of the variable (for example, abundance of a species or the size of an environmental variable). Graphics overlays use variously sized symbols (for continuous variables) or various symbols or colors (for categorical variables).

Many kinds of overlays are possible, for example:

- different symbols for different classes of points (Fig. 13.4).
- related points enclosed with loops (Fig. 13.5).
- sizes of symbols proportional to the magnitude of the variable used as an overlay (Fig. 13.6).

Other special kinds of overlays are described below, including joint plots, biplots, grids, and successional vectors.

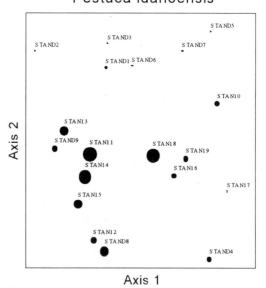

Figure 13.6. Overlay with sizes of symbols proportional to the magnitude of a variable, in this case abundance of the species *Festuca idahoensis*.

Correlations between axis scores and the other variables

The correlation coefficient provides a way of comparing positions of the sample units on the ordination axes with the abundance of species or other variables (often environmental variables). If a variable has any linear relationship with an ordination axis, it is expressed in the correlation coefficient. The squared values of the correlation coefficients express the proportion of variation in position on an ordination axis that is "explained" by the variable in question.

These correlations should primarily be used for descriptive purposes. It is both desirable and possible to resist the temptation to assign a *p*-value for the null hypothesis of no relationship between ordination scores and some other variable. Strictly speaking, the ordination scores are not independent of each other, violating the assumption of independence needed to believe a *p*-value. With even a modest sample size, however, this objection becomes rather moot, as the sample size overwhelms the partial dependence among points.

A more important consideration is that sample sizes are typically large enough that even a very small correlation coefficient is "statistically significant." The remedy is to set your own standards for how small an effect size, as indicated by the *r*-value or rank correlation coefficient, you are willing to interpret. In almost all cases, this threshold will be more conservative than one determined by the *p*-value. For these reasons, PC-ORD does not report *p*-values for correlations between variables and ordination axes. If they are desired, it is easy to look them up in a table of critical values (e.g., Rohlf & Sokal 1995).

Correlations of species with ordination axes present special problems. Species abundance can be shown as a scatterplot in relation to a gradient. In Figure 13.7, gradients in species composition are represented by two ordination axes. Data are based on Derr (1994) and are average lichen species cover on twigs in *Pinus contorta* bogs in southeast Alaska. Below and to the left of the ordination are two scatterplots showing abundance of a single species ("ALSA") against the ordination axis scores. Superimposed on each of these side scatterplots are two fitted lines. The envelope line was fit by PC-ORD as a smoothed line to include points falling within two standard deviations of a running mean of ALSA abundance along the gradient (PC-ORD flexibility parameter = 4). The straight lines are the least-squares regression lines. The correlation coefficients between

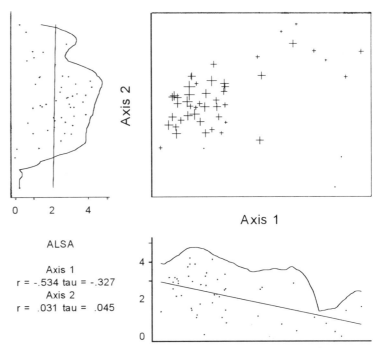

Figure 13.7. Scatterplot of abundance, measured as cover, of one species in relation to two ordination axes. Each point represents a plot. Symbol size is proportional to species abundance. Upper right: 2-D ordination of species composition. Lower right: scatterplot of abundance of "ALSA" (*Alectoria sarmentosa*) against score on Axis 1 (horizontal axis). Upper left: scatterplot of abundance of ALSA against Axis 2. Superimposed on the two abundance scatterplots are the least squares regression lines and a smoothed envelope. See text for interpretation of *r* and tau.

ALSA and ordination axes 1 and 2 are: -0.54 and 0.031, respectively. ALSA is a nearly ubiquitous species, but is more abundant in plots falling on the left side of axis 1 (Fig. 13.7). ALSA shows a unimodal pattern on axis 2, having low abundance in plots that scored very high or very low on that axis.

Is the correlation coefficient a reasonable descriptor of the relationship between species abundance and position of sample units on a gradient? Sometimes the answer is "yes," sometimes "no." If you use these correlation coefficients, it is important to evaluate this question for your own data. Remember that correlation coefficients express the *linear* (Pearson's *r*) and *rank* (Kendall's tau) relationships between the ordination scores and individual variables.

A table of correlation coefficients with ordination axes can be quite helpful in providing a quick interpretation of the ordination. Frequently, however, they can be misleading. Three common problems with interpreting an ordination just on basis of the correlation coefficients are:

1. Even a single outlier can have undue effect on the coefficients, often resulting in a very strong correlation that has little to do with the bulk of the data.
2. The coefficients will misrepresent nonlinear relationships (for example, if a species has a humped, unimodal distribution along an ordination axis).
3. These correlation coefficients are unsatisfactory with binary data.

Multiple regression and structural equations

Correlations between environmental variables and a community ordination can readily be extended to more than one environmental variable at once. Multiple regression can be used to express ordination scores as a function of a linear combination of two or more environmental variables. For example, Lee and Sampson (2000) regressed ordination scores representing a gradient in fish communities against a set of environmental variables and time. These independent variables can be selected using any of the usual methods of multiple regression.

When sufficient information is available to allow interpretation of causal relationships among multiple environmental variables, structural equation modeling allows you to represent those relationships (Ch. 30). You can develop and evaluate multivariate hypotheses about the relationships between community patterns and environmental conditions.

Joint plots

An increasingly popular method of showing the relationship between a set of variables (usually environmental variables) and ordination scores is a diagram of radiating lines (Fig. 13.8), often called a "joint plot." The angle and length of a line indicates the direction and strength of the relationship. These lines are often referred to as "vectors," but do not confuse them with successional vectors. Joint plots can be constructed for any ordination method.

Method 1 (PC-ORD).— The lines radiate from the centroid of the ordination scores. For a given variable, the line forms the hypotenuse (h) of a right triangle with the two other sides being r^2 values between the variable and the two axes (Fig. 13.9). The angle of the line relative to the horizontal axis is:

$$\alpha = \arccos(r_x \bullet |r_x|)$$

where r_x is the correlation of the variable with the horizontal axis. The length of the vector, h, is proportional to a function of the r^2 values with the two axes:

$$h \propto \sqrt{(r_x^2)^2 + (r_y^2)^2}$$

Although the r^2 values determine the relative scaling of the vectors, the absolute scaling is arbitrary.

This method was designed so that the components of a vector on an individual axis are proportional to the r^2 on that axis. This means that you can "see" the r^2 of the variable with each ordination axis, if you imagine a perpendicular from each axis to the tip of the vector. This design principle has the side effect that the lengths of vectors and angles between them can subtly change as the ordination is rotated (Ch. 15). The situation is further complicated by the fact that, as one rotates a point cloud, originally uncorrelated axes will

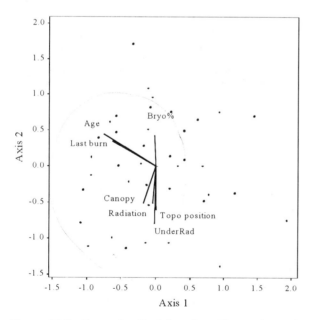

Figure 13.8. Example of a joint plot. The angles and lengths of the radiating lines indicate the direction and strength of relationships of the variables with the ordination scores.

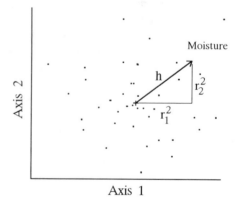

Figure 13.9. Method of calculating a joint plot vector in PC-ORD.

become correlated to some extent with each other, unless the point cloud is spherical. Regardless of how the ordination is rotated, however, the multiple r^2 for a variable with respect to the axes is constant.

Method 2 (Vector Fitting). — This method was first described by Dargie (1984). Kantvilas and Minchin (1989) introduced the term "vector fitting" for this procedure. Assume we have a k-dimensional ordination to which we are fitting a vector for variable V. Regress V against the ordination scores **X**:

$$V = b_0 + b_1 X_1 + b_2 X_2 + \ldots b_k X_k$$

The tip of the vector (Fig. 13.10) is then placed proportional to ($b_1, b_2, \ldots b_k$). Note that if $k > 2$ the apparent length of the vector is not necessarily proportional to the overall r^2 value for the regression. Using s as a scaling constant, the resulting geometry is shown in Figure 13.11. In a 3-D ordination each vector has unit length. In a 2-D projection, variables that are correlated with the third axis appear shorter than the unit circle. The variable "Disturbance" (Fig. 13.11), which projects out of the plane formed by axes 1 and 2, is only weakly related to those axes, but is strongly related to axis 3. The tip of a vector for a variable with no relationship to the third dimension (e.g., the variable "Moisture" in Fig. 13.11) will fall on the unit circle.

The statistical significance of these vectors can be tested with a randomization procedure (Faith and Norris 1989). The values of the variable V are randomly permuted and the linear correlation or r^2 is recalculated and recorded. A p-value is obtained by repeating this procedure many times and calculating the proportion of trials that yielded a test statistic as large or larger than the observed value.

Weak variables are usually not included in a joint plot. Strong variables are selected by applying either a significance criterion or by omitting variables with an r^2 smaller than a cutoff value.

With community data one normally shows sample units and environmental variables in a joint plot. But you can also show sample units and species simultaneously by using your species data for the radiating vectors. In PC-ORD you can do this by loading your species data in *both* your main and second matrices, then constructing a joint plot.

Biplots

A biplot is a way of representing sample units and variables in a single diagram. Exactly how this is done and how biplots are defined vary among authors. In its

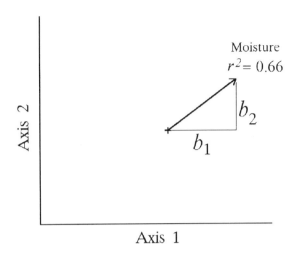

Figure 13.10. Finding the angle of the vector using vector fitting as described by Dargie (1984); $r^2 = 0.66$ between moisture and three ordination axes.

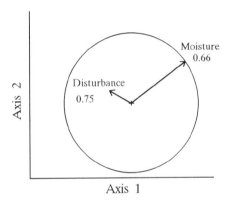

Figure 13.11. Scaling the length of the vector using vector fitting. Vectors were scaled to unit length in three dimensions then projected onto two dimensions. The numbers are r^2 values in three dimensions.

broad sense, biplots include joint plots. Gabriel (1971) coined the word "biplot" and applied the concept to Principal Components Analysis. Categorical variables can also be represented on a biplot, a point representing a particular level (or category) of a variable (Benzécri 1973, Gower & Hand 1996).

Gower and Hand (1996) represented continuous variables in ordination diagrams as additional axes. These are usually straight and intersect the centroid of the point cloud at various angles to the original coordinate system for the sample units. The biplot axes may also be curved or irregular in some applications.

Biplots were popularized in ecology through their adaptation to canonical correspondence analysis (ter Braak 1986). These could be considered "triplots" because they plot the sample units as points, the variables in the species matrix as an additional set of points, and the environmental variables as vectors of various lengths radiating from the centroid. As redefined by ter Braak (1986) and Jongman et al. (1995), biplots can be considered a special case of a joint plot where the positions of the tips of the vectors are calculated from the linear equations that define the ordination. By their definition, biplots follow a certain set of rules (Jongman et al. 1987). Both biplots and joint plots, however, are ordination diagrams that show two kinds of entities (often sample units and environmental variables). Ter Braak's method of constructing biplots is explained further under canonical correspondence analysis (Ch. 21).

Grid overlays

Grid overlays are specifically designed for artificial data sets based on a rectangular grid (e.g., Austin & Noy-Meir 1971, Fasham 1977, Kenkel & Orloci 1986, McCune 1994, Minchin 1987b). We typically use a grid for artificial two-dimensional data sets to see if ordination methods can reproduce the known underlying grid structure (Fig. 13.12).

The example in Figure 13.12 shows a grid overlay on an ordination of sample units in species space. Species responses were simulated in a 2-D environmental space, then sampled with a regular 10 × 10 grid. In this case, the ordination recovered the basic structure of the underlying grid but with some areas of distortion.

Grid overlays cannot be applied to normal field data. Rather, their use depends on a synthetic data set based on an underlying 2-D environment.

Successional vector overlay

Successional vectors can be used in any case where a number of sample units have been followed through time. For example, say you are recording species abundance in a number of plots and revisiting those plots every year. By connecting the data points in temporal sequence, you can see the "trajectory" of a sample unit in species space (Fig. 13.13; examples listed in McCune 1992; Philippi et al. 1998). Other applications are before/after data or paired samples.

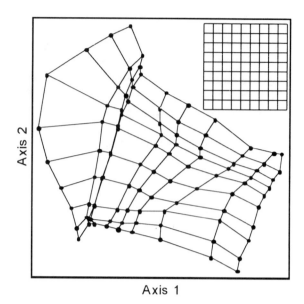

Figure 13.12. Grid overlay on an artificial data set. The grid represents an 2-D environmental gradient underlying noiseless species responses. Inset: the ideal result is a regular grid.

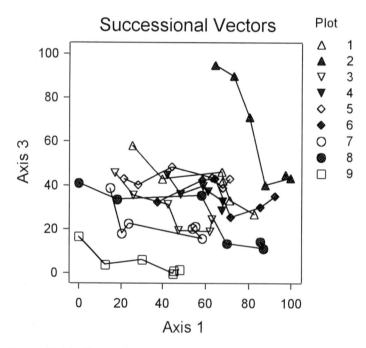

Figure 13.13. Successional vectors. Lines connect sequential measurements on permanent sample units.

Criteria for evaluating quality of ordination results

No single infallible criterion exists for evaluating the quality of an ordination. In practice, we use a combination of criteria, some statistical and some relying on prior knowledge. Gnanadesikan and Wilk (1969) wrote an insightful description of this process:

> The main business of statistical data analysis is the description and communication of the informational content of a body of data. This process requires summarization — perhaps in terms of a simple graph or perhaps using a complex mathematical model. The process also requires exposure — the presentation of the data in such a way as to facilitate the assessment of the summarization and to aid detection of the unanticipated. A simple example of this two-pronged process is that of fitting a straight line to y versus x data and plotting the residuals in a variety of ways — against x, against y, perhaps against values of some extraneous variable such as time, and probability plotting the residuals.
>
> Such processes of data analysis are directed toward description, communication, development of insight, and toward detecting unanticipated as well as anticipated peculiarities. Even for the problem of assessing the adequacy of a very specific model, the essential features are to summarize the data in terms of the model and to expose what is left in some cogent way as an indicator of adequacy.
>
> Though statistical and mathematical models, theories, and ideas are of key importance in this process, one cannot expect that formal theories of inference can subsume the direction of this flexible and iterative process.

Proportion of variance represented

The concept that ordination represents some fraction of the total variance in the original data, similar to a coefficient of determination (r^2) for regression, is appealing. Some methods, such as principal components analysis, have a built-in, logical way of evaluating this. Other methods, such as nonmetric multidimensional scaling and detrended correspondence analysis, are not based on partitioning variance and therefore do not yield a measure of variance explained.

One assessment, however, can be applied to all ordination methods: comparing how well the distances between points in the ordination diagram represent distances in the original, unreduced space. This method provides an after-the-fact evaluation of the quality of the data reduction, along with an assessment of how the variance explained is distributed among the primary axes. A further advantage of an after-the-fact assessment of variance explained is that any rotations or rescaling of the ordination axes are taken into account.

The distance-based evaluation calculates the coefficient of determination (r^2) between distances in the ordination space and distances in the original space (in PC-ORD, select *Graph Ordination | Statistics | Percent of variation in distance matrix*). One half of the distance matrix (not including the diagonal) is used in this calculation. If there are N entities, then there are $N*(N-1)/2$ pairs of values used to calculate r^2. The r^2 is commonly expressed as a percentage by multiplying by 100. Incremental and cumulative values can be calculated for each additional ordination axis (Table 13.1). These are commonly called the percentage of variance "explained," but "represented" is a more appropriate word. Data reduction by ordination is not so much an explanation as a summarization.

Table 13.1. Coefficients of determination for the correlations between ordination distances and distances in the original 40-dimensional space. For this example, the data matrix was 100 sample units × 40 species. The number of entity pairs used in the correlation = 4950.

	r^2	
Axis	Increment	Cumulative
1	0.371	0.371
2	0.199	0.570
3	0.146	0.716

Distances in the ordination space are *always* measured with Euclidean distance (because we perceive the graphic in a Euclidean way), whereas you have a choice of distance measure for the original space. You should select the distance measure most closely allied with the ordination method you are using (Table 13.2). For nonmetric multidimensional scaling, choose the distance measure that you used to construct the ordination. For PCA, the results from this method will not match the percentage of variance calculated from the eigenvalues. PCA is typically based on correlations among the columns rather than distances among the rows (Ch. 14).

Table 13.2. Suggested methods for calculating proportion of variance represented by ordination axes. The after-the-fact method is the r^2 between Euclidean distances in the ordination space and distances in the original space. The distances in the original space can be calculated with various distance measures, but this choice of distance measure is important. "Built-in" means the method is a part of the ordination method itself.

Ordination method	Suggested method for variance represented
Bray-Curtis	Built-in with residual distance over original distance
Canonical Correspondence Analysis	After-the-fact with relative Euclidean or chi-square distance
Correspondence Analysis	After-the-fact with chi-square distance
Detrended Correspondence Analysis	After-the-fact with relative Euclidean distance
Nonmetric Multidimensional Scaling	After-the-fact with same distance measure as used in analysis
Principal Components Analysis	Built-in with ratio of eigenvalues to total variance
Weighted Averaging	After-the-fact with relative Euclidean distance

The standard method for calculating total variance in CA and CCA was criticized by Økland (1999) as not accurately representing the effectiveness of these ordination methods at recovering ecological gradients. As an alternative, we suggest the after-the-fact evaluation of how well distances in the ordination space match the chi-square or relative Euclidean distances in the main matrix. Although both CA and CCA have an intrinsic basis in the chi-square distance measure, empirically we have found that the correlation between points in a CCA ordination space and the points in the original space is often higher when the latter is measured as relative Euclidean distance rather than as chi-square distance.

These r^2 values are by no means the whole story in evaluating the effectiveness of an ordination. For example, if a couple of outliers dominate the relationships in the data set, then a very high r^2 may result. This will give the appearance of an effective ordination, even though the ordination has been ineffective in describing the similarity relationships within the main cloud of points. Other cases of trivial structure in the data can cause similarly high r^2 values.

In some cases, the increment to r^2 as axes are added can be negative. This can happen for several reasons. The ordination method may be optimizing some criterion other than maximizing the percentage of variance extracted on each axis; for example, minimizing stress in NMS (nonmetric multidimensional scaling; Ch. 16). The results also depend on how the axes are scaled relative to each other. For example, rescaling axes to equal lengths ("Min to Max") usually weakens the relationship between ordination distance and original distance and may result in negative increments to r^2.

What is a "good" r^2? We offer no fixed answer, because our expectations and hopes should vary with the problem and the number of axes fitted to the data. Ordinations of data sets with very few variables should represent more variance than those with a large number of variables. As an extreme example, a principal components analysis of a data set with three variables will represent 100% of the variance in the first three axes. Very heterogeneous data sets may have more underlying complexity, requiring more dimensions to portray. On the other hand, the species variation in data sets from a narrow range of conditions may be mainly noise, defying a simple summarization and resulting in a low percentage of variance represented. Having qualified our answer, we now feel more comfortable stating that for data sets with 20 or more species, investigators are often pleased to explain more than 50% of the variation with two axes, though perfectly useful and interpretable ordinations commonly have 30-50% of the variation represented in two axes. Note, however, that unlike the r^2 from regression, we should not hold up 100% of the variation as a standard. Because one of the primary purposes of ordination is filtering a signal from noise (Gauch 1982b), one could say that an ordination representing 100% of the variation has accomplished nothing.

Interpretability

The criterion of interpretability may seem offensively circular and subjective, and certainly it has elements of those characteristics, but interpretability is actually a useful, nontrivial criterion. Its utility hinges on verifying whether the structure represented by the ordination is interpretable as an answer to the questions posed. For example, assume we seek to describe the relationship between an experimental treatment and community structure based on count data. If the range in counts is high, such that there are one or a few extreme values, then the ordination will express those extreme values and nothing else. The ordination is interpretable as a reaction to outliers, but it is not interpretable as a reaction of community structure to the treatments. A log transformation or elimination of the outliers may allow the bulk of the data to express itself in the ordination, greatly improving interpretability.

What if results appear to be interpretable but are actually spurious? Our best protection against this problem is to test the hypothesis that the observed structure is no stronger than expected by chance. We can do this with randomization tests (Ch. 22).

Strength of relationships to a second matrix

Evaluating the quality of an ordination requires a criterion or standard external to the ordination itself. The only situation in which the underlying pattern or process is completely known is when an artificial data set is generated mathematically (e.g., Austin & Noy-Meir 1972, Gauch & Whittaker 1972, McCune 1997, Minchin 1987). Nevertheless, with real data we often know something in advance about the relationships between communities and measured environmental factors. These factors can be useful in evaluating the quality of an ordination.

Ordination is usually a combination of confirming known relationships and seeking new ones. Often, the results describe the relative importance of several factors. All of the factors may be known to influence the communities, but the factors may be of uncertain importance in a particular context.

For example, consider a gradual gradient from a lake through a wetland to an upland. Community structure obviously varies in relation to vertical distance from the lake level. We can easily see this with our eyes, and the pattern is consistent with what we know about the importance of water to organisms. We probably do not, however, know the details of how species are related to each other, the rate of species change along the gradient, interactions between height above the lake and other factors (such as organic matter content of the soil), and how these interacting factors are related to the communities. In this case, ordination can provide an objective description of the community and its relationship to environment, including multiple, interrelated environmental factors, as well as insights into the relationships among species.

If an ordination of community structure does not reflect what seems obvious about the system, then we should reexamine both our assumptions and our methodology. Some of the main methodological culprits are outliers, failure to relativize to remove a grossly uneven emphasis among variables, and use of linear models to portray nonlinear relationships.

Do not push this logic too far by throwing every combination of options at a problem until you reach the best match to your preconceptions. One of the fun parts of science is discovering that you were wrong. A hard-won fact that pushes aside a preconception has a pleasant feeling of progress. Ordination can provide useful revelations if used parsimoniously and carefully.

Comparison with null model

One might suppose that comparing the strength of a structure in an ordination against a null model, ideally based on a randomization of the data, would provide a foolproof method for evaluating the quality of an ordination. In fact, it does not. One of the most common problems comes from outliers. For example, alter a community matrix such that two elements have much higher values (say 1000 times larger) than any other elements. Ordination of this matrix will describe the odd nature of the two rows that were altered. Almost any randomization of this matrix will have a structure of similar strength to the original data. No matter how it is shuffled, the basic nature of the matrix remains the same: two huge values contrast with a plethora of small, essentially equal values. In this case there may be nothing wrong with the representation in the ordination, but a randomization test would indicate a structure no stronger than expected by chance.

An ordination showing a relatively weak structure (based on randomization tests) can actually be more informative than one with a strong structure. Consider two versions of the same data set with 100 species, one relativized by species totals and one not relativized. If three intercorrelated species are much more abundant than the 97 others, randomization test statistics and the correlation between ordination distances and original distances will both be strong. Yet the structure represented by the ordination is trivial, since it is a 2-dimensional representation of an essentially 2- or 3-dimensional system. In this case, we are likely to learn little from the ordination. The relativized data set, however, poses a much more difficult problem of whether 100 species given equal influence can be represented in a few dimensions. Ordination of this data set is more likely to reveal something new and interesting, something that relates to the whole community. But a randomization test will almost certainly indicate a structure weaker than an ordination from the original, nonrelativized data set.

CHAPTER 14

Principal Components Analysis

"Where the relation between two species is studied, not for its own sake, but as part of an ordination procedure, the use of a linear model for essentially non-linear relations may lead to considerable obscurity." (Goodall 1953a, p. 127)

Background

Principal components analysis (PCA) is the first and most basic eigenvector method of ordination. It was first proposed by Pearson (1901) and Hotelling (1933). PCA was first used in ecology by Goodall (1954) under the name of "factor analysis." The object of the analysis is to represent a data set containing many variables with a smaller number of composite variables (components or axes). The most interesting and strongest covariation among variables emerges in the first few axes (components), hence "principal components."

The object of PCA is to reduce a data set with n cases (objects) and p variables (attributes) to a smaller number of synthetic variables that represent most of the information in the original data set. Thus, we reduce a data set of n objects in a p-dimensional space to n objects in a reduced k-dimensional space, where k is typically much smaller than p.

The following excerpt from Pearson's original (1901) paper nicely describes the differences between regression and a basic form of principal components analysis:

> In many physical statistical, and biological investigations it is desirable to represent a system of points in plane, three, or higher dimensional space by the "best-fitting" straight line or plane. Analytically this consists of taking
>
> $$y = a_o + a_1 x \quad \text{or} \quad z = a_o + a_1 x + b_1 y$$
> $$\text{or} \quad z = a_o + a_1 x_1 + a_2 x_2 + a_3 x_3 + \ldots + a_n x_n$$
>
> where $y, x, z, x_1, x_2 \ldots x_n$ are variables, and determining the "best" values for the constants $a_0, a_1, a_2, \ldots a_n, b_1$ in relation to the observed corresponding values of the variables. In nearly all cases dealt with in the text-books of least squares, the variables on the right side of our equations are treated as the independent, those on the left as the dependent variables. The result of this treatment is that we get one straight line or plane if we treat some one variable as independent, and a quite different one if we treat another variable as the independent variable [Fig. 14.1]. There is no paradox about this; it is, in fact, an easily understood and most important feature of the theory of a system of correlated variables. The most probable value of y for a given value of x, say, is not given by the same relation as the most probable value of x for a given value of y... The "best-fitting" lines and planes for the cases of z up to n variables for a correlated system are given in my memoir on regression (Phil. Trans.187:301 et seq.). They depend upon a determination of the means, standard-deviations, and correlation-coefficients of the system. In such cases the values of the independent variables are supposed to be accurately known, and the probable value of the dependent variable is ascertained.
>
> In many cases of physics and biology, however, the "independent" variable is subject to just as much deviation or error as the "dependent" variable... In the case we are about to deal with, we suppose the observed variables — all subject to error — to be plotted in plane, three-dimensioned or higher space, and we endeavour to take a line (or plane) which will be the "best fit" to such a system of points.

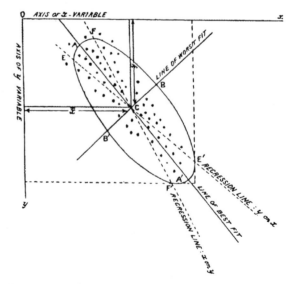

Figure 14.1. Comparison of the line of best fit (first principal component) with regression lines. Point C is the centroid (from Pearson 1901).

Of course the term "best fit" is really arbitrary; but a good fit will clearly be obtained if we make the sum of the squares of the perpendiculars from the system of points upon the line or plane a minimum [Fig. 14.2].

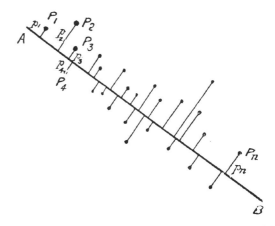

Figure 14.2. The basis for evaluating "best fit" (Pearson 1901). In contrast, least-squares best fit is evaluated from vertical differences between points and the regression line.

Pearson then demonstrated that "the best-fitting straight line for a system of points in a space of any order goes through the centroid of the system." He showed that the solution depends only on the knowledge of the means, standard deviations, and correlations of the variables. Although Pearson did not use the term "principal components," his best fitting line represents the first principal component, while the best-fitting plane is defined by the first and second principal components.

When to use it

PCA is an ideal technique for data with approximately linear relationships among variables. In some cases, transformation of a data set with nonlinear relationships can markedly improve its representation by PCA. Ecological community data are, however, rarely amenable to PCA. This was demonstrated as early as Swan (1970). The typical multivariate "dust bunny" distribution (Ch. 5) found in species data forces very awkward fits to a linear model. This poor fit has been documented repeatedly in the literature and is summarized by Beals (1984). See Gnanadesikan and Wilk (1969) for extended examples of other cases where PCA cannot recover nonlinear structures.

PCA of heterogeneous ecological community data has a telltale pattern: a horseshoe or arch of points in the ordination space (Fig. 19.2). The most extreme ends of gradients are pulled toward each other in the ordination space. This occurs because PCA interprets shared zeros as an indication of a positive relationship. The structure of the ordination is based solely on the matrix of correlations among variables.

With relatively homogenous community data sets, PCA can be an effective tool for ordinating community data. But even moderately heterogeneous data sets will be severely distorted by PCA. It is remarkable that even recent papers in our most reputable journals sometimes still use PCA on community data with no justification for the linear model (e.g., Bellwood & Hughes 2001).

Sample size

Tabachnik and Fidell (1989) proposed a rule of thumb of at least five cases (sample units) for each observed variable. Taken literally, the rule says we should analyze only elongate matrices. This should not apply when the data are not considered a sample (e.g., a matrix of species by traits). Also, if we are equally interested in relationships among rows as among columns, why shouldn't the matrix be square?

Tabachnik and Fidell (1996, p. 640) changed their recommendation to using hundreds of SUs: "If there are strong and reliable correlations and a few, distinct factors, a smaller sample size is adequate." Pillar (1999b) determined through bootstrapped ordinations that, for a given strength of correlation structure, increasing the number of variables improved the consistency of the principal components.

The bottom line is that if you are going to map a set of objects based on the correlation structure among descriptors of those objects, then you want good estimates of that correlation structure. The weaker the correlation structure, the larger the sample size needed to describe it reliably.

Normality

If statistical inference is made, normality is important; PCA requires multivariate normality. If your goals are descriptive, this assumption is relaxed.

Tabachnik and Fidell (1989) stated that "Normality of all linear combinations of variables is not testable." In their 1996 edition, they stated that the various schemes for evaluating multivariate normality tend to be too conservative — finding significant deviations from normality even when the violations are not severe.

Normality of single variables can be assessed by **skewness** (asymmetry) & **kurtosis** (peakiness). If the distribution is normal, then skewness = 0 and kurtosis = 0. With species data, skewness is almost always positive, meaning the right tail is too long. If kurtosis > 0, then the curve is more peaky than the normal distribution. In general, skewness and kurtosis are best used descriptively. You can test an hypothesis of skewness = 0, but with a large sample size, skewness will always differ significantly from zero. Our personal rule of thumb when using parametric multivariate analyses, such as PCA, is to try to have |skew| < 1. Kurtosis will usually be much improved if skewness is remedied.

Linearity

The quality of ordination by PCA is completely dependent on how well relationships among the variables can be represented by straight lines. In other words, if the cross-products matrix represents relationships well, then the principal components may represent the data structure well. If relationships among variables cannot be well represented by linear functions, then PCA results will be poor.

Note that linearity is related to multivariate normality, in that a multivariate normal data set will be reasonably approximated by linear relationships. Appropriateness of fitting linear relationships among your variables can be checked with bivariate scatterplots, selecting various pairs of variables.

Outliers

Because outliers can influence correlation coefficients greatly, they can influence the results of PCA. In Figure 14.3, the correlation coefficient is 0.92 with the outlier included, but -0.96 with the outlier excluded. An outlier can so strongly influence the correlation matrix that the first axis of PCA will be devoted to describing the separation of the outlier from the main cloud of points.

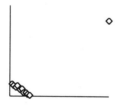

Figure 14.3. Outliers can strongly influence correlation coefficients.

How it works

PCA basics

1. The object of PCA is to express covariation in many variables in a smaller number of composite variables.

2. PCA accomplishes this by seeking the strongest linear correlation structure among variables.

3. The basic results are variance explained by each axis (eigenvalues) and linear equations (eigenvectors) that combine the original variables. Eigenvalues are found as roots of a polynomial. Eigenvectors are solutions to simultaneous equations.

4. Each object is projected into an ordination space using the eigenvector and the data matrix.

Step by step

1. From a data matrix **A** containing n sample units by p variables, calculate a **cross-products matrix**:

$$\mathbf{S} = \mathbf{B'B} \quad \text{where } b_{ij} = a_{ij} - \bar{a}_j$$

The dimensions of **S** are p rows × p columns. This is known as the "R-route," referring to the first step of calculating a cross-products matrix among the columns.

This form of **S**, the cross-products matrix, is a **variance/covariance matrix**. It contains covariances in the triangles and variances in the diagonal. More commonly, the cross-products matrix is a **correlation matrix**, containing Pearson (the usual) correlation coefficients in the triangles and ones in the diagonal. The equation for a correlation matrix is the same as above except that each difference is divided by the standard deviation, s_j. With a covariance matrix, variables contribute to the results in proportion to their variance. With a correlation matrix, each variable contributes equally to the total variance. The total variance is the sum of the diagonal. With correlation coefficients in the cross-products matrix, the total variance is simply the number of elements in the diagonal, which is equal to the number of variables, p.

2. Find the **eigenvalues**. Eigenvalues are roots of a polynomial. Each eigenvalue represents a portion of the original total variance, that portion corresponding to a particular principal component (or axis). The largest eigenvalue goes with the axis explaining the most variance. The size of each eigenvalue tells us how much variance is represented by each axis. Each eigenvalue (= latent root) is a lambda (λ) that solves:

$$|\mathbf{S} - \lambda \mathbf{I}| = 0$$

I is the identity matrix (App. 1). This is the "characteristic equation."

There are p values of λ that will satisfy this equation. Each λ is a root of a polynomial of degree p. The coefficients in the polynomial are derived by expanding the determinant (see example calculations below).

$$a\lambda^p + b\lambda^{p-1} + c\lambda^{p-2} + \ldots + constant = 0$$

[For a broader view of the context for finding roots of a polynomial, see Pan (1998). Exact roots cannot be found algebraically – instead we find approximations using iterative computational algorithms.]

3. Then find the **eigenvectors**, **Y**. For every eigenvalue λ_i there is a vector **y** of length p, known as the eigenvector. Each eigenvector contains the coefficients of the linear equation for a given component (or axis). Collectively, these vectors form a $p \times p$ matrix, **Y**. To find the eigenvectors, we solve p equations with p unknowns:

$$[\mathbf{S} - \lambda \mathbf{I}]\mathbf{y} = \mathbf{0}$$

Again we use an iterative computational algorithm to solve the problem.

Each element of the eigenvectors represents the contribution of a given variable to a particular component. Each eigenvector has zero correlation with the other eigenvectors (i.e., the eigenvectors are orthogonal).

4. Then find the **scores** for each case (or object) on each axis: Scores are the original data matrix postmultiplied by the matrix of eigenvectors:

$$\underset{n \times p}{\mathbf{X}} = \underset{n \times p}{\mathbf{A}} \quad \underset{p \times p}{\mathbf{Y}}$$

Y is the matrix of eigenvectors

A is the original data matrix

X is the matrix of scores on each axis (component)

We interpret the first k principal components of **X** (typically $k < 4$). Note that each component or axis can be represented as a linear combination of the original variables, and that each eigenvector contains the coefficients for the equation for one axis. For eigenvector 1 and entity i ...

$$\text{Score1 } x_i = y_1 a_{i1} + y_2 a_{i2} + \ldots + y_p a_{ip}$$

This yields a linear equation for *each* dimension. Note this is the same form as other linear models, such as those derived from multiple linear regression.

5. The $p \times k$ matrix of correlations between each variable and each component is often called the principal components **loading matrix**. These correlations can be derived by rescaling the eigenvectors or they can be calculated as correlation coefficients between each variable and scores on the components.

Geometric analog

1. Start with a cloud of n points in a p-dimensional space.

2. Center the axes in the point cloud (origin at centroid)

3. Rotate axes to maximize the variance along the axes. As the angle of rotation (θ) changes, the variance (s^2) changes.

$$\text{Variance along axis } v = s^2_v = \underset{1 \times p}{\mathbf{y'}} \underset{p \times p}{\mathbf{S}} \underset{p \times 1}{\mathbf{y}}$$

At the maximum variance, all partial derivatives will be zero (no slope in all dimensions). This is another way of saying that we find the angle of rotation θ such that:

$$\frac{\delta \ln s^2}{\delta \cos \theta} = 0$$

for each component (the lower case delta (δ) indicates a partial derivative).

As we rotate the axis through the point cloud, the variance changes (Fig. 14.4). Visualize the line rotating through the cloud of points, using the centroid as the pivot. When the line is aligned with the long axis of the cloud of points (Fig. 14.5), the variance along that axis is maximized.

Now do the same for subsequent axes. In the example with only two variables (Fig. 14.5) there is only one more axis. In practice, all of the axes are solved simultaneously.

Example calculations, PCA

Start with a 2 × 2 correlation matrix, **S**, that we calculated from a data matrix of $n \times p$ items where $p = 2$ in this case:

$$\mathbf{S} = \begin{bmatrix} 1.0 & 0.4 \\ 0.4 & 1.0 \end{bmatrix}$$

We need to solve for the eigenvalues, λ, by solving the characteristic equation:

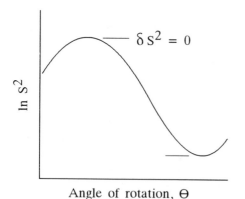

Figure 14.4. PCA rotates the point cloud to maximize the variance along the axes.

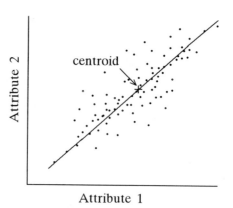

Figure 14.5. Variance along an axis is maximized when the axis skewers the longest dimension of the point cloud. The axes are formed by the variables (attributes of the objects).

$$|S - \lambda I| = 0$$

Substituting our correlation matrix, S:

$$\left| \begin{bmatrix} 1 & 0.4 \\ 0.4 & 1 \end{bmatrix} - \begin{bmatrix} \lambda & 0 \\ 0 & \lambda \end{bmatrix} \right| = 0$$

$$\begin{vmatrix} 1-\lambda & 0.4 \\ 0.4 & 1-\lambda \end{vmatrix} = 0$$

Now expand the determinant:

$$\lambda^2 - 2\lambda + 0.84 = 0$$

$$(1-\lambda)(1-\lambda) - 0.4(0.4) = 0$$

Note that we now have a polynomial of degree 2. We then solve this polynomial for the values of λ that will satisfy this equation. You probably remember that to find the roots of a quadratic equation (polynomial of degree 2) we can apply the following equation:

$$\lambda = \frac{-b \pm \sqrt{b^2 - 4ac}}{2a}$$

Since $a = 1$, $b = -2$, and $c = 0.84$, then

$$\lambda = \frac{2 \pm \sqrt{4 - 3.36}}{2}$$

Solving for the two roots gives us $\lambda_1 = 1.4$ and $\lambda_2 = 0.6$.

The percentage of variance explained by the two corresponding principal components is found by the following equation (remember that the total variance is simply p).

$$\% \text{ of variance}_i = \frac{100 \cdot \lambda_i}{p}$$

Now find the eigenvectors, Y. For each λ there is a y:

$$[S - \lambda I] y = 0$$

For the first root we substitute $\lambda_1 = 1.4$, giving:

$$\begin{bmatrix} 1 - 1.4 & 0.4 \\ 0.4 & 1 - 1.4 \end{bmatrix} \begin{bmatrix} y_1 \\ y_2 \end{bmatrix} = 0$$

Multiplying this out gives two equations with two unknowns:

$$-0.4 y_1 + 0.4 y_2 = 0$$
$$0.4 y_1 + -0.4 y_2 = 0$$

In this case, it is easy to solve these simultaneous equations ($y_1 = 1$ and $y_2 = 1$). But consider the general case that you will have p equations each with p unknowns! Setting up and solving the equations for the second eigenvector yields $y_1 = 1$ and $y_2 = -1$.

We now normalize the eigenvectors, rescaling them so that the sum of squares = 1 for each eigenvector. In other words, the eigenvectors are scaled to unit length. The scaling factor k for each eigenvector i is

$$k_i = \frac{1}{\sqrt{\sum_{q=1}^{p} y_{qi}^2}}$$

So for the first eigenvector,

$$k_1 = \frac{1}{\sqrt{1^2 + 1^2}} = 0.707$$

Then multiply this scaling factor by all of the items in the eigenvector:

$$\begin{bmatrix} 1 \\ 1 \end{bmatrix} 0.707 = \begin{bmatrix} 0.707 \\ 0.707 \end{bmatrix}$$

The same procedure is repeated for the second eigenvector, then the eigenvectors are multiplied by the original data matrix to yield the scores (**X**) for each of the entities on each of the axes (**X** = **A Y**).

What to report

- ☐ Form of the cross-products matrix (correlation coefficients or variance-covariance?).
- ☐ If applied to community data, justify using a linear model for relationships among species.
- ☐ How many axes were interpreted, proportion of variance explained by those axes.
- ☐ Tests of significance of principal components (not necessary for many applications).
- ☐ Principal eigenvectors if model is to be used to assign scores to new data points.
- ☐ If and how solution was rotated.
- ☐ Standard interpretive aids (overlays, correlations of variables with axes).

Significance of principal components

The eigenvalues form a descending series with the first component (axis) explaining the most variance, the second component the next most, etc. Even with random data, the eigenvalues will form a descending series. If there is a significant correlation structure on a given component, then the eigenvalue will be higher than expected by chance alone. You can evaluate this by comparing the observed series of eigenvalues with a series generated by a null model: either by repeatedly shuffling the data or by some other null model.

Many different parametric methods have been used to test for significant correlation structure in eigenanalysis of multivariate data. One common example is Bartlett's test of sphericity (this and other techniques reviewed by Grossman et al. 1991 and Jackson 1993). This test was found by Jackson (1993) to be erratic.

Jackson found that one of the easiest and most meaningful guides for determining the number of significant axes was comparison of the eigenvalues to those produced by a broken-stick model. He called this approach "heuristic" rather than statistical. If the variance is divided randomly among the components, it should follow a broken-stick distribution. If the broken-stick eigenvalue for axis k (b_k) is less than the actual eigenvalue for that axis, then that axis contains more information than expected by chance and should be considered for interpretation. The broken stick eigenvalue is easily calculated (Jackson 1993, Frontier 1976):

$$b_k = \sum_{j=k}^{p} \frac{1}{j}$$

where p is the number of columns and j indexes axes k through p.

Another common criterion is that the eigenvalue for an axis should be greater than one (the Kaiser-Guttman criterion). Most people feel that this criterion is not conservative enough. Few people are willing to interpret all of the principal components with eigenvalues above 1. The Kaiser-Guttman criterion overestimates the number of nontrivial axes (Jackson 1993).

A Monte Carlo test for determining significant eigenvalues, "parallel analysis," was proposed by Franklin et al. (1995). They analyzed an equal-sized matrix of normally distributed random variables, then compared the eigenvalues with those from the actual data matrix. If the variables are normally distributed, this approach might be reasonable. But for most data sets it can be much improved by randomizations that preserve the distributions of the original variables.

Pillar (1999b) proposed a method of parallel bootstrapping of a real and randomized data sets (see Chapter 22). In contrast to Franklin et al. (1995), Pillar's method randomly reassigns elements within columns (variables), thus preserving the distributions of individual variables but destroying the correlation structure. Pillar's method reliably detected the underlying dimensionality of artificial data sets, as long as the correlation structure was sufficiently strong. Parallel bootstrapping is, therefore, a promising method for determining how many dimensions are stronger than expected by chance.

Examples

Unfortunately, most uses of PCA in the literature of community ecology exemplify inappropriate use. Nevertheless, the consequences of using PCA versus other ordination techniques are not severe when beta diversity is low. Be sure to consider this before dismissing a paper that used PCA on community data.

Some special circumstances invite the use of PCA. In a study of temporal change in species abundance, the transformation by differences between consecutive samples (Ch. 9) will yield both positive and negative numbers and often a mean near zero. This prevents use of proportionate distance measures (e.g., Sørensen distance) and methods with an explicit or implicit chi-square distance (CA, DCA, CCA). Allen and Koonce (1973) applied PCA to differences between successive samples of phytoplankton.

Experimental studies on relatively abundant organisms can also be amenable to PCA. With a matrix filled with nonzero values, a linear approximation to the relationships among species abundances (or log abundances) may be reasonable.

Some molecular markers used as "species" have fairly linear relationships among them, making PCA an appropriate method for community analysis. For example, Ritchie et al. (2000) examined frequency distributions of abundance of markers based on both length-heterogeneity PCR and fatty acid methyl esters. They concluded that the variables were amenable to PCA. Frequency distributions and bivariate scatterplots should be examined and reported for each new kind of marker.

Similarly, analyses of sole-source carbon use profiles of microbial communities, determined by BIOLOG plates, should report the properties of the data and relationships among variables, at least until the nature of the data is better understood. Typically, methods based on linear models, such as PCA, have been applied to BIOLOG data without reporting how well the data meet distributional assumptions (e.g., Zak et al., Myers et al. 2001). In at least one case however, (K. Strauss, unpublished, 2002), the data appear to be amenable to PCA.

Community analysts will find PCA a useful tool for relating other kinds of data to community data. You can use PCA to distill suites of intercorrelated variables into one or a few variables (e.g., Hamann et al. 1998). For example, many plant demographic or morphological traits are related to each other because they respond similarly to plant vigor. A PCA-derived synthetic variable describing plant vigor can be used to show how performance of a target species varies with community composition. Similarly, PCA is useful for combining climatic or other environmental variables (see example for soil variables in Chapter 30).

Variations

Other eigenanalysis techniques are covered in later sections, but one commonly used technique in the social sciences, **factor analysis**, is not. The term "factor analysis" has various uses. Some authors consider PCA to be a special case of factor analysis. Others consider these terms mutually exclusive.

The difference between PCA and factor analysis begins with the content of the diagonal of the cross-products matrix. PCA has ones in the diagonal of **S**. Each variable contributes unit variance to the pool of variance being redistributed, so you analyze as much variance as there are variables. All of the variance is redistributed to the components. In factor analysis, on the other hand, there are smaller values in the diagonal. These values are the proportion of variance shared by a given variable with other variables. These values are called "communalities," and they must range between zero and one. You can think of factor analysis as weighting variables by the strength of their relationship with other variables in the analysis, whereas PCA gives them equal weight.

A key question in deciding on PCA vs. factor analysis is: Are you interested in analyzing the total variance or only the intercorrelated portion of the variance? If you choose the latter, then factor analysis is the appropriate technique.

Principal coordinates (PCoA; also known as Gower's ordination or principal axes analysis; see Gower 1966, Kenkel & Bradfield 1981, Williamson 1978) can be thought of as a more general form of PCA that allows the use of a wide array of distance measures, including non-Euclidean measures. Since the use of correlation in PCA is the largest part of the poor performance of PCA with community data, PCoA can give a marked improvement over PCA.

In PCoA, distances (**D**) are converted to elements of the crossproducts matrix (**S**) by the formula

$$s_{ij} = -d_{ij}^2 / 2$$

In effect, this changes the distances to similarities, "pretending" that the distances are Euclidean. The distances can either be among the attributes of the matrix, followed by an R-route eigenanalysis, or the distances can be among the sample units, followed by a Q-route eigenanalysis.

At least one of the resulting eigenvalues is always negative, since the sum of the eigenvalues is always zero (remember that the diagonal of the crossproducts matrix contains zeros because when $d = 0$ then $s = 0$). If a non-Euclidean distance measure is used, additional eigenvalues will be negative. Negative eigenvalues and their associated eigenvectors are ignored.

One variant, "**reduced-rank regression**," combines PCA with multiple regression. It is essentially multivariate regression; multiple response (dependent) variables are regressed at the same time on multiple independent variables. It is called "reduced-rank" because a restriction is imposed on the rank of the matrix of regression coefficients, yielding a model in which the response variables react to the regressors through a small number of "latent variables." The results can be expressed in an ordination graph (ter Braak and Looman 1994). Parameter estimates can be obtained from a canonical correlation analysis or from a PCA of the fitted values for the response variables.

If PCA is constrained by a multiple regression on a second matrix (usually environmental variables in community ecology), it is called **redundancy analysis** (RDA). See ter Braak (1994) and Van Wijngaarden et al. (1995) for introductions to RDA. Despite its increasing popularity, the underlying linear model for RDA is generally inappropriate for community analysis.

Noncentered PCA skips the centering step (subtraction of the means) while generating the cross-products matrix. While this method has had sporadic proponents over the decades (e.g., Noy-Meir 1973a, Feoli 1977, Carleton 1980), there is no convincing evidence that community analysts gain anything by noncentering. The first axis from a noncentered PCA typically reflects the difference between the centroid and the origin. Because this differences contributes to the total variance being explained in noncentered PCA, the variance represented by the first axis is often higher than with centered PCA. This higher variance explained should not be considered a better result, because it is simply an artifact of noncentering, rather than representing any biologically stronger result. See further discussion by Orloci (1976b).

CHAPTER 15

Rotating Ordination Axes

Ordination solutions are often rotated to improve ease of interpretation and to make explanation of the results more straightforward. The general procedure is to obtain an initial ordination, select a small number of axes (or components), then rotate the axes through the point cloud to optimize some criterion other than that on which the ordination was originally based.

Orthogonal (rigid) rotation

Orthogonal (or rigid) rotations are simply a way of viewing an existing constellation in k dimensions from another angle.

Rigid rotations **change** the following:
- correlations of variables with ordination axes and, in the case of eigenanalysis-based ordinations, loadings of variables onto the eigenvectors and
- variance represented by an individual axis.

Rigid rotations **do not change** the following:
- geometry of the constellation of points in the ordination space and
- cumulative variance represented by the axes being rotated.

Eigenanalysis-based rotations

Rotations based on eigenanalysis are also known as "analytical rotations." The $n \times k$ configuration is rigidly rotated by eigenanalysis of a $k \times k$ matrix. The contents of this matrix depend on the exact type of rotation. See examples in Ayyad and El-Ghareeb (1982), Enright (1982), Ivimey-Cook and Proctor (1967), Noy-Meir (1971), Oksanen (1985), Tabachnik and Fidell (1996), and Wiegleb (1980).

Varimax rotation (Fig. 15.1) is the most commonly used rotation (though much more so in the social sciences than in ecology). The $k \times k$ matrix contains the variance of the loadings squared; this quantity is maximized by the eigenanalysis. The method is described in detail by Mather (1976). It tends to finds groups of species and sample units that correspond, and is thus a reasonable method for detecting clusters of sample units. Noy-Meir (1971, 1973b) applied this to community data, calling it "nodal component analysis." After varimax rotation, axes tend to shoot through clusters of sample units and variables. High correlations of the individual variables tend to be higher and low correlations tend to be lower.

Quartimax rotation maximizes the fourth power of the loadings. It seeks to isolate variables on individual axes.

Rotation to maximize correspondence with respect to another matrix can also be done. This can be done with canonical correlation between the axis scores and a second matrix. When the main matrix contains species data and the second matrix contains environmental data, the method attempts to maximize the correlation of points in the reduced species space with the positions of points in environmental space. For example, Kenkel and Burchill (1990) rotated axes from nonmetric multidimensional scaling by canonical correlation with an environmental matrix.

Procrustes rotation. Procrustes rotation has been used in social sciences to compare two ordination spaces (Gower 1975, Harries & Macfie 1976, Schönemann 1966, Ten Berge 1977). (By the way, Procrustes was the legendary robber in ancient Greece who would cut off peoples feet or stretch the people to make them the right length for his bed. The multivariate applica-tion is much more reasonable.) Procrustes rotation is typically a rigid, orthogonal rotation that best matches two ordinations. In addition, the point clouds can be dilated or contracted to maximize the match. You may have observed the directional instability that can occur with slight variants in ordination — sometimes a gradient may be expressed with from wet-to-dry and sometimes from dry-to-wet. Procrustes rotation eliminates these simple reversals as well as differences in the angle of the gradients through the ordination space.

Other rotations

Rigid rotations not based on eigenanalysis are also known as graphical or visual rotation. Note that these methods, like the eigenanalysis methods, change the correlations of individual variables with the ordination axes and the distribution of variance explained on each axis. These statistics should, therefore, be recalculated after a rotation, to update them to the new configuration.

Rotating Ordinations

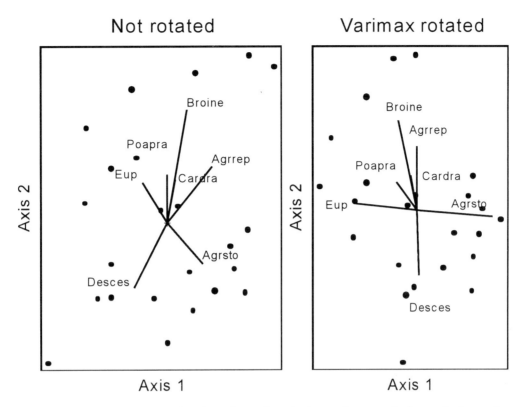

Figure 15.1. Comparison of ordination of sample units in species space before and after varimax rotation. Dots represent sample units. The vectors show correlations of species with the ordination space. Note the improved alignment of the vectors with the ordination axes in the rotated ordination. The vectors are calculated with the PC-ORD method (Ch. 13).

When comparing different ordinations of the same data set, it is often desirable to rotate one of them to improve the correspondence with the other. This can be done analytically with Procrustes rotation (above), or it can readily be done visually. See the example in Figure 15.2.

One can load a single variable onto a single axis or rotate to find the maximum correspondence of an environmental variable with points in the ordination space (Dargie 1984). When the analyst desires maximum correspondence with a single variable (usually a variable from a second matrix that is external to the ordination), this can be done by rotating axes such that the selected variable has a correlation coefficient of zero with all axes but one. Rotating to maximize loading of a single environmental variable is logical when there is a single variable of primary interest. It is also useful when one wishes to remove the variation associated with a given environmental variable. For example, assume we are interested in community variation that is independent of stand age (succession). One could rotate the configuration to load stand age completely on axis 1 and then examine factors related to axes 2 and 3.

In PC-ORD, rotation is accomplished through the "rotate" option in the graphing routine. The ordination can be rotated to align a selected variable with the horizontal axis.

One can also choose an angle of rotation by eye or trial and error. By examining a joint plot, showing correlational vectors radiating from the centroid, one might see that the major vectors can be approximately aligned with the axes by rotating the axes. Simply choose a trial angle and inspect the results.

Oblique rotation

Oblique rotation allows individual axes to be rotated independently. This results in axes that are correlated to some degree with each other. This is a heavy-handed manipulation that alters the constellation of points in the ordination space. Oblique rotation has been advocated as improving the cluster-seeking properties as compared with orthogonal rotations (Carleton 1980). Examples of its application in ecology are Carleton (1980) and Oksanen (1985).

You can visualize the effects of an oblique rotation by considering a two-dimensional ordination. One axis can be rotated while the other is fixed, such that the axes are no longer orthogonal. When the points are replotted on new perpendicular axes, the original space will have had an angular compression in some parts, with an angular expansion in other parts. Although oblique rotations distort two-dimensional graphs, they may be useful for structural models of intercorrelated ecological factors (see Chapter 31).

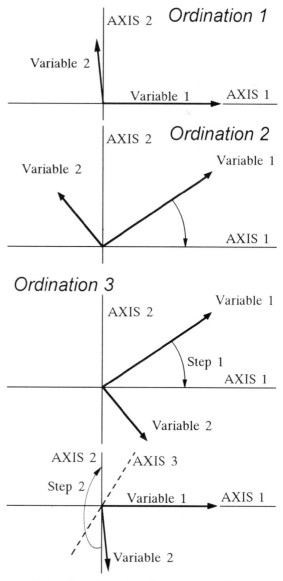

Figure 15.2. Rotation of ordinations to align patterns from separate ordinations. In Ordination 1, the point cloud has been rotated to maximize the loading of Variable 1 onto Axis 1. In Ordination 2, the same dominant trends were found but at an angle to those found in Ordination 1. Therefore, Ordination 2 can be rotated through an angle (indicated by arrow) so that it too aligns Variable 1 with Axis 1. In Ordination 3, the configuration is not only at an angle to Ordination 1, but it is also a mirror image of it. To align Ordination 3 with Ordination 1, Axes 1 and 2 are first rotated to align Axis 1 with Variable 1 (Step 1). In Step 2, the ordination is rotated by 180° in the plane of Axis 2 and 3, such that both Variables 1 and 2 are now similar in orientation to Ordination 1.

CHAPTER 16

Nonmetric Multidimensional Scaling

Background

Ordination provides views into a high-dimensional space by seeking and displaying the strongest structure. Methods such as PCA and CCA see only the portion of the configuration that fits a limited perspective, as specified by the particular underlying model. Nonmetric multidimensional scaling, on the other hand, can "see" a much wider range of structures. The future of ordination is in exploiting computational power with iterative optimization methods such as nonmetric multidimensional scaling.

Nonmetric multidimensional scaling (= nonmetric MDS, NMDS, NMS) differs *fundamentally* in design and interpretation from other ordination techniques (Kruskal and Wish 1978, Clarke 1993). Shepard (1962a, b) proposed the method and Kruskal (1964a, b) refined it, making it workable and generally effective. Kruskal and Carroll (1969), Kruskal and Wish (1978), and Mather (1976) contain good descriptions of the method and examples of its application to a variety of problems.

NMS has performed very well with simulated gradients, even when beta diversity is high (Fasham 1977, Kenkel & Orloci 1986, Minchin 1987a, McCune 1994) or gradient strengths unequal (Fasham 1977). The technique is increasingly used in community ecology and is currently one of the most defensible techniques during peer review.

We agree with Clarke (1993) that

> ...non-metric MDS is often the method of choice for graphical representation of community relationships... principally because of the flexibility and generality bestowed by: (1) its dependence only on a biologically meaningful view of the data, that is, choice of standardization, transformation and similarity coefficient appropriate to the hypotheses under investigation; (2) its distance-preserving properties, that is, preservation of the rank order of among-sample dissimilarities in the rank order of distances.

When to use it

Nonmetric multidimensional scaling is an ordination method well suited to data that are nonnormal or are on arbitrary, discontinuous, or otherwise questionable scales. For this reason, NMS should probably be used in ecology more often than it is. This method can be used both as an ordination technique and as a method for assessing the dimensionality of a data set.

See Clarke (1993) for an excellent summary of the advantages and uses of NMS. If you are unfamiliar with NMS, be sure to read the precautions discussed below.

The main **advantages** of NMS are:
- It avoids the assumption of linear relationships among variables.
- Its use of ranked distances tends to linearize the relationship between distances measured in species space and distances in environmental space. This relieves the "zero-truncation problem," a problem which plagues all ordinations of heterogeneous community data sets.
- It allows the use of any distance measure or relativization.

The historical **disadvantages** of NMS were (1) failing to find the best solution (minimum stress) because of intervening local minima and (2) slow computation with large data sets. Advances in computing power have, however, virtually eliminated these disadvantages.

Nonmetric multidimensional scaling is the most generally effective ordination method for ecological community data and should be the method of choice, unless a specific analytical goal demands another method.

How it works

NMS is an iterative search for the best positions of n entities on k dimensions (axes) that minimizes the stress of the k-dimensional configuration. The calculations are based on an $n \times n$ distance matrix calculated from the $n \times p$-dimensional data matrix, where n is the number of rows (entities) and p is the number of columns (attributes) in the data matrix. "Stress" is a measure of departure from monotonicity in the relationship between the dissimilarity (distance) in the original *p*-dimensional space and distance in the reduced *k*-dimensional ordination space. Details are given below.

Definitions

X = coordinates of *n* sample units or entities in a *k*-dimensional space. The element x_{il} is the coordinate of sample unit *i* on dimension (axis) *l*.

Δ = matrix of dissimilarity coefficients from the original data. Any distance measure can be used. We recommend using the quantitative version of the Sørensen coefficient for most ecological community data.

δ_{ij} = elements of Δ.

D = matrix of interpoint distances in the *k*-space. This matrix of $n \times n$ Euclidean distances is calculated from **X**.

d_{ij} = elements of **D**.

iteration = in this context, one small step of adjusting the positions of the entities in the ordination space.

run = in this context, the process of minimizing stress from a starting configuration, achieved through some number of iterations.

Basic procedure

1. Calculate dissimilarity matrix Δ.

2. Assign sample units to a starting configuration in the *k*-space (define an initial **X**). Usually the starting locations (scores on axes) are assigned with a random number generator. Alternatively, the starting configuration can be scores from another ordination.

3. Normalize **X** by subtracting the axis means for each axis *l* and dividing by the overall standard deviation of scores:

$$\text{normalized } x_{il} = \frac{x_{il} - \overline{x}_l}{\sqrt{\sum_{l=1}^{k} \sum_{i=1}^{n} (x_{il} - \overline{x}_l)^2 / (n \bullet k)}}$$

4. Calculate **D**, containing Euclidean distances between sample units in *k*-space.

5. Rank elements of Δ in ascending order.

6. Put the elements of **D** in the same order as Δ.

7. Calculate **D̂** (containing elements $\hat{d}_{ij}$, these being the result of replacing elements of **D** which do not satisfy the monotonicity constraint). The method of calculating the replacement values is shown graphically in Figure 16.1. The point *d* disrupts the otherwise monotonic pattern. (A monotonically increasing series has successive values that increase or stay the same, but never decrease.) That point is moved horizontally (see arrow) until monotonicity is achieved, at point $\hat{d}$. The amount of movement required to achieve monotonicity is measured by the difference between *d* and $\hat{d}$ (Fig. 16.1). The sum of these squared differences is the basis for evaluating stress (the opposite of "fit"). An example with real data points but without the monotonic line is in Figure 16.2. Stress can be visualized as the departure from monotonicity in the plot of distance in the original *n*-dimensional space (dissimilarity) vs. distance in the ordination space (Fig. 16.2). The closer the points lie to a monotonic line, the better the fit and the lower the stress.

8. Calculate raw stress, S^*:

$$S^* = \sum_{i=1}^{n-1} \sum_{j=i+1}^{n} (d_{ij} - \hat{d}_{ij})^2$$

Note that S^* measures the departure from monotonicity. If $S^* = 0$, the relationship is perfectly monotonic.

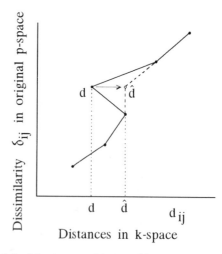

Figure 16.1. Moving a point to achieve monotonicity in a plot of distance in the original *p*-dimensional space against distances in the *k*-dimensional ordination space.

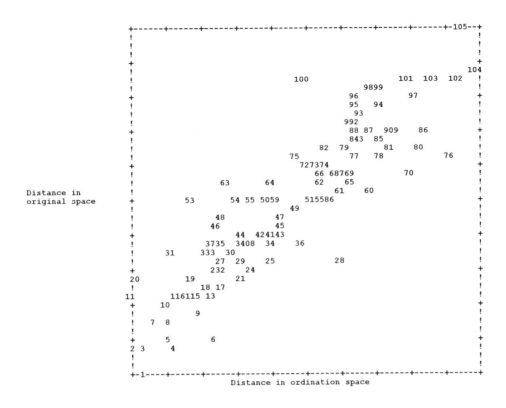

Figure 16.2. Plot of distance in ordination space (d_{ij}, horizontal axis) vs. dissimilarity in original p-dimensional space (δ_{ij}, vertical axis), analogous to Fig. 16.1. Points are labeled with the ranked distance (dissimilarity) in the original space.

9. Because raw stress would be altered if the configuration of points were magnified or reduced, it is desirable to standardize ("normalize") stress. This can be done in several ways, but PC-ORD follows Mather (1976) in using Kruskal's "stress formula one:"

$$S = S^* / \sum_{i=1}^{n-1} \sum_{j=i+1}^{n} d_{ij}^2$$

Finally, PC-ORD reports stress as S_R, the square root of the scaled stress, analogous to a standard deviation, then multiplied by 100 to rescale the result from zero to 100:

$$S_R = 100\sqrt{S}$$

This rescaling follows Mather (1976, p. 409). In some versions of NMS the standardized stress is expressed on a scale from 0-1.

An alternative scaling recommended by Kruskal and Carroll (1969) as "stress formula two" is:

$$S = S^* / \sum_{i=1}^{n-1} \sum_{j=i+1}^{n} (d_{ij} - \bar{d})^2$$

Kruskal and Carroll stated that stress formula two typically yields a larger value than stress formula one, perhaps twice as large in many cases. The two formulas yield very similar configurations.

10. Now we try to minimize S by changing the configuration of the sample units in the k-space. This is done with the steepest-descent minimization algorithm. Other algorithms are available that accomplish the same thing. For the method of steepest descent, we first calculate the "negative gradient of stress" for each point i (the direction of movement of each point i in the lth dimension of the k-space in which stress will decrease fastest). The gradient vector (a vector of length k containing the movement of each point h in each dimension l) is calculated by:

$$g_{hl} = S \sum_{i=1}^{n} \sum_{j=1}^{n} \left(\Delta^{hi} - \Delta^{hj}\right) \left[\frac{d_{ij} - \hat{d}_{ij}}{S^*} - \frac{d_{ij}}{\sum_{i,j} d_{ij}^2}\right] \frac{\left(x_{il} - x_{jl}\right)}{d_{ij}}$$

The letter i indexes one of the n points, j indexes another of the n points, l indexes a particular dimension, and h indexes a third point, the point of interest which is being moved on dimension l. So g_{hl} indicates a shift for point h along dimension l. The Δ^{hi} and Δ^{hj} are Kroekner deltas which have the value 1 if i and h or h and j indices are equal, otherwise they have the value 0. If the strict lower triangles of the distance matrices have been used then the Kroekner deltas can be omitted.

An alternative to the method of steepest descent is "simulated annealing." The method is slower than steepest descent but more effectively finds the global minimum stress. See Press et al. (1986, p. 326) for intuitive and mathematical descriptions of simulated annealing.

11. The amount of movement in the direction of the negative gradient is set by the step length, α, which is normally about 0.2 initially. The step size is recalculated after each step such that the step size gets smaller as reductions in stress become smaller. The manner in which the step size is adjusted is complex and need not concern most users. See Mather (1976) for more details.

12. Iterate (go to step 3) until either (a) a set maximum number of iterations is reached or (b) a criterion of stability is met. The ideal maximum number of iterations depends on the speed of your computer, the number of entities being ordinated, your patience, and whether or not you provide a starting configuration. More iterations are better, unless a stable configuration has been found, in which case additional iterations are pointless.

In versions 3 and lower of PC-ORD, the number of iterations was fixed by the user. In version 4, the maximum number of iterations is used only if a predefined criterion of stability is not met. Instability is calculated as the standard deviation in stress over the preceding 10 iterations. If the instability is less than the cutoff value set by the user, iterations are stopped and the solution is considered final. Instabilities of 0.005 are usually achieved rapidly, while instabilities of 0.00001 are usually reached slowly or not at all. Note that an instability of zero can never be reached unless the final stress is zero. Zero instability is not achieved because NMS always nudges the configuration in the direction of the gradient vector. So even when a local minimum is found, NMS is exploring for better solutions by making small adjustments in the configuration. The degree of stability that can be attained is therefore set by the algorithm that determines the magnitude of the gradient vector.

Using an instability criterion rather than a fixed number of iterations has several beneficial effects:

- The computer doesn't waste time calculating many additional iterations that are not improving the solution. The time saved can be used more profitably to explore starting configurations and perform more precise Monte Carlo tests.
- Tracking instability allows the program and the analyst to readily detect and avoid unstable solutions, where stress oscillates over a small number of iterations.
- The analyst can tune the thoroughness of the search for good solutions to the goals of the analysis, the available time, and the speed of the computer. The method allows relatively rapid assessments with large data sets or slow, exhaustive searches if time allows.

At the stopping point, the stress should be acceptably small. One way to assess this is by calculating the relative magnitude of the gradient vector **g**:

$$\textit{magnitude of } \mathbf{g} = \sqrt{\sum_{h,l} g_{hl}^2 / n}$$

Mather suggested that $mag(\mathbf{g})$ should reach a value of 2-5% of its initial value for an arbitrary configuration. This is often not achievable.

Landscape analogy for NMS

Imagine that you are placed at a random location in a p-dimensional landscape (your "starting configuration"). Your task is to find the lowest spot (with the minimum stress) in the landscape as efficiently as possible. Because you can see only those parts of the landscape near you, you don't have a perfect knowledge of where the lowest point is located. You proceed by using your hyperspace clinometer, a device for measuring slope, and you choose to go in the direction of steepest descent (= the gradient vector). After a prescribed distance (= step length), you stop and reassess the direction of steepest descent. The steeper the slope, the larger the step you take. You repeat this a maximum number of times (= number of iterations). Usually you will find yourself in the bottom of a valley well before you reach the maximum number of steps, so that the curve of elevation vs. iteration flattens out. The worriers among you will be concerned that there are other lower places in the landscape, perhaps in a different drainage (i.e., that you are in a local minimum). Worriers should be dropped by helicopter into several random locations where they can repeat the process until they are convinced that they have found the lowest point in the landscape (= the global minimum).

Local minima

Although NMS has performed very well in comparative studies of ordination methods, it has not been as widely adopted as deserved. The main reasons for this seem to be that (a) historically it was very time consuming and expensive to run, being computationally intensive, and (b) authors inevitably express concern about the dangers of local minima. Both of these reasons are no longer germane, now that we have more computational power sitting in a small box on our desks than we used to have filling up a room. It is easy to compare solutions from many runs from different starting configurations, to be sure that the method is converging on the best solution.

Some local minima produce ordinations with strong, regular, geometric patterns. For example, the sample units may be arranged in a circle, concentric circles, or pinwheel patterns. The weaker the correlation structure in the data (using "correlation" in the broad sense), the more likely is this kind of local minimum. These geometric patterns obvious when inspecting the ordination diagram. The problem is easily avoided by requesting more random starts, more iterations, and a more stable solution.

Starting configuration

The starting configuration can be produced with a random number generator or you can supply the starting configuration. If you use random starting coordinates, it is essential to try many to ensure as much as possible that the minimum of stress is not a local minimum. In our experience, the method usually finds repeatedly a stable consistent configuration, differing only in minor details. Expect to find final configurations that are simple reflections or rotations of the same basic configuration. These are unimportant differences because the solution can be reflected or rigidly rotated at will (Ch. 15).

On the other hand, some data structures are relatively challenging for NMS. These are usually expressed by unstable solutions. See "Avoid Unstable Solutions" below.

To speed up the convergence on a minimum stress and avoid local minima, some authors recommended that you supply a starting configuration rather than using a series of random starts. For example, in PC-ORD you would follow these steps to use a starting configuration:

1. Run an ordination on your data set using a method other than NMS.
2. Save the coordinates of the ordination in a new file.
3. Run NMS.
4. Select the option that allows you to supply your own starting configuration. You will be asked to indicate the file containing the starting coordinates from step 2.

While this approach is often effective, we have found that a very large number of random starting configurations will often provide an ordination with lower stress than does starting with another ordination.

Monte Carlo test

You can evaluate whether NMS is extracting stronger axes than expected by chance by performing a randomization (Monte Carlo) test. You select the number of runs used to conduct the test. Because NMS is computationally intensive, you will find that it is time consuming to request even 20 Monte Carlo runs for a large data set.

Compare the stress obtained using your data set with the stress obtained from multiple runs of randomized versions of your data (Table 16.1). The data are randomly shuffled within columns (variables) after each run. In the example in Table 16.1, we can conclude that the best of the 1-D to 5-D solutions provide significantly more reduction in stress than expected by chance, accepting a probability of Type I error < 0.05. You can also compare the observed stress to the range and mean of stresses obtained in the randomized runs.

The p-value is calculated as follows. If n is the number of randomized runs with final stress less than or equal to the observed minimum stress and N is the number of randomized runs, then $p = (1+n)/(1+N)$.

Table 16.1. Stress in relation to dimensionality (number of axes), comparing 20 runs on the real data with 50 runs on randomized data. The *p*-value is the proportion of randomized runs with stress less than or equal to the observed stress (see text). This table is the basis of the scree plot shown in Figure 16.3.

Axes	Stress in real data			Stress in randomized data			p
	Minimum	Mean	Maximum	Minimum	Mean	Maximum	
1	37.52	49.00	54.76	36.49	48.70	55.54	0.0199
2	15.30	18.10	27.27	19.93	27.12	43.35	0.0050
3	10.96	12.33	19.88	13.90	19.01	32.06	0.0050
4	7.62	10.09	15.34	10.23	15.00	28.87	0.0050
5	5.73	6.06	6.39	7.27	13.28	30.44	0.0050
6	6.96	8.65	10.56	5.42	12.17	33.82	0.0796

Table 16.2. A second example of stress in relation to dimensionality (number of axes), comparing 15 runs on the real data with 30 runs on randomized data, for up to 4 axes.

Axes	Stress in real data			Stress in randomized data			p
	Minimum	Mean	Maximum	Minimum	Mean	Maximum	
1	42.80	49.47	56.57	43.18	50.39	56.52	0.0323
2	27.39	29.01	30.69	27.75	29.68	31.99	0.0323
3	20.30	20.48	21.27	19.87	20.81	22.30	0.3548
4	15.76	16.02	16.97	14.89	16.03	16.96	0.3548

Table 16.3. Thoroughness settings for autopilot mode in NMS in PC-ORD version 4.

Parameter	Thoroughness setting		
	Quick and dirty	Medium	Slow and thorough
Maximum number of iterations	75	200	400
Instability criterion	0.001	0.0001	0.00001
Starting number of axes	3	4	6
Number of real runs	5	15	40
Number of randomized runs	20	30	50

The second example (Table 16.2) illustrates results for a data set with weaker structure. We tried 15 different random starts with real data, versus 30 randomizations, for 1-D to 4-D solutions. The best 2-D solution has fairly poor fit (minimum stress = 27.39), but this is lower stress than all of the 2-D randomized runs. Although the 3-D solution reduced the stress considerably, it is no better than the solutions obtained with the randomized data. In this case, the 2-D solution is best, but the high stress gives us little confidence in the results.

NMS autopilot

In version 4 of PC-ORD, an "autopilot" mode was added to NMS to assist the user in making multiple runs, choosing the best solution at each dimensionality, and testing for significance. Everything that is done in autopilot mode can also be done by manually adjusting the parameter settings and options. When you select autopilot mode all options are set automatically (Table 16.3) except for the distance measure, which you must select. For community data, we recommend Sørensen (Bray-Curtis) distance.

Autopilot in NMS uses the following procedure:

1. Multiple runs with real data. Perform x real runs, each with a random starting configuration. A "run" consists of a series of solutions, stepping down in dimensionality from the highest number of axes to one axis.

For each dimensionality, PC-ORD saves to disk the best starting configuration from the x real runs. These files are named config1.gph, config2.gph, etc., for the 1-dimensional; 2-dimensional, etc. configurations respectively

2. Multiple runs with randomized data. Perform y runs with randomized data. Before each run, the data from the main matrix are shuffled within columns. Each run uses a different random starting configuration. Statistics on the final stress for each dimensionality are accumulated.

3. Select the best solutions. The best solution is defined by a particular starting configuration and number of dimensions. PC-ORD selects the best solution for each dimensionality as that with the lowest final stress from a real run.

4. Select the dimensionality. PC-ORD selects the dimensionality by comparing the final stress values among the best solutions, one best solution for each dimensionality. Additional dimensions are considered useful if they reduce the final stress by 5 or more (on a scale of 0-100). PC-ORD selects the highest dimensionality that meets this criterion. At that dimensionality, the final stress must be lower than that for 95% of the randomized runs (i.e., $p \leq 0.05$ for the Monte Carlo test). If this criterion is not met, PC-ORD does not accept that solution or it chooses a lower-dimensional solution, providing that it passes the Monte Carlo test.

Choosing the best solution

The basic principles for choosing the best solution from multiple NMS runs are summarized below. We discuss each of these principles, then prescribe a general procedure for running NMS.

Basic principles for choosing the best NMS solution:

   1. Select an appropriate number of dimensions.
   2. Seek low stress.
   3. Use a Monte Carlo test.
   4. Avoid unstable solutions.

1. Select an appropriate number of dimensions

The number of axes or dimensions (k) is critical. This value is important because axis scores with NMS depend on the number of axes selected. NMS is unlike most other ordination method in that, as dimensions are added to the solution, the configuration on the other dimensions changes. In particular, note two features of how NMS deals with multiple dimensions:

1. The first dimension in a 2-D solution is not necessarily the same as the first dimension in a 3-D solution or a 1-D solution. For a given number of dimensions, the solution for a particular axis is unique. By contrast, with other ordination methods such as PCA, the scores of sample units on the second component (axis) are independent of whether you look at 2 or 20 dimensions.

2. Axis numbers are arbitrary, so that the percent of variance on a given axis does not necessarily form a descending series with increasing axis number. However, the variance represented (using an after-the-fact distance-based evaluation) usually increases and the final stress decreases as axes are added to the solution.

Kruskal & Wish (1978) pointed out that there is no firm fixed statistical criterion for selecting an

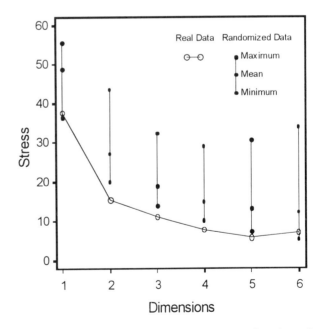

Figure 16.3. A scree plot shows stress as a function of dimensionality of the gradient model. "Stress" is an inverse measure of fit to the data. The "randomized" data are analyzed as a null model for comparison. See also Table 16.1.

appropriate number of dimensions: "Even if a good statistical method did exist for determining the 'correct' or 'true' dimensionality, this would not in itself suffice for assessing what dimensionality to use. Since MDS is almost always used as a descriptive model for representing and understanding the data, other considerations enter into decisions about the appropriate dimensionality, e.g., interpretability, ease of use, and stability."

An appropriate number of dimensions (axes) should be determined by plotting (a scree plot; Fig. 16.3) or tabulating (Tables 16.1 & 16.2) final stress vs. the number of dimensions. Choose a number of axes beyond which reductions in stress are small. Look for an "elbow" in the scree plot. In Figure 16.3 the first two axes provide far greater reductions in stress than later axes, so that we conclude that a 2-D solution is best for this data set. For comparison, you may wish to use a randomized version of your data set as a null model. This can be done by shuffling the contents of each column. Then run NMS on the randomized data set and plot stress versus number of dimensions. If there is appreciable correlation structure in the data ("correlation" in the broad sense), then the curve for the observed data should be lower than the curve for the randomized data. In this case (Fig. 16.4), the observed stress is lower than expected by chance for one to five dimensions (see also Table 16.1).

To quote the report from which Figure 16.3 was taken, "Two major gradients captured most of the variance in the lichen communities, the first two dimensions containing 27.5% and 31.5%, respectively, of the information in the analytical data set (cumulative = 59%). Higher dimensions improved the model very little."

Using too many dimensions can have a bad effect on interpretability of the results. The important variation will be spread over all of the axes, defeating the primary purpose of ordination: expressing the variation in as few dimensions as needed to express the covariation among as many attributes as possible.

On the other hand, analysts frequently are indecisive between two adjoining dimensionalities, usually 2 versus 3, less often 1 versus 2. In these cases, it is probably safer to chose the higher dimension. Doing so will often result in a strong gradient on a diagonal through the ordination space. This configuration can easily be rotated to facilitate interpretation and discussion. Rotation of the 3-D configuration can often result in the factors of interest being displayed in a convenient 2-D representation, with less interesting or minor variation relegated to a third axis that need not be displayed.

2. Seek low stress

How low is "low"? Both Kruskal (1964a) and Clarke (1993) gave rules of thumb for evaluating the final stress, using Kruskal's stress formula 1. Kruskal (1964a) and PC-ORD (McCune & Mefford 1999) multiply this stress by 100, expressing it as a percentage. Multiplying Clarke's (1993) rule of thumb by 100, the two scales can easily be compared (Table 16.4). Because different software packages rescale the stress in different ways, results may not be comparable.

Table 16.4. Rules of thumb for interpreting final stress from nonmetric multidimensional scaling, using Kruskal's stress formula 1 multiplied by 100.

Stress	
	Kruskal's rules of thumb
2.5	Excellent
5	Good
10	Fair
20	Poor
	Clarke's rules of thumb
< 5	An excellent representation with no prospect of misinterpretation. This is, however, rarely achieved.
5-10	A good ordination with no real risk of drawing false inferences
10-20	Can still correspond to a usable picture, although values at the upper end suggest a potential to mislead. Too much reliance should not be placed on the details of the plot.
> 20	Likely to yield a plot that is relatively dangerous to interpret. By the time stress is 35-40 the samples are placed essentially at random, with little relation to the original ranked distances.

In our experience, most ecological community data sets will have solutions with stress between 10 and 20. Values in the lower half of this range are quite satisfactory, while values approaching or exceeding 20 are cause for concern.

Clarke (1993) cautioned against over-reliance on these guidelines: "These guidelines are over-simplistic.

For example, stress tends to increase with increasing sample size. Also, it makes a difference to the interpretation if contributions to the stress arise roughly evenly over all points or if most of the stress results from difficulty in placing a single point in the two-dimensional picture."

Kruskal and Wish (1978) stated: "Never forget that the interpretation of stress values depends somewhat on the number of objects, I, and the dimensionality, R. As long as I is large compared to R, this effect is slight, and the exact values of I and R do not make much difference. As a rule of thumb, if $I > 4R$, the interpretation of stress is not sensitive to I and R. However, as I gets close to R, great changes occur."

We illustrate this by removing sample units at random from a community data set of 50 sample units by 29 species. The data contain abundances of epiphytic lichens in forests of coastal Alaska. We fit 2-D solutions, 100 runs of 500 iterations each, with a stability criterion of 0.00001. The lowest final stress for each subsample was plotted against the size of the subsample. As the number of sample units decreased, final stress declined gradually at first, then precipitously, reaching zero stress with six sample units remaining (Fig. 16.4).

Final stress also depends on the number of original variables. If species are deleted at random, the final stress is rather inconsistent, because some species are more influential in controlling the pattern than others. If, however, only rare species are deleted, then the final stress will normally be quite similar to that in the whole data set. We show this for an example data set by progressively deleting species in order of increasing frequency, then running NMS (same conditions as for example in Fig. 16.4) on the reduced data sets. Final stress was remarkably uniform until only species occurring in more than half of the sample units remained (Fig. 16.5). After that point, further removals of species steadily reduced the final stress.

3. Use a Monte Carlo test

The Monte Carlo test described above is helpful for selectin dimensionality but not foolproof. The most common problems are:

A. **Strong outliers** caused by one or two extremely high values can result in randomizations with final stress values similar to the real data.

B. The same can be true for data dominated by a **single super-abundant species**.

C. With **very small data sets** (e.g., less than 10 SUs), the randomization test can be too conservative.

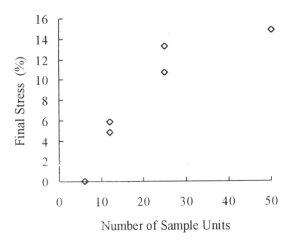

Figure 16.4. Dependence of stress on sample size, illustrated by subsampling rows of a matrix of 50 sample units by 29 species.

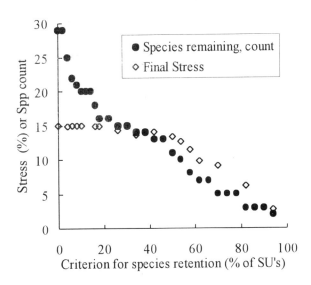

Figure 16.5. Dependence of final stress on progressive removal of rare species from a data set of 50 sample units and 29 species. For example, when only species occurring in 19% or more of the sample units were retained, 16 species remained in the data set and the final stress was about 15%.

Randomizations in this case are very likely to produce configurations for which NMS finds solutions with zero stress (e.g. Fig. 16.4).

D. If a data set contains **very many zeros**, then most randomizations will produce multiple empty sample units, making applications of many distance measures impossible.

These effects are often so strong that they will emerge from every single set of runs; in other words, they cannot be overcome by using a large number of random starts. See Chapter 22 (Reliability of Ordination Results) for more on the limitations of randomization tests.

4. Avoid unstable solutions

The stability of the solution should be examined with a plot of stress versus iteration number. Figure 16.6 shows the usual decline in stress as the iterations proceed. Occasionally the curve will continue to fluctuate erratically rather than settling on an even stress. In that case the outcome of the ordination will depend on the exact number of iterations chosen (say 80 vs. 81). This undesirable situation can be caused by overfitting the data (i.e., using a dimensionality that is too high for the number of sample units) or where two or more local minima are equal "competitors" for the global minimum. (Imagine a ball rolling back and forth between two valleys separated by a small hill.)

The following examples illustrate common patterns of changes in stress and stability as iterations (steps) proceed in NMS. Figure 16.7 shows the ideal case. Stress drops quickly and stabilizes smoothly. Instability (standard deviation in stress over the preceding x iterations; in this case $x = 10$) drops to extremely low levels (almost 10^{-6}) in less than 200 steps. In the second example (Fig. 16.8), NMS cannot find a stable solution. This would also be evident in a low-resolution plot (a more erratic tail on the curve than in Fig. 16.6), and a high final instability value. For the unstable case, stress, step length, and the magnitude of the gradient vector continue to fluctuate wildly, with no evidence of a gradual dampening. The resulting ordination would be unacceptable. In the third example (Fig. 16.9), NMS finds a fairly stable solution, interrupted by persistent but low-level fluctuations in stress. Despite the minor instability, the resulting ordination would be acceptable, if no better result were available, because the final stress is low and the final instability is about 10^{-3}.

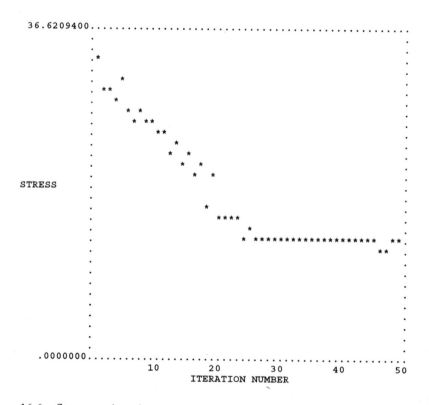

Figure 16.6. Stress vs. iteration (step) number, low-resolution plot from PC-ORD. Check such a plot to be sure the curve stabilizes at a relatively low level. The minor instability near the final iterations suggests that more iterations are needed.

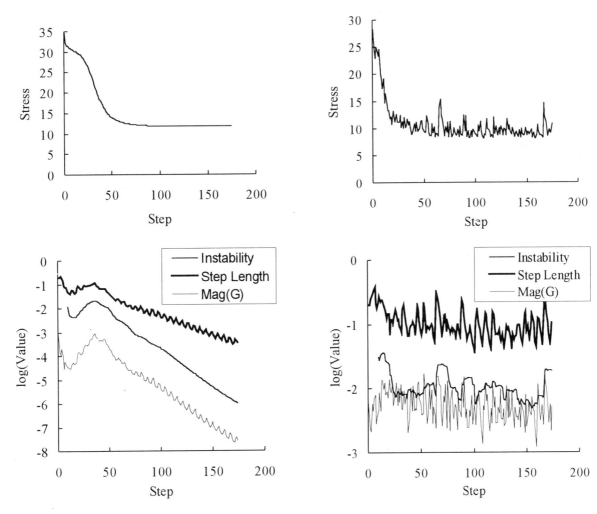

Figure 16.7. NMS seeks a stable solution. Instability is the standard deviation in stress over the preceding 10 steps (iterations); step length is the rate at which NMS moves down the path of steepest descent; it is based on *mag*(**g**), the magnitude of the gradient vector (see text).

Figure 16.8. NMS finds an unstable solution; normally this would be unacceptable instability. In this case it was induced by overfitting the model (fitting it to more dimensions than required). The plotted variables are described in Figure 16.7.

A general procedure for NMS

To use NMS you should either use an automated search through a variety of solutions or manually go through a series of steps. In PC-ORD, the automated search is called *autopilot mode* (McCune & Mefford 1999; see above). The following procedure determines appropriate dimensionality and statistical significance and seeks to avoid local minima.

1. **Adjust the data** as needed with transformations (monotonic transformations or relativizations). Unlike PCA and the CA family of ordinations, NMS has no built-in relativization. Students commonly wonder whether data transformations are important with NMS, since it is based on ranked distances. The answer is "yes" — remember that the values in the distance matrix are ranked, not the original data. For example, consider a monotonic transformation, such as a log transformation. Monotonic transformations do not change the ranks of the observations in the data set. But these transformations *generally do* have an effect on NMS, because even monotonic transformations usually alter the ranks of *distances* calculated from the data.

Chapter 16

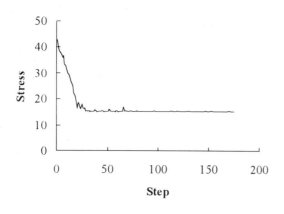

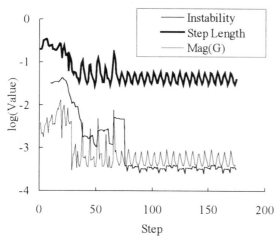

Figure 16.9. NMS finds a fairly stable solution, ending with a periodic but low level of instability. The instability is so slight that the stress curve appears nearly flat after about 45 steps. The plotted variables are described in Figure 16.7.

2. **Preliminary runs**: Request a 6-D solution, stepping down to a 1-D solution, an instability criterion of 0.0005 (or smaller), 200-500 iterations, 10-50 runs with your real data, and 20-50 runs with randomized data. The latter provide the basis for a Monte Carlo test of significance at each dimensionality.

3. **Scree plot**. Plot the final stress vs. the number of dimensions (Fig. 16.3). Do this in PC-ORD by selecting *Graph | NMS Scree Plot* just after running NMS.

4. **Select a number of dimensions** beyond which additional dimensions provide only small reductions in stress. Check for a better-than-random solution by using the results of the Monte Carlo test. Note that the first axis with randomized community data is often nearly as strong or stronger than the real data, even when the pattern in the real data is strong. This is because the randomization scheme creates rows with unequal abundances – some rows can result that have much higher or lower totals than any rows in the real data. Thus a 1-D NMS solution from the shuffled data tends to describe variation in row totals.

5. **Check stability** of the solution using either (a) the plot of stress vs. iteration or (b) the final instability value for the chosen solution, as listed in the numerical output from NMS. Strive for instability $< 10^{-4}$.

6. **Rerun NMS** requesting the following:

the number of dimensions you decided on,

the starting configuration that worked best (note: you will need to have saved either this starting configuration or the random number "seed" if you want to recreate the same ordination later),

no step-down in dimensionality,

one real run, and

no Monte Carlo test (no randomized runs).

What to report

☐ Distance measure used.

☐ Algorithm and/or the software used (If using PC-ORD you can cite Mather (1976) and Kruskal (1964) for the method).

☐ Whether you used random starting configurations, or, if you supplied a starting configuration, how that was done.

☐ Number of runs with real data.

☐ How you assessed the dimensionality of the data set.

☐ Number of dimensions in the final solution.

☐ Monte Carlo test result: number of randomized runs and probability that a similar final stress could have been obtained by chance.

☐ How many iterations for the final solution.

☐ How you assessed the stability of the solution.

☐ Proportion of variance represented by each axis, based on the r^2 between distance in the ordination space and distance in the original space.

☐ Standard interpretive aids (overlays, correlations of variables with axes). Remember to "put a face on it" if you are ordinating community data. Many people composition, instead focusing on interpretation in terms of environmental factors or experimental treatments.

Examples

Examples in the literature

Some examples of uses of NMS in community ecology are Anderson (2001), Clymo (1980), Dargie (1984), Field et al. (1982), Kantvilas and Minchin (1989), MacNally (1990), McCune et al. (1997a,b), Neitlich and McCune (1997), Oksanen (1983), Peterson and McCune (2001), Prentice (1977), Sprules (1980), Tuomisto et al. (1995), Waichler (2001), and Whittington and Hughes (1972).

The essence of NMS

We already gave mini-examples of various aspects of NMS, including Monte Carlo test results, dependence of stress on sample size, and evaluation of stability. One final example graphically illustrates the core of the method: gradual adjustments of points in the ordination space in an attempt to minimize stress.

The data set is very small so that you can follow along with coins on a table or people standing in a room, using these as an ordination space. Begin with a data set of five sample units and six species (Table 16.5). Calculate a distance matrix (Table 16.6). Assign positions to the sample units (people or coins) at random, then make a series of small steps to improve the relationship between Table 16.6 and distances in the ordination space, all objects moving simultaneously. The series of steps taken by the computer for one run of NMS are summarized in Figure 16.10.

Table 16.5. Abundance of six species in each of five sample units.

	Species					
SU	sp1	sp2	sp3	sp4	sp5	sp6
1	1	2	3	4	5	5
2	1	3	2	4	6	6
3	0	3	0	1	0	1
4	1	2	2	2	3	4
5	5	2	1	3	5	6

Table 16.6. Sørensen distances among the five sample units from Table 16.5.

	SU				
SU	1	2	3	4	5
1	0				
2	0.0952	0			
3	0.6800	0.6296	0		
4	0.1765	0.2222	0.5789	0	
5	0.1905	0.1818	0.7037	0.2778	0

Variations

Global vs. local NMS

Global NMS is the basic form as presented above. Local NMS (Sibson 1972; Prentice 1977, 1980; Minchin 1987a) uses a different criterion. For each sample unit, the rank order of distances to all other sample units in the ordination space is matched as well as possible with the n-dimensional dissimilarities between that sample unit and the other sample units. This allows the possibility that the rate of increase in distance in species space with environmental distance may differ in different parts of the ordination space. The consequences of the choice of local vs. global NMS have not been fully explored in the literature.

Constrained ordination

NMS can be constrained by multiple regression on environmental variables (ter Braak 1992), similar to constraining a correspondence analysis (reciprocal averaging) by regression in canonical correspondence analysis (Ch. 21). The theory and econometric applications are given by Heiser and Meulman (1983). The method is apparently untested in ecology.

Isomap and locally linear embedding

Two new, related methods of ordination have great promise for multivariate analysis of ecological communities. Isometric Feature Mapping (Isomap) and Locally Linear Embedding (LLE) were published side-by-side in Science (Tenenbaum et al. 2000, Roweis & Saul 2000). Although ecologists have so far not used these methods (to our knowledge), they probably should, because the methods have demonstrated an ability to extract nonlinear structures from hyperspace. Extracting nonlinear structures is exactly the problem in detecting and describing gradients in the multivariate dust-bunny distribution.

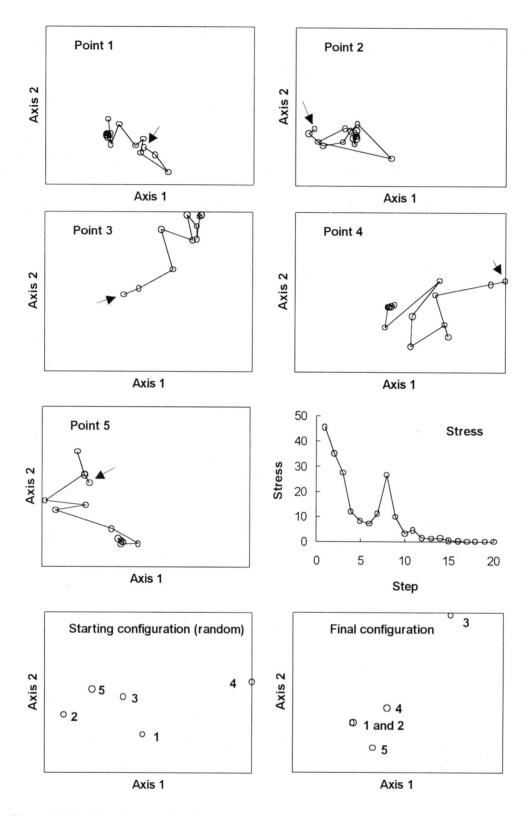

Figure 16.10. Migration of points from the starting configuration through 20 steps to the final configuration using NMS. Each point in the ordination represents one of five sample units. The first five graphs show the migration of individual points (arrows indicate starting points), followed by the graph of the decline in stress as the migration proceeds.

Both methods seek a global structure by examining local structures. Isomap builds on multidimensional scaling, while LLE sets up an eigenanalysis problem where only local distances contribute.

The key to Isomap is estimating long distances based on accumulating local distances between points, rather than a direct calculation. Tenenbaum et al. called this a "geodesic" distance. This method should alleviate the loss of sensitivity of our usual distance measures with increasing distances in species space.

Isomap uses three basic steps (Tenenbaum et al. 2000). (1) Determine the neighbors for each point. This can be determined by finding nearest neighbors or by searching for other points within a fixed radius. (2) Estimate the geodesic distances among all pairs of points by finding the shortest geodesic paths. (3) Apply classical multidimensional scaling to the geodesic distances.

Locally Linear Embedding (LLE; Roweis & Saul 2000) is more akin to PCA than to NMS. LLE uses a locally linear model to construct an overall nonlinear structure. The position of each point is reconstructed from linear coefficients relating it to its local neighborhood. They defined the neighborhood as the a fixed number of nearest neighbors, based on a Euclidean distance measure. The error in this reconstruction is measured by a "cost function," the sum of weighted squared distances between the data points and their reconstructed positions. The weights control the contribution of particular points to the reconstructions (zero for points outside the neighborhood, proportions summing to one for points within the neighborhood). Optimal weights are sought with a least-squares method. Once the weights are defined, eigenanalysis is used to minimize the cost function. This results in a set of coordinates defining the positions of the points in a low-dimensional space.

The success of both Isomap and LLE hinge on exactly how we define the local neighborhood. This can be viewed either as a drawback (another arbitrary decision is needed) or as an advantage (if we systematically vary the definition of the neighborhood, we can explore scaling effects in ecological spaces). In any event, both Isomap and LLE deserve thorough testing and exploration with ecological data.

Principal curves

Principal curves (PC; Hastie & Stuetzle 1989, De'ath 1999b) is an iterative, nonlinear ordination technique. PC minimizes squared distances to curves, while PCA minimizes squared distances to straight lines. The concept is very appealing, considering the convoluted paths of environmental gradients through the dust-bunny distribution in species space (Chapter 5). De'ath (1999b) demonstrated the effectiveness of principal curves for one-dimensional ordination with ecological data sets.

An initial curve is specified (a set of SU scores, e.g., from another ordination method). Imagine a line bending through a multidimensional species space. The data points are then projected onto the line, finding the point on the line closest to each data point. The location of the line in species space is then adjusted to reduce the distances from the data points to the line, using a smoothing algorithm. These steps are repeated until the solution stabilizes.

Numerous decisions affect whether the solution will be useful or not; most importantly, the choice of the initial line and the type and flexibility of the smoother (De'ath 1999b). Widespread implementation in ecology will require some automation of these decisions, based on the characteristics of the data, because few ecologists will invest the time necessary to assure that the resulting gradient model is optimal. Extension to more than one dimension will greatly increase the versatility of this method.

Predicting scores from an NMS model

A criticism of ordination in general is that practitioners never take it seriously, as evidenced by the rarity of its use to predict anything or estimate scores for new data points (Peters 1991). While predictions are easily made with equations resulting from eigenanalysis (e.g., the eigenvectors resulting from PCA), prediction is more difficult with some other methods, such as NMS.

Scores for a sample unit that was not part of the initial data set can be calculated with a prediction algorithm for NMS. This is not prediction in the sense of forecasting but rather statistical prediction in the same way that we use multiple regression to estimate a dependent variable. NMS Scores in PC-ORD calculates scores for new items based on prior ordinations.

New points are fitted to an existing ordination without altering the positions of the original points. Conceptually, it is easiest to think of prediction as using a model to estimate certain outputs, given a set of inputs. In regression, the model is the linear equation, the input is a set of values for the independent variables, and the output is an estimate of the dependent variable. In NMS Scores, the model is defined by a set of ordination scores for one to three axes, along with the original data used to derive the scores. The input data are one or more new sample units in the main matrix. These must share some or all of the variables with the original data set (for community data this means a set of overlapping but not necessarily the same species). The output consists of ordination scores (positions on the ordination axes) for each of the new sample units.

The model is applied by an iterative search for the position of best fit for each of the new data points, one at a time. Trial positions are selected in successively finer focus on the region of the axis where they best fit (Fig. 16.11). The data points in the original model are held fixed in the ordination space, rather than being adjusted as in regular NMS.

For examples of predictive-mode NMS, see McCune et al. (1997a,b). In those cases the model was a two dimensional NMS ordination related to a regional climatic gradient and air quality. The model was used to assign scores to new data points that were not part of the calibration data set.

Two methods are available for finding scores for new points:

1. Simultaneous axes: For a k-dimensional solution, we seek the position (score) that will minimize stress in the overall k-dimensional configuration. The search is hierarchical, sharpening the focus in three steps. The basic method is detailed below under "one axis at a time," except that trial scores are sought from all combinations of positions on the k axes. This makes it slower to seek positions on three axes at once than on three axes one at a time. The computation time increases with the square of the number of axes, while with "one axis at a time," the computation time increases linearly with the number of axes. Choose "simultaneous axes" if you wish to optimize the overall position of the fitted points in the ordination space. Choose "one axis at a time" if you seek the best fit with respect to a single axis. For example, if one axis in particular is of interest for predictions, and the other axes are irrelevant to your specific purpose, select "one axis at a time."

2. One axis at a time: For each axis, we seek the position (score) that will minimize stress in the overall one-dimensional configuration. The search is hierarchical, increasing the focus in three steps. In the first step, the axis plus a 20% margin on both ends of the axis is divided into 29 equally spaced trial points (20 intervals on the axis + 4 intervals at each end). The overall stress is calculated for each of the 29 possible positions for the trial point. The position with the lowest stress is selected, then used as the basis for refining the position of best fit (Fig. 16.11). The interval is then narrowed by dividing the previous interval by five (the focus factor). Five trial points spaced by this new interval are examined on each side of the previous best point. Again, the best point is found, the interval is divided by the focus factor, and the final best point is found.

Example of successive focusing: Assume an axis is 80 units long, margin = 20%, focus factor = 5, number of focus levels = 3. The focusing parameters are calculated in Table 16.7.

Distance measure. Any distance measure can be used. Normally you should choose the same distance measure as used for the ordination of the calibration data set.

Flag for poor fit. Two kinds of poor fit for new SUs are screened by PC-ORD: (1) points that fall beyond an **extrapolation limit** and (2) final configurations exceeding a final stress criterion. In the first case, a flag = 2 (score too low) or flag = 3 (score too high) is written into the output file (Table 16.8; Fig. 16.12).

The default extrapolation limit in PC-ORD is 5% of the axis length (minimum = 0%, maximum = 50%). For example, if the calibration scores range from 0-100 and the extrapolation limit is 5%, then points that score greater than 105 are flagged with a 3 in the output file. Scores less than -5.0 are flagged with a 2 in the output file. Flagged SUs should be considered for exclusion from the results, because they are foreign to the model. In some cases they will represent data recording or data entry errors.

Table 16.7. Focusing parameters for fitting new points to an existing NMS ordination. The units for interval width and range are always in the original axis units.

Focus level	Range examined	Intervals	Points	Interval Width
1	$80 + 2(0.2 \times 80) = 112$	$20 + 4 + 4 = 28$	29	$112/28 = 4$
2	$2 \times 5 \times 4/5 = 8$	$2 \times 5 = 10$	11	$4/5 = 0.8$
3	$2 \times 5(0.8/5) = 1.6$	$2 \times 5 = 10$	11	$0.8/5 = 0.16$

Nonmetric Multidimensional Scaling

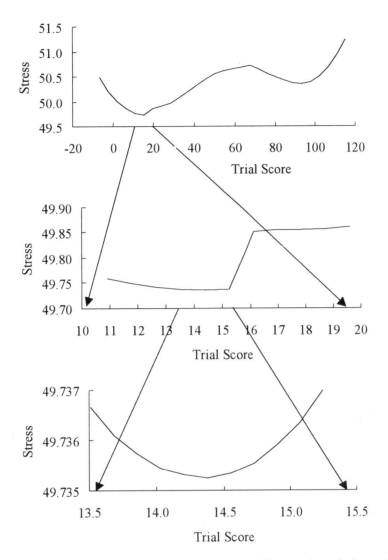

Figure 16.11. Successive approximation of the score providing the best fit in predictive-mode NMS. The method illustrated is for one axis at a time. Multiple axes can be fit similarly by simultaneously varying two or more sets of coordinates.

Table 16.8. Flags for poor fit of new points added to an existing NMS ordination.

Flag	Meaning
0	Criteria for acceptability were met
1	Final stress is too high
2	Score is too far off low end of gradient
3	Score is too far off high end of gradient

```
The following options were selected:
    Distance measure     = Sørensen
    Method for fitting   = one axis at a time
    Flag for poor fit    = autoselect (2 stand.dev. > mean stress)
    Extrapolation limit, % of axis length:    5.000

Calibration data set: D:\FHM\DATA\94\SEModel.wk1

Contents of data matrices:
    85 Plots    and     87 Species in calibration data set
    34 plots    and    140 species in test data set
    85 scores from calibration data set
Title from file with calibration scores: SEModel.wk1,
    rotated 155 degrees clockwise

    63 species  in test data set are not present in calibration data

AXIS SUMMARY STATISTICS
-------------------------------------------------------------------
                                          Axis:       1          2
Mean stress                                       49.833     50.251
Standard deviation of stress                       0.351      0.391
Stress used to screen poor fit (flag = 1)         50.536     51.033
Cutoff score, low end-of-axis warning flag (2)    -6.473    -12.895
Cutoff score, high end-of-axis warning flag (3)  106.473    112.895
Minimum score from calibration data                6.559      1.619
Maximum score from calibration data               93.441     98.381
Range in scores from calibration data             86.881     96.761
-------------------------------------------------------------------


BEST-FIT SCORES ON EACH AXIS
-------------------------------------------------------------------
           Axis:    1                       2
Item (SU)  Score    Fit    Flag    Score    Fit    Flag
-------------------------------------------------------------------
3008281    25.673   49.713   0     90.253   50.085   0
3008367    44.440   50.058   0     74.190   50.582   0
3008372    36.099   50.141   0     80.964   50.437   0

etc.

3108526     2.563   49.505   0      8.199   50.007   0
3108564    50.000   49.953   0     87.543   50.491   0
3108651     0.825   49.327   0    114.830   49.714   3
3108684     4.995   49.328   0    111.347   50.148   0
3108732   121.416   48.633   3    -17.539   48.957   2
3208432    14.379   49.840   0     16.714   50.243   0
3208436    56.429   49.988   0     66.062   50.480   0
```

Figure 16.12. Example output (PC-ORD) from "one axis at a time" method of fitting new points in an NMS ordination.

CHAPTER 17

Bray-Curtis (Polar) Ordination

Background

Bray-Curtis ordination (Bray & Curtis 1957) is also commonly known as polar ordination because it arranges points in reference to "poles" or endpoints (also known as reference points). The method selects endpoints either subjectively or objectively, then uses a distance matrix to position all other points relative to the endpoints. Beals (1984) discussed the method in detail, comparing it with many other ordination methods.

Bray and Curtis sought a method for extracting ecological gradients that did not depend on the assumption of linear relationships among species. As field ecologists, they sought a method that could be practically applied to hundreds of field plots with no more computing power than a mechanical calculator. In fact, they performed their first ordinations by manual geometry, using a compass and ruler, following hand calculation of a distance matrix based on presence-absence data.

Their first attempt was flawed in the method of choosing the endpoints, resulting in excessive sensitivity to outliers. A series of improvements (Table 17.1) resulted in a method that is still useful today, despite a checkered history (Box 17.1).

When to use it

Historically, Bray-Curtis ordination was used as a fast, effective, multipurpose ordination technique. The speed is no longer much of an advantage, except for a quick look at a large data set. Other methods, such as principal components analysis and reciprocal averaging, are just as fast but force a particular distance measure on the analyst. In contrast, Bray-Curtis ordination can be applied to a matrix containing any distance measure, including non-Euclidean semimetrics such as Sørensen (Bray-Curtis) distance.

Table 17.1. Development and implementation of the most important refinements of Bray-Curtis ordination (from McCune & Beals 1993).

Stage of Development	Implementation
Basic method (Bray's thesis 1955, Bray & Curtis 1957)	Ordination scores found mechanically (with compass)
Algebraic method for finding ordination scores (Beals 1960)	BCORD, the Wisconsin computer program for Bray-Curtis ordination, ORDIFLEX (Gauch 1977), and several less widely used programs developed by various individuals
Calculation of matrix of residual distances (since 1970 at Wisconsin; published by Beals 1973), which also perpendicularizes the axes; given this step, the methods for perpendicularizing axes by Beals (1965) and Orloci (1966) are unnecessary.	BCORD
Variance-regression method of reference point selection (in use since 1973, first published in Beals 1984)	BCORD

Box 17.1. Do ecologists select quantitative tools logically? Consider the case of Bray-Curtis ordination. The following paragraphs are updated from McCune and Beals (1993).

While ordination as an approach to data analysis gained acceptance in the 1960's, the Bray-Curtis method came under attack beginning with Austin and Orloci (1966), and it quickly fell into general disfavor among ecologists worldwide, as new methods of ordination were introduced and championed. Only ecologists trained at the University of Wisconsin persisted in using Bray-Curtis, not out of blind loyalty, but because it generally gave ecologically more interpretable results than did the newer or more sophisticated methods. But reviewers and editors were sharply critical and were reluctant to publish the results of such ordinations. It was out of frustration, combined with his experience of the superior performance of Bray-Curtis, that Beals wrote his 1984 paper. Today, resistance by editors and reviewers to Bray-Curtis ordination continues to some extent.

Papers that compared Bray-Curtis with other methods (Gauch et al. 1977, Gauch & Scruggs 1979, del Moral 1980, McCune 1984, Robertson 1978) all provided evidence that Bray-Curtis may perform better than the others, though the authors do not always make that clear! See Beals (1984, pp 10, 15) for that evidence. Other comparative studies of ordination methods have simply ignored Bray-Curtis, including the exhaustive investigation by Minchin (1987). Some recent books on ordination (Legendre & Legendre 1998, Pielou 1984) also ignored Bray-Curtis ordination.

It is amazing, given the strength of its performance, that the Bray-Curtis method suffered such an ignominious decline. We believe there are two primary reasons for this. First were the criticisms of the method, some of which were justified, some not (Beals 1984). The most outspoken critic was Orloci (1966, 1975, 1978). To some extent his views were echoed by others (e.g., Whittaker & Gauch 1973, Robertson 1978). Second, there was a general but false impression that Bray-Curtis was simply an approximation of more "rigorous" techniques (Lambert & Dale 1964, Greig-Smith 1964, Goodall 1970, Whittaker & Gauch 1973), with the implication that it may have been acceptable before the widespread availability of computers, but was no longer.

The second major cause for the unpopularity of Bray-Curtis was probably a practical one: the lack of access to good software. Program BCORD, which was in use at Wisconsin, was not made generally available. Ecologists at Cornell University developed and distributed easy-to-use programs (ORDIFLEX in 1977 and DECORANA in 1979) for their preferred methods of ordination. The Bray-Curtis method in ORDIFLEX was inadequate because the only automatic method of endpoint selection was the original method. We suggest that reciprocal averaging (in ORDIFLEX) became so popular in the late 1970's and detrended correspondence analysis (DECORANA) in the 1980's, because of easy availability and built-in standardizations, rather than any superior performance of those methods.

Bray-Curtis ordination has, however, increasingly been used by non-Wisconsin ecologists (e.g., Puroleit & Singh 1985; Schoenly & Reid 1987, 1989; Wunderle et al. 1987). Another development was the discovery that a new approach to ordination based on fuzzy set theory could produce results identical to Bray-Curtis ordination (Roberts 1986).

Two recent books have been kind to Bray-Curtis. Causton (1988), without apparent knowledge of Beals' arguments and evidence (1973, 1984), concluded after analyzing simulated data sets, "Perhaps the biggest surprise to emerge from this study was the robustness of Bray & Curtis Ordination." He went on to describe its strengths over other methods, based on his results (pp. 220-222). Ludwig and Reynolds (1988), in their text on statistical ecology, stated, "we agree with Beals (1984) that PO [= Bray-Curtis ordination] is a valuable method."

Today, the computational power on everyone's desktop makes nonmetric multidimensional scaling increasingly appealing, and its performance is often superior to Bray-Curtis ordination (e.g., McCune 1994). Bray-Curtis remains, however, a useful tool when the object is a multivariate comparison to reference points.

The main attraction remaining for Bray-Curtis ordination is for evaluating problems that have conceptual reference points. Placing points in reference to two other points per axis reduces the amount of information used on each axis. This is a disadvantage if one seeks a general-purpose description of community structure. It can be an advantage, however, allowing the user to describe community variation relative to a specific hypothesis, reference condition, or experimental design (example below). Bray-Curtis ordination can be used to examine specific hypotheses by subjectively selecting endpoints.

The original Bray-Curtis method was sensitive to outliers because of their method of endpoint selection. This is an advantage for outlier analysis (example below), in effect forcing a description of the community relative to an outlier. This sensitivity is still a problem, however, if one wants to avoid undue influence of outliers, and the software does not offer an alternative method of endpoint selection.

How it works

1. Select a distance measure (usually Sørensen distance) and calculate a matrix of distances (**D**) between all pairs of N points.

2. Calculate sum of squares of distances for later use in calculating variance represented by each axis.

$$SS_{TOT} = \sum_{i=1}^{N-1} \sum_{j=i+1}^{N} D_{ij}^2$$

3. Select two points, A and B, as reference points for first axis. This is the most crucial step of the ordination, because all points are placed in relationship to these reference points. The original method chose the most distant points. This and other methods of endpoint selection are described later.

4. Calculate position (x_{gi}) of each point i on the axis g. Point i is projected onto axis g between two reference points A and B (Fig. 17.1). The equation for projection onto the axis is:

$$x_{gi} = \frac{D_{AB}^2 + D_{Ai}^2 - D_{Bi}^2}{2 D_{AB}} \quad \text{Eqn. 1}$$

The basis for the above equation can be seen as follows. By definition,

$$\cos A = x_{gi} / D_{Ai} \quad \text{Eqn. 2}$$

By the law of cosines,

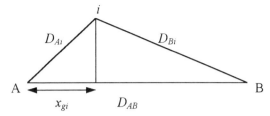

Figure 17.1. Geometry of positioning point i on the Bray-Curtis ordination axis between endpoints A and B, based on distances (D) between all pairs of A, B, and i.

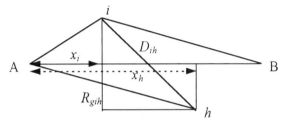

Figure 17.2. Geometry behind the calculation of residual distances from a Bray-Curtis ordination axis between endpoints A and B. Point i is positioned on the axis at distance x_i from A. The residual distance between points i and h, R_{gih}, is calculated from the original distance between those points (D_{ih}) and scores of points i and h on the axis (x_i and x_h).

$$\cos A = \frac{D_{AB}^2 + D_{Ai}^2 - D_{Bi}^2}{2 D_{Ai} D_{AB}} \quad \text{Eqn. 3}$$

Then substitute cos(A) from Equation 2 into Equation 3.

5. Calculate residual distances R_{gih} (Fig. 17.2) between points i and h where f indexes the g preceding axes.

$$R_{gih} = \sqrt{D_{ih}^2 - \sum_{f=1}^{g} (x_{fi} - x_{fh})^2} \quad \text{Eqn. 4}$$

6. Calculate variance represented by axis k as a percentage of the original variance ($V_k\%$). The residual sum of squares has the same form as the original sum of squares and represents the amount of variation from the original distance matrix that remains.

$$SS_{RESID} = \sum_{i=1}^{N-1} \sum_{h=i+1}^{N} R_{ij}^2$$

$$\text{Cumulative } V_k\% = 100\left(1 - \frac{SS_{RESID}}{SS_{TOT}}\right)$$

$$V_k\% = \text{cumulative } V_k\% - \text{cumulative } V_{k-1}\%$$

7. Substitute the matrix **R** for matrix **D** to construct successive axes.

8. Repeat steps 3, 4, 5, and 6 for successive axes (generally 2-3 axes total).

Selecting endpoints

The **original method** (Bray & Curtis 1957) selected the two points having maximum distance. This tends to select outliers, resulting in a pile of points on one side of the ordination and one or two isolated opposing points.

The **variance-regression** method (Beals 1984) is preferred. The first endpoint has highest variances of distances to other points. This finds a point at the long end of the main cluster in species space. This ignores outliers because their distances will be consistently higher, resulting in a lower variance. Then the distances between the first endpoint and every other point D_{1i} are regressed against distances between a trial endpoint and every other point D_{2i}^*. This is repeated with each remaining point as a trial endpoint.

Endpoint 2 is selected to minimize a (find the most negative number) in fitting simple linear regressions to

$$D_2 = aD_1 + b.$$

This procedure selects the second endpoint at the edge of the main cloud of points, opposite to endpoint 1.

In the example (Fig. 17.3), the first endpoint has already been chosen. Distances are shown from points 8 and 9 to the first endpoint and the trial second endpoint. In this case, the trial second endpoint will not be selected as the second endpoint because other points (those at lower left in the graph) will yield a more negative regression coefficient. Table 17.2 illustrates the basis for the regression.

Table 17.2. Basis for the regression used in the variance-regression technique. Distances are tabulated between each point i and the first endpoint D_{1i}, and between each point and the trial second endpoint D_{2i}^*.

point i	D_{1i}	D_{2i}^*
1	0.34	0.88
2	0.55	0.63
.	.	.
.	.	.
n	0.28	0.83

Choosing a **centroid of a number of points** as an endpoint diminishes the problem of using only a small amount of information (i.e., that in two entities) to place the remaining points on the axes. Our experimentation with synthetic grid datasets suggests that this does not work very well, but it needs more study and experimentation.

Choice of endpoints can also be **subjective**: you can choose which entities should be used as endpoints. Subjective endpoints are appropriate in some cases. For example if you wish to describe an outlier, that point can be forced as an endpoint, allowing the selection of the other endpoints by objective criteria of disturbance.

Subjective endpoints can also be used to contrast an outlier with the remaining points. The ordination will give a visual description of the degree and nature of the outlier. Overlays and correlations can be used to help describe the variables related to the outlier.

Sometimes there are "natural" subjective endpoints. For example, if you have sampled a transect along an environmental gradient, you may wish to choose the ends of the gradient as the endpoints and see how the intermediate sample units are arrayed in the ordination space.

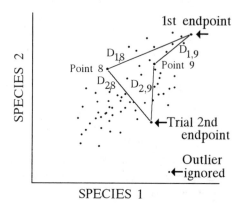

Figure 17.3. Example of the geometry of variance-regression endpoint selection in a two-dimensional species space.

What to report

- Rationale for choosing a reference-point method (state why the reference point concept was useful).
- Algorithm and/or the software used.
- Distance measure used.
- Method of endpoint selection.
- Proportion of variance represented by each axis.
- Standard interpretive aids (overlays, correlations of variables with axes).

Examples

Reference point in an experimental design.

One or more subjective endpoints are also appropriate for representing a control point or reference point in an experimental design (e.g., response to different degrees of disturbance). You could array the treatments on the first axis by specifying the control as one endpoint and the most extreme treatment as the other endpoint. The other treatments are arrayed between those endpoints, then subsequent axes describe the trajectories of the sample units through species space after the treatments were imposed. In this example (Fig. 17.4), the trajectories head back toward the control, suggesting gradual convergent return by various paths toward the control.

Outlier analysis

We applied Bray-Curtis ordination to the 25 sample unit × 16 species data set used to illustrate multivariate outliers in Chapter 7. This data set contains a single strong outlier. We used the original method of endpoint selection to force the outlier as an endpoint for the first axis, then allowed subsequent endpoints to be chosen with the variance-regression method, based on Sørensen distances. The result was rotated clockwise 20° to maximize the contrast between the outlier and the long axis of the main cloud of points (Fig. 17.5). We then overlaid species in a biplot (Ch. 13), which clearly indicated the abundance of species 3 and the lack of species 6 contributed to the unusual nature of the outlying point (Fig. 17.5).

Variations

The most important, yet largely untested, variations in Bray-Curtis ordination are non-Euclidean options for projecting points onto axes and for

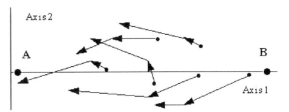

Figure 17.4. Using Bray-Curtis ordination with subjective endpoints to map changes in species composition through time, relative to reference conditions (points A and B). Arrows trace the movement of individual SUs in the ordination space.

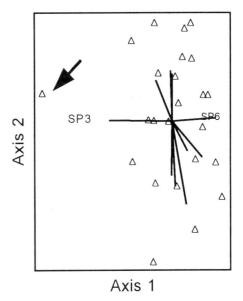

Figure 17.5. Use of Bray-Curtis ordination to describe an outlier (arrow). Radiating lines are species vectors. The alignment of Sp3 and Sp6 with Axis 1 suggests their contribution to the unusual nature of the outlier.

calculating residual distances (Table 17.3). Experimentation with city-block methods for calculating scores and residual distances, based on data sets with known underlying structure, suggests that, at least in some circumstances, the city-block projection geometry and residual distances perform better than the Euclidean method. In particular, distortions (mainly compression) around the edges of the ordination may be relieved.

Table 17.3. Comparison of Euclidean and city-block methods for calculating ordination scores and residual distances.

Operation	Euclidean (usual) method	City-block method
Calculate scores x_i for item i on new axis between points A and B.	$x_i = \dfrac{D_{AB}^2 + D_{Ai}^2 - D_{Bi}^2}{2 D_{AB}}$	$x_i = \dfrac{D_{AB} + D_{Ai} - D_{Bi}}{2}$
Calculate residual distances R_{ij} between points i and j.	$R_{ij} = \sqrt{D_{ij}^2 - (x_i - x_j)^2}$	$R_{ij} = \left| D_{ij} - |x_i - x_j| \right|$

CHAPTER 18

Weighted Averaging

Background

One of the simplest but still effective methods of ordination is by weighted averaging (Whittaker 1967). This has been reinvented numerous times for various purposes and under various names (below), but the essential operation is the same: a set of previously assigned species weights (or weights for species groups) is used to calculate scores for sites. The calculation is a weighted averaging for species or species groups actually present in a sample unit. Several examples illustrate the many incarnations of this method.

Curtis and McIntosh (1951) sought a quantitative description of forest succession in Wisconsin. They assigned "climax adaptation values" to tree species. The climax adaptation values weight species on a continuum from pioneer species (weight = 1) to climax species (weight = 10). These values are used as weights to calculate the "continuum index" for each sample unit. The continuum index is a score on a gradient from early successional to late successional species composition.

LeBlanc and de Sloover (1970) constructed an "index of atmospheric purity" (IAP) to describe air quality based on bioindicators. The IAP was calculated for each sample unit from presence of epiphytic lichen species. Each species was assigned a weight indicating its degree of tolerance to air pollution. They then mapped air quality in and around Montreal based on this index.

The most influential ordination method in the history of ecology, in terms of altering human impacts on ecosystems and economic impact, is in the Federal Wetlands Manual (Federal Interagency Committee for Wetland Delineation 1989). It prescribes a one-dimensional weighted averaging, but most users of this manual do not realize they are doing an ordination! The Federal Wetlands Manual assigns to each plant species a five-point numerical value of plant tolerance to hydric soils. The ranks are then used to calculate scores for individual sites. The purpose of these scores is to give a quantitative statement of the degree of dominance by wetland-demanding plants (e.g., obligate wetland plants, facultative wetland plants, facultative plants, facultative upland plants, and upland plants). These categories are weighted 1 through 5, respectively (the "EI values"). The resulting score for each plot is called a "prevalence index," but we can also think of it as a one-dimensional ordination in species space. If the prevalence index is less than three, the site is considered to have hydrophytic vegetation.

Aquatic invertebrates have been used to score lakes and streams with a variety of weighted averaging ordination techniques (although rarely explicitly recognized as such). The Benthic Quality Index (BQI) weights the proportions of counts in a given taxon or indicator group by a series of weights from 1 to 5, according to pollution tolerance (Wiederholm 1980; Johnson 1998). Similar approaches in Britain are the Biological Monitoring Working Party score (BMWP; Wright et al. 1993), the Benthic Response Index (Smith et al. 2001), and the Average Score Per Taxon (ASPT; Armitage et al. 1983). The BMWP score is a weighted sum, while the ASPT divides the weighted sum by the number of nonzero items contributing to the sum, converting the index to a weighted average and improving its performance. The Index of Biotic Integrity (IBI) uses a weighted average of indicator values that attempt to incorporate species sensitivity to several different stressors. An IBI has been constructed using fish (Karr 1991) and invertebrates (Kerans & Karr 1994). Lougheed and Chow-Fraser (2002) used a weighted average for their wetland zooplankton index (WZI), an indicator of wetland quality.

When to use it

Much of the appeal of weighted averaging comes from its simplicity. It is easy-to-use, understand, and communicate. Witness, for example, its adoption in the Federal Wetlands Manual (Federal Interagency Committee for Wetland Delineation 1989). It is both an advantage and disadvantage that it focuses results on a single gradient. When a single gradient is of concern and the results must be communicated to nonscientists, weighted averaging may be the method of choice. Focus on a single gradient is based on the way that species weights are assigned.

On the other hand, restriction of weighted averaging to a single axis (or dimension) seriously limits exploratory analyses in ecology. Few research problems patently concern one and only one compositional or environmental gradient.

Use of weighted averaging requires previous knowledge of species weights or indicator values. This can be a useful application of existing knowledge but a disadvantage for more open-ended data explorations.

In one of the few evaluations of weighted averaging as a basis for describing species responses, Oksanen et al. (1988) calculated pH optima and tolerances of diatoms. They compared weighted averaging, least-squares (curvilinear) regression, and maximum likelihood methods. In this case, the evaluation concerned the derivation of species weights from environmental and community data, rather than the application of a priori species weights. The main drawback to the weighted averaging method for solving this problem was that when the true optimum lay outside the observed ecological range, the weighted average was an inadequate indication of the ecological optimum. In their test data set, however, weighted averaging estimates yielded better predictions than either the maximum likelihood or least squares methods.

Ter Braak and Looman (1986) compared Gaussian logistic regression and weighted averaging for simulated presence-absence data. In this case, the goal was to describe the central tendency of the species response function on a one-dimensional gradient, rather than to construct a weighted averaging ordination of sites. They found that the performance of weighted averaging was as good as Gaussian logistic regression in most situations. The major exception was when the weighted average was biased by truncation (the species optimum occurring outside the sampled range) or by strongly uneven sampling along the gradient.

How it works

Begin with a matrix **A** containing abundances of p species in n sample units and a vector of p species weights, w_j. The weights might, for example, indicate tolerance for soil moisture. Calculate an ordination score v_i for each plot i:

$$v_i = \frac{\sum_{j=1}^{p} a_{ij} w_j}{\sum_{j=1}^{p} a_{ij}}$$

Although not part of most weighted averaging ordinations, the procedure can be reversed to refine the species weights. A new set of species weights can be calculated from the site scores:

$$w_j = \frac{\sum_{i=1}^{n} a_{ij} v_i}{\sum_{i=1}^{n} a_{ij}}$$

Repeating this procedure back and forth, between calculating new species weights and new sample unit scores, we have the basis of reciprocal averaging ordination. It is not, however, quite that simple. In reciprocal averaging ordination the data are first standardized, the scores are rescaled, and the method is applied to more than one axis (see next chapter).

What to report

☐ Source or derivation of the original species weights.

☐ Species weights.

☐ Standard interpretive aids (overlays, correlations of variables with the ordination axis).

Example

This example calculates scores on a moisture axis, based on previously assigned values indicating each species' tolerance for wet to dry sites. Begin with a matrix containing abundances of 3 species in 2 sample units (Table 18.1).

Table 18.1. Example data matrix with 2 plots × 3 species.

	sp 1	sp 2	sp 3	sum
plot A	4	1	0	5
plot B	0	2	4	6
sum	4	3	4	

Based on prior information, we assigned values to a vector of species weights, w_j, indicating tolerance for wet to dry sites (Table 18.2).

Table 18.2. Weights assigned to each species, the weights indicating tolerance of wet to dry sites.

Sp 1	Sp 2	Sp 3
Dry site indicator	Medium site indicator	Wet site indicator
weight = 0	weight = 50	weight = 100

Calculate a moisture score v_i for each plot i as an average of the abundances weighted by the moisture indicator values. For example, plot B gets a score of 83.3.

$$v_B = \frac{0(0) + 2(50) + 4(100)}{6} = 500/6 = 83.3$$

This high value indicates that plot B is relatively wet, as we would have expected from the prevalence of the wet site indicator, species 3. The same calculation for plot A yields a moisture score of 10, indicating a relatively dry site. In this way, we have constructed a one-dimensional ordination of plots in species space (Fig. 18.1). If the species weights are considered scores on the same gradient, we also have a species ordination, but in this case it is a priori (not derived from the data).

For example, Hutchinson (2001) used weighted averaging to calculate climatic affinities for a set of plots, based on the species composition within those plots. She calculated a score for each plot on each of three axes of climatic affinity: oceanic, suboceanic, and continental. Weights for each species were one or zero, indicating whether that species had that particular climatic affinity or not. If both the weights and species data are binary, the score for a given plot is simply the fraction of species in that plot belonging to a particular climatic affinity. If the abundance data are quantitative, then the plot scores for climatic affinities are weighted by species abundance. Climatic affinities of the flora were then related to environmental variables by scatterplots and correlation coefficients.

Weighted averaging to produce species scores

It has been widely overlooked that *any* ordination of sample units in species space can be supplemented by species as points in the same ordination space. This can be done by a *single* weighted averaging step, following the ordination. For example, assume we have an NMS ordination of SUs in species space. The scores for each SU on an given axis in species space can be considered the vector **v** in the above equations. Then the species scores **w** are calculated for each axis by weighted averaging of the abundances a_{ij} of each species j in each sample unit i.

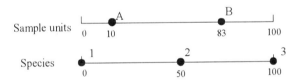

Figure 18.1. Example of ordination by weighted averaging.

Variations

Creative design of species weights

Weighted averaging can be applied to a much wider range of problems than just describing compositional or environmental gradients. One-dimensional ordinations can be produced for any attribute of species, including functional groups, behaviors, life history characteristics, and geographic distributions.

CHAPTER 19

Correspondence Analysis

(Reciprocal Averaging)

Background

Correspondence analysis (CA) was first popularized in ecology under the name reciprocal averaging (RA). Reciprocal averaging refers to the repeated weighted averaging of site scores to yield species scores and vice versa. The name "reciprocal averaging" is a helpful reminder of the fundamental nature of the technique. Correspondence analysis is an eigenvector technique (see PCA, Ch. 14) that accomplishes the same end as RA: a simultaneous ordination of both rows and columns of a matrix.

Correspondence analysis is often used on contingency table data (Greenacre 1984, 1988) but has been used in ecology primarily on abundance data (cover, frequency, or counts; see Legendre & Legendre 1998, p. 463, for the conceptual link between community data and contingency tables). It enjoyed a burst of popularity shortly after its introduction in ecology (Hatheway 1971, Hill 1973b), in part because of good performance on simulated one-dimensional gradients, as compared to PCA (Gauch et al. 1977).

In particular, with 1-D simulated data sets with hump-shaped species responses to environmental gradients (Fig. 19.1), PCA contorts the gradient into a horseshoe shape that turns in on the ends (Fig. 19.2). In this example, 11 species have evenly spaced peaks along the gradient and noiseless responses to the underlying environmental gradient. The gradient was "sampled" with 11 evenly spaced SUs. Although PCA successfully aligns the major axis with the underlying gradient, the order of SUs on Axis 1 is distorted by the curvature (Fig. 19.3). On the other hand, CA results in the same order of SUs on Axis 1 as on the underlying gradient (Figs. 19.2 & 3). NMS performed similarly to CA, but with slightly less stretching of distances in the middle of Axis 1.

A few examples of the use of CA in ecology are Marks and Harcombe (1981), Moral (1978), and Westman (1981). It is a logical extension from the weighted averaging technique used by Whittaker (1956). Whittaker assigned an indicator value to a species on an environmental gradient by averaging the values of that environmental variable for those sample units in which a species occurred. If the data are quantitative, a weighted average is taken, using the abundance values as weights for the environmental values. The resulting indicator values can then be used to locate each sample unit on the environmental gradient by performing another set of weighted averages, this time of the indicator values for species occurring in a sample unit, weighted by their abundance values.

If the weighted averaging is repeated, going back and forth from the species scores to the sample unit scores, the scores will quickly stabilize, producing a unique and invariant solution. This solution is a 1-D ordination of both sample units and species.

One of the most appealing characteristics of CA (=RA) was that it yielded both normal and transpose ordinations automatically and superimposed the two sets of points on the same coordinate system. Because a given data matrix has only one CA solution, there are few options to consider. As with other ordination methods, however, you should attend to the question of appropriate data transformations before performing the analysis.

Today we realize that CA has some serious faults for ecological community data:

- While the first axis may be a good representation of a single dominant underlying gradient, second and later axes are usually quadratic distortions of the first. This is an undesirable artifact, producing an arching scatter of points that can be seen in almost every 2-D CA scatterplot (see Examples).

- Distances tend to be compressed at the ends of the axes and stretched in the middle, even if the sample units are equally spaced on a series of unimodal species curves along an environmental gradient (Figures 19.1 & 19.2).

- The underlying distance measure (chi-square distance where samples are weighted by their totals) exaggerates the distinctiveness of samples containing several rare species.

Correspondence Analysis

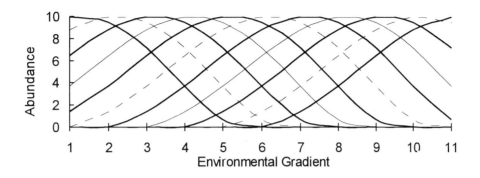

Figure 19.1. A synthetic data set of eleven species with noiseless hump-shaped responses to an environmental gradient. The gradient was sampled at eleven points (sample units), numbered 1-11.

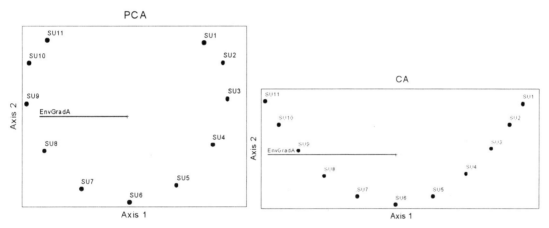

Figure 19.2. Comparison of PCA and CA of the data set shown in Figure 19.1. PCA curves the ends of the gradient in, while CA does not. The vectors indicate the correlations of the environmental gradient with the axis scores.

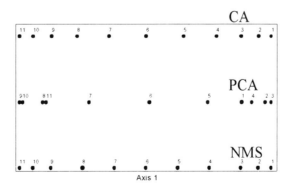

Figure 19.3. One-dimensional ordinations of the data set in Figures 19.1 and 19.2, using nonmetric multidimensional scaling (NMS), PCA, and CA. Both NMS and CA ordered the points correctly on the first axis, while PCA did not.

Despite these faults, it is worth acquiring some familiarity with CA because it forms the basis for several other techniques:

- detrended correspondence analysis (DCA; the brute-force approach to correcting the first two problems listed above; Ch. 20; Hill and Gauch 1980),
- two-way indicator species analysis (TWINSPAN; Ch. 12; a divisive classification method based on dividing CA axes; Hill 1979b), and
- canonical correspondence analysis (CCA; Ch. 21; additional steps introducing a weighted least squares regression of site scores on environmental data; ter Braak 1986).

When to use it

Because of the faults described above, there should be no regular application of CA to ecological community data. It is, however, conceivably useful if (a) you need to superimpose species space and sample unit space, (b) your system is essentially one-dimensional, and (c) you consider chi-square distances conceptually appropriate (see below). Usefulness of CA, as compared to distribution-free techniques such as NMS, has not been argued convincingly in the literature.

How it works

Axes are rotated simultaneously through species space and sample space with the object of maximizing the correspondence between the two spaces. This produces two sets of linear equations, $\mathbf{X}$ and $\mathbf{Y}$, where

$$\mathbf{X} = \mathbf{A}\mathbf{Y} \quad \text{and} \quad \mathbf{Y} = \mathbf{A}'\mathbf{X}$$

$\mathbf{A}$ is the original data matrix with n rows (henceforth sample units) and p columns (henceforth species). $\mathbf{X}$ contains the coordinates for the n sample units on k axes (dimensions). $\mathbf{Y}$ contains the coordinates for the p species on k axes. Note that both $\mathbf{Y}$ and $\mathbf{X}$ refer to the same coordinate system.

The goal is to maximize the correspondence, R, defined as:

$$R = \frac{\sum_{i=1}^{n}\sum_{j=1}^{p} a_{ij} x_i y_j}{\sum_{i=1}^{n}\sum_{j=1}^{p} a_{ij}}$$

under the constraints that

$$\sum_{i=1}^{n} x_i^2 = 1 \quad \text{and} \quad \sum_{j=1}^{p} y_i^2 = 1$$

Major steps — eigenanalysis approach

1. Calculate weighting coefficients based on the reciprocals of the sample unit totals and the species totals. The vector $\mathbf{v}$ contains the sample unit weights and $\mathbf{w}$ contains the species weights, where a_{i+} is the total for sample unit i and a_{+j} is the total for species j.

$$v_i = \frac{1}{a_{i+}} \quad \text{and} \quad w_j = \frac{1}{a_{+j}}$$

The square roots of these weights are placed in the diagonal of the two matrices $\mathbf{V}^{\frac{1}{2}}$ and $\mathbf{W}^{\frac{1}{2}}$, which are otherwise filled with zeros. If there are n sample units and p species, then $\mathbf{V}^{\frac{1}{2}}$ has dimensions $n \times n$ and $\mathbf{W}^{\frac{1}{2}}$ has dimensions $p \times p$.

2. Weight the data matrix $\mathbf{A}$ by $\mathbf{V}^{\frac{1}{2}}$ and $\mathbf{W}^{\frac{1}{2}}$:

$$\mathbf{B} = \mathbf{V}^{1/2} \mathbf{A} \mathbf{W}^{1/2}$$

In other words,

$$b_{ij} = \frac{a_{ij}}{\sqrt{a_{i+} \, a_{+j}}}$$

This is a simultaneous weighting by row and column totals. The resulting matrix $\mathbf{B}$ has n rows and p columns.

3. Calculate a cross-products matrix: $\mathbf{S} = \mathbf{B}'\mathbf{B} = \mathbf{V}^{\frac{1}{2}}\mathbf{A}\mathbf{W}\mathbf{A}'\mathbf{V}^{\frac{1}{2}}$. The dimensions of $\mathbf{S}$ are $n \times n$. The term on the right has dimensions:

$$(n \times n)(n \times p)(p \times p)(p \times n)(n \times n)$$

Note that $\mathbf{S}$, the cross-products matrix, is a variance-covariance matrix as in PCA except that the cross-products are weighted by the reciprocals of the square roots of the sample unit totals and the species totals. CA also differs from PCA in the scaling of the scores for sample units and species.

4. Now find eigenvalues as in PCA. Each eigenvalue (latent root) is a lambda (λ) that solves:

$$|\mathbf{S} - \lambda \mathbf{I}| = 0$$

This is the "characteristic equation." Note that it is the same as that used in PCA, except for the contents of $\mathbf{S}$.

5. Also find the eigenvectors **Y** ($p \times k$) and **X** ($n \times k$) for each of k dimensions:

$$[S - \lambda I]x = 0$$

and

$$[W^{½}A'VAW^{½} - \lambda I]y = 0$$

using the same set of λ in both cases. For each axis there is one λ and there is one vector **x**. For every λ there is one vector **y**.

6. At this point, we have found **X** and **Y** for k dimensions such that:

$$\underset{n \times k}{X} = \underset{n \times p}{A} \; \underset{p \times k}{Y}$$

and

$$\underset{p \times k}{Y} = \underset{p \times n}{A'} \; \underset{n \times k}{X}$$

where

Y contains the species ordination,

A is the original data matrix, and

X contains the sample ordination.

As in PCA, note that each component or axis can be represented as a linear combination of the original variables, and that each eigenvector contains the coefficients for the equation for one axis. For eigenvector 1 (the first column of **Y**):

Score1 $x_i = y_1 a_{i1} + y_2 a_{i2} + ... + y_p a_{ip}$ for entity i

The sample unit scores are scaled by multiplying each element of the SU eigenvectors, **X**, by

$$SU \; scaling \; factor = \sqrt{a_{++} / a_{i+}}$$

where a_{++} is the grand total of the matrix **A**. The species scores are scaled by multiplying each element of the SU eigenvectors, **Y**, by

$$Species \; scaling \; factor = \sqrt{a_{++} / a_{+j}}$$

Major steps — reciprocal averaging approach

1. Arbitrarily assign scores, **x**, to the n sample units. The scores position the sample units on an ordination axis.

2. Calculate species scores as weighted averages, where a_{+j} is the total for species j:

$$y_j = \frac{\sum_{i=1}^{n} a_{ij} x_i}{a_{+j}}$$

3. Calculate new site scores by weighted averaging of the species scores, where a_{i+} is the total for sample unit i:

$$x_i = \frac{\sum_{j=1}^{p} a_{ij} y_j}{a_{i+}}$$

4. Center and standardize the site scores so that

$$\sum_{i=1}^{n} a_{i+} x_i = 0 \quad \text{and} \quad \sum_{i=1}^{n} a_{i+} x_i^2 = 1$$

5. Check for convergence of the solution. If the site scores are closer than a prescribed tolerance to the site scores of the preceding iteration, then stop. Otherwise, return to step 2.

The underlying distance measure

Minchin (1987a) cleared up some confusion in the ecological literature about the expression of compositional distance in CA and related techniques. Hill and Gauch (1980) claimed that correspondence analysis "makes no use of the concept of compositional distance." Minchin (1987a) pointed out that this is not true; correspondence analysis can be viewed as an eigenanalysis technique where compositional dissimilarity is measured using the chi-square distance measure and samples are weighted by their totals (see also Chardy et al. 1976). Correspondence analysis results in an ordination space in which distances between sample points are approximately proportional to their chi-square distance values.

The chi-square distance measure gives high weight to species whose total abundance in the data matrix is low. This can result in exaggeration of the distinctiveness of samples containing several rare species (Faith et al. 1987; Minchin 1987a). Unlike proportionate city-block distance measures, chi-square distance does not have a constant maximal value for pairs of sample units with no species in common, but fluctuates according to variations in the abundance of species with high or low total abundances (Minchin 1987a).

Ter Braak (1985) showed that under four conditions, the chi-square distance represented by CA is an approximate solution to representing gradients of unimodal species responses. The four conditions are:

1. Sample units are homogeneously distributed along the gradients.
2. Species have equal environmental tolerances (species abundances rise and fall at about the same rate along the gradient), or species tolerances are independent of the position of the species on the gradient.
3. Species abundances at their optima are either equal to each other or species abundances are independent of the position of species optima.
4. Species optima are homogeneously distributed along the gradients.

Each of these conditions is always violated to some varying degree in actual community data. The consequences of those violations are unknown. Suffice it to say that CA has performed poorly with synthetic data sets of known underlying structure in two or more dimensions (example below).

Differences between PCA and CA

CA can be considered a variant of PCA. PCA preserves the Euclidean distances among the SUs, while CA preserves the chi-square distances among the SUs and among the variables. Remember that chi-square distances are similar to Euclidean distances except that they have been doubly weighted by the row and column totals.

PCA analyzes linear relationships among the variables, while CA analyzes linear relationships among the variables after doubly weighting by row and column totals. The difference in the resulting ordinations can be slight or profound. The difference is slight when row totals are fairly homogeneous, column totals are fairly homogeneous, and few zeros are present in the data.

Although strictly speaking, both PCA and CA fit a linear model to the data, the use of chi-square distances in CA imparts a distinct change in the emphasis of the ordination. If we assume that species are responding to an underlying environmental gradient, then, in effect, CA uses the central tendencies of the species on the gradient to locate the position of the sample units and vice-versa. In contrast, PCA uses the linear relationships between species to locate sample units.

What to report

☐ If applied to community data, a justification of why you consider appropriate the implicit chi-square distance measure.

☐ How many axes were interpreted and proportion of variance represented by those axes.

☐ If and how solution was rotated.

☐ Standard interpretive aids (overlays and correlations of variables with axes).

Examples

Simple 1-D example

If weighted averaging ordination is taken to its logical conclusion by reciprocal averaging, we obtain a set of refined scores for species and sample units (Fig. 19.4). Compare this one-dimensional CA ordination with the one derived from one-step weighted averaging (Fig. 18.1). In this case, the relative positions of sample units and species based on CA scores were fairly similar to those from the weighted averaging. Many are surprised to learn that the CA outcome is always the same, regardless of the initial species weights. Iteration of the reciprocal averaging forces convergence on a single solution.

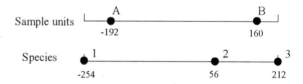

Figure 19.4. One dimensional CA ordination of the same data set used in the weighted averaging example in the previous chapter (Fig. 18.1). Scores were standardized to unit variance, then multiplied by 100.

2-D example

The most illuminating way of probing the quality of ordination spaces is to experiment with artificial data sets having a known underlying structure. The data set should selectively include various features of real data sets.

We generated a data set with smooth, noiseless, unimodal species responses to a strong primary gradient and to an independent, much weaker, secondary gradient. The species responses were sampled with a 10 × 3 grid of 30 sample units, with the long axis of the grid corresponding to the stronger environmental gradient.

Examine the distortion of the underlying grid (Fig. 19.5). For clarity, we have connected only the points along the major gradient. The CA (RA) ordination is shown, along with nonmetric multidimensional scaling (NMS) and a principal components analysis (PCA).

Table 19.1. Comparison of correspondence analysis CA, nonmetric multidimensional scaling (NMS), and principal components analysis (PCA) of a data set with known underlying structure.

	CA	NMS	PCA
Proportion of variance represented*			
Axis 1	0.473	0.832	0.327
Axis 2	0.242	0.084	0.194
Cumulative	0.715	0.917	0.521
Correlations with environmental variables			
Axis 1			
Gradient 1	-0.982	0.984	-0.852
Gradient 2	-0.067	0.022	-0.397
Axis 2			
Gradient 1	-0.058	0.102	-0.204
Gradient 2	-0.241	0.790	0.059

* Eigenvalue-based for CA and PCA; distance-based for NMS

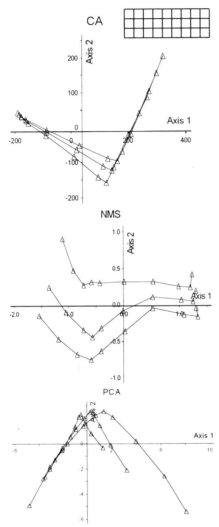

Figure 19.5. Comparison of 2-D CA (RA), nonmetric multidimensional scaling (NMS), and principal components analysis (PCA) of a data set with known underlying structure. The lines connect sample points along the major environmental gradient. The minor secondary gradient is nearly orthogonal to the major gradient. In the perfect ordination, the points would form a horizontally elongate grid. Inset: the ideal result is a regular 3 × 10 grid.

CA recovered the first gradient, as shown by the strong correlation between Gradient 1 and scores on Axis 1 (Table 19.1). The method failed, however, to recover the second gradient. Note the lack of separation between the lines (Fig. 19.5) and the absence of strong correlations with Gradient 2 (Table 19.1). Combined, the first two axes represented about 71% of the variation in the data (Table 19.1). The inverted arch in the CA ordination is quite typical. One can usually spot CA ordinations without even reading the methods because they typically scatter points along an arch or two sides of a triangle.

NMS recovered both gradients, the first gradient equally well with CA and the second gradient much more clearly than either of the other methods. Note the separation into parallel lines (Fig. 19.5) and the strong correlation between axis two scores and Gradient 2 ($r = 0.79$; Table 19.1). Combined, the first two axes represent 92% of the variation in the data.

As expected, the first two PCA axes express its struggle to fit a linear model to the nonlinear relationships among species. PCA must skewer the dust-bunny distribution with a straight line, also producing an arch. PCA failed to separate the two gradients (Fig. 19.5, Table 19.1) and expressed only 52% of the variation in the data.

Examples from the literature

All uses of CA in the literature of community ecology that we have seen are either inappropriate or inadequately justified, or both. Nevertheless, the consequences of using CA versus other ordination techniques are not severe when beta diversity is low. Be sure to consider this before dismissing a paper that used CA on community data.

CHAPTER 20

Detrended Correspondence Analysis

Background

Detrended correspondence analysis (DCA; Hill and Gauch 1980) is an eigenvector ordination technique based on correspondence analysis (CA or RA). DCA is geared to ecological data sets and the terminology is based on sample units and species.

DCA is frequently referred to by the name of the original computer program for this analysis, DECORANA, which originated in the Cornell Ecology Program series (Hill 1979a). The method is explained in detail in accessible terms by Gauch (1982a).

Like CA (Ch. 19), DCA ordinates both species and sample units simultaneously. The arch that is almost inevitable with more than one dimension in CA is squashed with DCA. The arch is squashed by dividing the first axis into segments, then setting the average score on the second axis within each segment to zero.

Another fault of CA, the tendency to compress the axis ends relative to the middle, is also corrected with DCA. This is done by rescaling the axis to equalize as much as possible the within-sample variance of species scores along the sample ordination axis. The rescaling also tends to equalize the species turnover rate along the gradient.

Like CA, its parent technique, DCA implicitly uses a chi-square distance measure. This was not, however, recognized by the creators of DCA (Hill and Gauch 1980), who claimed that correspondence analysis "makes no use of the concept of compositional distance." Although DCA and CA do not require an explicit calculation of a matrix of chi-square distances, CA is equivalent to an eigenanalysis of distances in a space defined by the chi-square distance metric (Chardy et al. 1976; also see "underlying distance measure" in the preceding chapter).

DCA has been quite popular in ecology. But it has come under attack on a number of fronts, one of the most penetrating of these being Minchin's (1987) comparison of DCA with nonmetric multidimensional scaling (NMS) and other methods.

Minchin (1987) found lack of robustness and erratic performance of DCA as compared to NMS using a wide variety of simulated data sets based on a range of assumptions about the underlying distributions of species along environmental gradients. He attributed this poor performance to two factors: properties of the implicit dissimilarity measure and the behavior of the detrending and rescaling processes. Minchin, as well as others (e.g., Beals 1984, Fewster & Orloci 1983, Ter Braak 1986), have also expressed reservations about the desirability of detrending and rescaling all ordinations to fit a single mold. This has been expressed in several ways:

Beals (1984): "...with detrending, getting from the original spatial model to the final ordination involves intense manipulations that obscure the direct relation between that model and its simplified representation, the ordination. CA generates information (the arch) that is not wanted, so that information is obliterated by drastic measures. It is rather like taking a hammer to pound out an unwanted bulge. The question of why CA causes severe curvature if not given such special prophylaxis is avoided rather than answered: can information lost by compression onto the first axis ever be retrieved along later axes? The theoretical as well as practical ramifications of such treatment need to be evaluated. Furthermore, a real sociological curvature, due to bimodality of some species, will be eliminated by this method, and hence useful ecological information may not be detected."

Jackson and Somers (1991): "...we believe that multidimensional configurations obtained with DCA may be unstable and potentially misleading. ...We have found that the choice of axis segmentation may substantially affect the interpretation and hence the utility of higher dimensions generated with DCA."

Legendre and Legendre (1999): "Present evidence indicates that detrending should be avoided, except for the specific purpose of estimating the lengths of gradients; such estimates remain subject to the assumptions of the model being true. In particular, DCA should be avoided when analysing data that represent complex ecological gradients."

Minchin (1987): "Not all curvilinear structures which may appear in an ordination are distortions, arising from the non-linear relationship between dissimilarity and environmental distance. DCA has no way of distinguishing between 'horse-shoe' or 'arch' distortions and features of the environmental configuration which happen to be nonlinear. There is a danger that DCA will introduce new distortions of its own."

Ter Braak (1986): "Detrending, however, also attempts to impose such a homogeneous distribution of scores on the data where none exists."

Besides these problems with the underlying concept of DCA, an instability in the results was induced by the algorithm for solving DCA (an algorithm is a specific series of steps and rules used by the computer to solve a problem). Tausch et al. (1995) found that scores in DCA depended on the order of sample units in the data set. While the instability was small on the first axis, later axes had considerable instability, sometimes enough to move a point to the opposite end of an axis! Standard deviation values among site scores in reordered data sets "generally increased through the axes, reaching the maximum with axis 4. On axis 2, some plots were jumping from one end of the axis to the other. This increased on axis 3 and by axis 4 the relative positions of the plots [for one data set] changed so much from randomization to randomization that they gave the appearance of being shuffled like a deck of cards" (Tausch et al. 1995). They also noted that this variation diminished with the use of presence-absence data.

Oksanen and Minchin (1997) showed two causes of instability in DCA (DECORANA): (1) lax criteria for stability of the solutions and (2) a bug in the smoothing algorithm of DECORANA. Note that this bug was present in the original version of DECORANA from Cornell as well as all its descendants, including CANOCO and PC-ORD. The problems were fixed with PC-ORD version 2.11 and higher and CANOCO version 3.15 and higher. PC-ORD applied Oksanen and Minchin's "super strict" criteria of tolerance = 10^{-7} and maximum number of iterations = 999. The bug in the rescaling algorithm was also corrected.

When to use it

DCA unnecessarily imposes assumptions about the distribution of sample units and species in environmental space. Other methods, especially nonmetric multidimensional scaling, perform as well or better without making those assumptions. There is no need to use DCA.

How it works

There is only one basic algorithm for DCA, DECORANA (Hill & Gauch 1980), although it has been adapted to various software packages. The algorithm is complex and has never been fully documented apart from the source code (computer instructions in FORTRAN) itself. To sketch the basic steps:

1. Solve the **eigenanalysis** problem for correspondence analysis (CA) (see preceding chapter).

2. **Detrend.** An axis is sliced into an arbitrary number of segments and scores on the next axis are adjusted so that the mean score within each segment is zero (Fig. 20.1). This is analogous to slicing the ordination into strips, then sliding them vertically to equalize the mean within each strip. The default number of segments is 26. The results are sensitive to

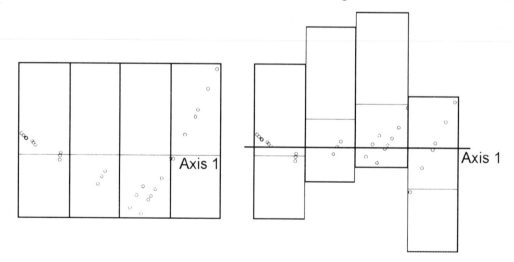

Figure 20.1. Detrending by segments. In this simplified example, a CA ordination axis is divided into an arbitrary number of segments (four in this case), then the sample unit scores on the vertical axis are centered on zero, within each segment, analogous to sliding the segments with respect to each other.

this parameter (Jackson & Somers 1991).

3. **Rescale**. The objective is to have species turn over at a uniform rate along the gradient. "For technical reasons this goal is best achieved by an indirect, somewhat complex procedure" (Gauch 1980). Individual segments of each axis are expanded or contracted to equalize the within-sample variation of species scores. The rescaling calculations are made on the species ordination (Fig. 20.2). The average local standard deviation is calculated (this is a pooled within-segment standard deviation). Then the width of each segment is adjusted such that a half change in species composition spans about 1-1.4 standard deviations. This is "Hill's scaling." Often the standard deviation units are multiplied by 100.

Output

List of residuals. The eigenvectors are found by iteration and the residuals indicate how close a given vector is to an eigenvector. A vector is deemed to be an eigenvector if it has a residual of less than a specified very small value (the "tolerance"). The residual is calculated from the sum of squared differences between the current estimate of the eigenvector and that estimate from the previous step. If this value is not reached in a set number of iterations, the current trial vector is taken to be the eigenvector and a warning is printed that the residual exceeds the desired tolerance.

Eigenvalue. The eigenvalue is produced from the CA part of the method. Beware that, after detrending and rescaling, the meaning of the eigenvalue is corrupted so that it no longer represents a fraction of the variance represented by the ordination. Thus, eigenvalues in DCA cannot be interpreted as proportions of variance represented. To evaluate the effectiveness of a DCA ordination, one can use an after-the-fact coefficient of determination between relative Euclidean distance in the unreduced species space and Euclidean distance in the ordination space.

Length of segments. These lengths indicate the degree of rescaling that occurred. For example, small values for the middle segments suggest that a gradient was compressed at the ends before rescaling. The rescaling routine equalizes the length of the segments, and these new segment lengths are then printed.

What to report

☐ If applied to community data, a justification of why you consider acceptable the assumptions behind the rescaling and detrending.

☐ How many axes were interpreted, proportion of variance represented by those axes, as measured by the correlation between Euclidean distances among sample units in the ordination and relative Euclidean distances in the original space.

☐ If and how solution was rotated.

☐ Standard interpretive aids (overlays, correlations of variables with axes).

Examples

One strong and one minor gradient

We illustrate DCA with the same synthetic data set as in the example under correspondence analysis (CA; see Fig. 19.5 & Fig. 20.3). This data set has smooth, noiseless, unimodal species responses to a strong primary gradient and an independent, much weaker, secondary gradient. The species responses were sampled with a 10 × 3 grid of 30 sample units, with the long axis of the grid corresponding to the stronger environmental gradient.

Figure 20.2. Segmenting a species ordination as the basis for rescaling in DCA. The arrows indicate the boundaries of segments. Circles are species. For each segment, DCA calculates the within-sample variance for species whose points occur within that segment. The lengths of the segments are then stretched to equalize those within sample variances. After rescaling, species tend to rise and fall (full turnover) over four standard deviations.

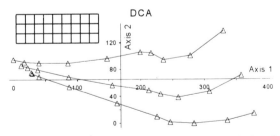

Figure 20.3. Detrended correspondence analysis of a data set with known underlying structure. The lines connect sample points along the major environmental gradient. The minor secondary gradient is nearly orthogonal to the major gradient. In the perfect ordination, the points would form a horizontally elongate rectilinear grid (inset).

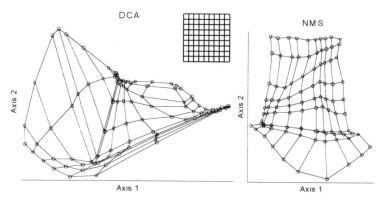

Figure 20.4. Comparison of DCA and NMS ordinations of a data set with known underlying structure, containing two strong gradients. DCA crumpled and folded the grid and while NMS more successfully extracted the underlying structure. Inset: the ideal result is a regular 10 × 10 grid.

In this case, DCA performed nearly as well as NMS and much better than CA (previous chapter; Fig. 19.5). The arch was effectively flattened (Fig. 20.3). Both the major and minor environmental gradients were detected, the correlation of scores on Axis 1 with the major gradient (r = -0.984) and Axis 2 with the minor gradient (r = 0.787). The correlation between distances in the ordination space and the relative Euclidean distances in the original space was 0.76 and 0.05 for Axes 1 and 2, respectively, for a cumulative variance represented of 0.81.

Two equally strong gradients

The case of one strong gradient is relatively easy for DCA to express. More complex structures, such as two strong gradients, are more challenging. An artificial data set was prepared with 40 species responding in smooth hump-shaped distributions to two environmental gradients. The response curves were sampled with a 10 × 10 grid of sample points.

Ideally, ordination methods should extract a rectilinear grid. In this case, however, DCA folded and crumpled the grid, altering both the overall and local structure of the grid (Fig. 20.4). The same data set was ordinated with NMS using Sørensen distance, selecting the lowest stress solution from 100 runs. NMS successfully extracted the overall structure of the grid and introduced relatively minor local distortions.

Discontinuity in one gradient

The assumptions required for detrending and rescaling introduce concerns about how DCA handles discontinuities in the sampling or in species response. We examined this problem by introducing a discontinuity into the data set of the previous example, eliminating the two middle sample points along that gradient. This simulated lack of sampling in the middle portion of a gradient. Ideally, an ordination method should show two rectilinear grids separated by a gap (Fig. 20.5, inset).

The discontinuity in sampling is distorted in the DCA ordination (Fig. 20.5) as a band of larger polygons, starting with the two largest polygons in the upper left of the graph then turning sharply to the right in the lower portion of the graph. Neither the overall structure of the grid nor the discontinuity are revealed clearly. NMS, on the other hand, clearly showed the discontinuity while preserving the overall structure of the grid (Fig. 20.5).

Variations

Detrending of CA can be done by segments (described above) or by polynomials (Hill & Gauch 1980; Jongman et al. 1987, 1995). Scores on axis 2 and higher can be replaced by residuals from the regression of ordination scores of that axis against quadratic or higher degree polynomials of earlier axes. This seems no more desirable than detrending by segments. Although it avoids an arbitrary decision on number of segments, it does not correct the compression at the ends of the axes.

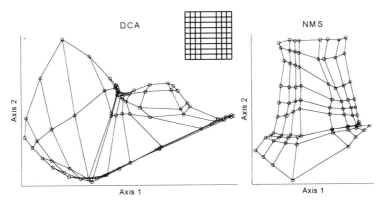

Figure 20.5. Comparison of DCA and NMS ordinations of a data set with known underlying structure, containing two strong gradients, one with a discontinuity in the sample. The data set is the same as for Figure 20.4 except for the introduction of this discontinuity. The discontinuity is barely recognizable in DCA while NMS successfully showed it, visible as relatively broad segments roughly parallel to Axis 1. Inset: the ideal result is a regular grid with a gap down the middle.

CHAPTER 21

Canonical Correspondence Analysis

Background

Canonical correspondence analysis (CCA; ter Braak 1986, 1994, 1995; Palmer 1993) constrains an ordination of one matrix by a multiple linear regression on variables in a second matrix. The ordination step is correspondence analysis (CA; Ch. 19), which has the same result as reciprocal averaging (RA). In community ecology, the usual form of CCA is an ordination of sample units in environmental space. A matrix of sample units × species is paired with a matrix of sample units × environmental variables.

CCA ignores community structure that is unrelated to the environmental variables. In contrast, performing an ordination on just the community data, and then secondarily relating the ordination to the environmental variables, allows an expression of pure community gradients, followed by an independent assessment of the importance of the measured environmental variables. See Økland (1996) for a discussion of the contrasting goals of CCA and most other ordination techniques.

When to use it

Multivariate analysis is a way of getting messages from a high-dimensional world that we cannot see directly. Using CCA is like getting those messages through a narrow mail slot in a door. The edges of the slot are defined by the measured environmental variables. Messages that do not fit the slot are either deformed to push them through or just left outside. Nonmetric multidimensional scaling (Ch. 16) is a more open, accepting approach. We accept all the messages, no matter their shape and size, sort them by their importance, then evaluate the important ones against the rulers of our environmental variables. Important messages from the community are retained, whether or not they relate to our environmental variables.

The following two questions can be used to decide whether CCA is appropriate: (1) Are you interested only in community structure that is related to your measured environmental variables? (2) Is a unimodal model of species responses to environment reasonable? If, for a specific problem you answer yes to both of these, then CCA might be appropriate (Fig. 21.1).

The basic questions appropriate for CCA are different than those for indirect ordination methods (Table 21.1). Nevertheless, there is widespread indiscriminate use of CCA.

CCA is best suited to community data sets where: (1) species responses to environment are unimodal (hump-shaped), and (2) the important underlying environmental variables have been measured. Note that the first condition causes problems for methods assuming linear response curves (such as RDA; see Variations below) but causes no problems for CCA. The second condition results from the environmental matrix being used to constrain the ordination results, unlike most other ordination technique (exceptions: Canonical Correlation (CC) and Redundancy Analysis (RDA)). Because CCA uses data on environment to structure the community analysis, and CCA plots points in a space defined by environmental variables, CCA can be considered a method for "direct gradient analysis".

CCA is currently one of the most popular ordination techniques in community ecology. It is, however, one of the most dangerous in the hands of people who do not take the time to understand this relatively complex method. The dangers lie principally in several areas: (1) Because CCA includes multiple regression of community gradients on environmental variables, CCA is subject to all of the hazards of multiple regression (Ch. 30). These are well documented in the statistical literature but are often not fully appreciated by newcomers to multiple regression. (2) As the number of environmental variables increases relative to the number of observations, the results become increasingly dubious: species-environment relationships appear to be strong even if the predictors are random numbers. (3) Statistics indicating the "percentage of variance explained" are calculated in several ways, each for a different question, but users frequently confuse these statistics when reporting their results.

Given the above cautions, it may seem somewhat paradoxical that, as the number of environmental variables approaches the number of sites, CCA becomes very similar to CA — i.e., the direct ordination becomes equivalent to an indirect ordination (ter Braak 1994, p. 129). This is because as more environmental variables are used, the constraints on the axes

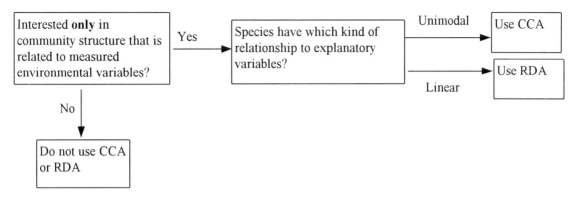

Figure 21.1. Decision tree for using CCA for community data. Assume that we have a site × species matrix and a site × environment matrix and that chi-square distances are acceptable. RDA is a constrained ordination method based on a linear model (see "Variations" below).

Table 21.1. Questions about the community (**A**) and environmental or experimental design (**E**) matrices that are appropriate for using CCA.

Matrix relationships	Questions for which CCA is OK	Questions for which CCA is not OK
A only	Not applicable.	What are the strongest gradients in species composition?
H_o: no linear relationship between **A** ← → **E**	Are any aspects of community structure related to these environmental variables?	Are the strongest community gradients related to these environmental variables?
Describe **A** ← → **E**	How is the community structure related to these environmental variables?	How are the strongest gradients in community structure related to these environmental variables?

become weaker and weaker until, when the number of environmental variables is one less than the number of sites, the constraints are effectively absent. But despite this lack of constraint by the environmental variables, the statistic for the species-environment correlation will equal 1.0!

CCA is available in the software CANOCO (ter Braak 1988, 1990) and PC-ORD (McCune and Mefford 1999). The CANOCO version includes a number of options, such as inclusion of covariates and detrending, which are not available in the PC-ORD version of CCA.

Nomenclature

By this point you may be a bit confused by the names of multivariate methods. It might seem as though methods are named by recombining the same few words, such as "canonical" and "correspondence."

We have canonical analysis (= discriminant analysis), correspondence analysis (= reciprocal averaging), canonical correspondence analysis, canonical variate analysis (= discriminant analysis), and canonical correlation.

The word "canonical" refers to the canonical form, the simplest and most comprehensive form of a mathematical structure (Legendre & Legendre 1998). The canonical form of a covariance matrix is the matrix of eigenvalues. Statisticians adopted the term "canonical" for the simultaneous analysis of two or more related data matrices. Historically "canonical analysis" has been used in a number of ways, including as a synonym of discriminant analysis.

Two methods

CCA can be performed either by an iterative, reciprocal-averaging method (steps 1-7 on p. 1169 in ter Braak 1986) or by eigenanalysis (Appendix A in ter Braak 1988). Apart from rounding errors, the results are identical.

CCA does not explicitly calculate a distance matrix. But, like CA and DCA, CCA is implicitly based on the chi-square distance measure where samples are weighted according to their totals (Chardy et al. 1976, Minchin 1987). This gives high weight to species whose total abundance in the data matrix is low, thus exaggerating the distinctiveness of samples containing several rare species (Faith et al. 1987, Minchin 1987).

How it works

The basic method

The iterative method is summarized below (after ter Braak 1986). The species data matrix **Y** contains nonnegative abundances, y_{ij}, for $i = 1$ to n sample units and $j = 1$ to p species. The notation y_{+j} and y_{i+} indicates species totals and sample unit (site) totals, respectively. The environmental matrix **Z** contains values for n sites by q environmental variables.

1. Start with arbitrary but unequal site scores, **x**.

2. Calculate species scores, **u**, by weighted averaging of the site scores:

$$u_j = \lambda^{-\alpha} \sum_{i=1}^{n} y_{ij} x_i / y_{+j}$$

The constant α is a user-selected scaling constant as described below. It can be ignored when studying the logic of the weighted averaging process.

3. Calculate new site scores, **x***, by weighted averaging of the species scores:

$$x_i^* = \lambda^{\alpha-1} \sum_{j=1}^{p} y_{ij} u_j / y_{i+}$$

These are the "**WA scores**" of Palmer (1993).

4. Obtain regression coefficients, **b**, by weighted least-squares multiple regression of the sites scores on the environmental variables. The weights are the site totals stored in the diagonal of the otherwise empty, $n \times n$ square matrix **R**.

$$\mathbf{b} = (\mathbf{Z'RZ})^{-1}\mathbf{Z'Rx}^*$$

5. Calculate new site scores that are the fitted values from the preceding regression:

$$\mathbf{x} = \mathbf{zb}$$

These are the "**LC scores**" of Palmer (1993), which are linear combinations of the environmental variables.

6. Adjust the site scores by making them uncorrelated with previous axes by weighted least squares multiple regression of the current site scores on the site scores of the preceding axes (if any). The adjusted scores are the residuals from this regression.

7. Center and standardize the site scores to a mean = 0 and variance = 1 (see "Axis scaling" below).

8. Check for convergence on a stable solution by summing the squared differences in site scores from those in the previous iteration. If the convergence criterion (detailed below) has not been reached, return to step 2.

9. Save site scores and species scores, then construct additional axes as desired by going to step 1.

Axis scaling

Step 7 above has several alternative possibilities. The two most commonly used rescalings can be called Hill's Scaling (Hill 1979a) and Centered with Unit Variance. The early versions of CANOCO (2.x) used Hill's Scaling as a default. CCA in PC-ORD and later versions of CANOCO (ter Braak 1990) use Centered with Unit Variance as a default. These are described below.

Centered with Unit Variance. The site scores are rescaled such that the mean is zero and the variance is one. This is accomplished in three steps:

$$\bar{x} = \sum_i w_i^* x_i$$

$$s^2 = \sum_i w_i^* (x_i - \bar{x})$$

$$x_i^* = \frac{(x_i - \bar{x})}{s}$$

where x_i^* is the new site score and w_i^* is the weight for site i, the weight being a ratio of the sum of species in site i (y_{i+}) to the grand total for the matrix (y_{++}).

With this standardization, the ordination diagram can be interpreted as a biplot. Fitted species abundances can be estimated for any point on the diagram, based on the position of the point. If this standardization is not used, the resulting ordination diagrams can be interpreted only in the sense of relative positions of the objects in the ordination space. Jongman et al. (1987, 1995) defined joint plots vs. biplots in this way: joint plots are interpreted in a relative way, while biplot scores have a more precise mathematical meaning, as defined above. In practice this distinction makes little difference, since ordinations are usually interpreted by the relative positioning of objects in the ordination space. This kind of interpretation is compatible with either scaling.

Hill's Scaling. The site scores are rescaled such that the standard deviation of species abundance ("tolerance" of ter Braak) along the axis is on average about 1 unit, and a species' response curve can be expected to rise and decline over an interval of about 4 units (± 2 standard deviations). It has been argued that this scaling makes the resulting ordination more interpretable ecologically, hence its use as the default scaling in early versions of CANOCO. The chief drawback is that this rescaling prevents interpretation of the resulting ordination as a biplot in the strict sense (Jongman et al. 1987, 1995).

Hill's scaling standardizes the scores such that:

$$\sum_{i,k} y_{ij}(x_i - u_j)^2 = y_{++}$$

In CCA, Hill's scaling is accomplished by multiplying the scores by a constant based on $\lambda^\alpha / 1-\lambda$ (see below). Thus it is a *linear* rescaling of the axis scores. In contrast, Hill's scaling in DECORANA is applied within segments of the ordination, such that the scaling applies both within segments and across the whole ordination. In DECORANA, the original scores do not have a linear relationship to the new scores.

Optimizing species or sites

The relative scaling of species versus sites (sample units) is controlled by the parameter α (Table 21.2). One can optimize the representation of sample units or species but not both at the same time.

If **site scores are weighted mean species scores** (α = 1, default with CANOCO version 2), then the configuration of the sites in the ordination is optimized, such that the distances between sites approximates their chi-square distances (Jongman et al. 1987, 1995).

If **species scores are weighted mean site scores** (α = 0, default with CANOCO 3.1 and CCA in PC-ORD), then the configuration of species in the ordination is optimized, such that the distances between species best approximates their chi-square distances. If both species points and biplot points for environmental variables are present in the ordination, then this choice allows a direct spatial interpretation of the relationship between environmental and species points. The biplot approximates weighted averages of species with respect to environmental variables.

If a **compromise** is chosen, then α = 0.5 in the transition formulas.

Table 21.2. Constants used for rescaling site and species scores in CCA. Combining the choices for axis scaling and optimizing species or sites results in the following constants used to rescale particular axes. Lambda (λ) is the eigenvalue for the given axis. Alpha (α) is selected as described in the text.

	Biplot scaling	Hill's scaling
Constant for rescaling species scores	$\dfrac{1}{\sqrt{\lambda^\alpha}}$	$\dfrac{1}{\sqrt{\lambda^\alpha(1-\lambda)}}$
Constant for rescaling site scores	$\sqrt{\lambda^\alpha}$	$\sqrt{\dfrac{\lambda^\alpha}{1-\lambda}}$

Interpreting output

1. Correlations among explanatory variables. As in multiple regression, examining the simple correlations among the explanatory variables (Table 21.3) will often help you understand your results.

2. Iteration report. Convergence on a stable solution is evaluated in relationship to a specified "tolerance" value. The test statistic that is checked for tolerance is the weighted mean square difference between current site scores and those of the previous iteration. Each term in the summation is weighted by the fraction of the total abundance that is present at that site. Historically in PC-ORD and CANOCO this tolerance was 10^{-10} with a maximum of 20 iterations. Oksanen and Minchin (1997) showed that both this tolerance and the maximum number of iterations were unnecessarily lax. Beginning with PC-ORD ver. 2.11, the tolerance was tightened to 10^{-13}. The maximum number of iterations was increased from 20 to 999.

The results are stated either as having reached this tolerance in a certain number of iterations or as having failed to reach this tolerance after the maximum number of iterations. Even if this tolerance was not reached the results may be reasonable. But if the difference from the previous iteration was large (ter Braak (1988) suggests a limit of 0.02) you should consider discarding the results from that axis.

3. Total variance in the species data. The total variance (or "inertia" in the terminology of ter Braak, 1990) in the species data is a statement of the total amount of variability in the community matrix that could potentially be "explained". Because of the rescaling involved in CCA, the total variance is calculated based on proportionate abundances. It can be thought of as the sum of squared deviations from expected values, which are based on the row and column totals. Let

e_{ij} = the expected value of species j at site i

y_{+j} = total for species j,

y_{i+} = total for site i, and

y_{++} = community matrix grand total.

Then

$$e_{ij} = \frac{y_{i+} \, y_{+j}}{y_{++}}$$

The variance of species j, var(y_j), is

$$\text{var}(y_j) = \frac{\sum_{i=1}^{n}(y_{ij} - e_{ij})^2 / e_{ij}}{y_{+j}}$$

and the total variance is

$$\text{total variance} = \sum_{i=1}^{n}\left(y_{+j} / y_{++}\right) \text{var}(y_j)$$

4. Axis summary statistics (Table 21.4).

Eigenvalues. The eigenvalue represents the variance in the community matrix that is attributed to a particular axis. The eigenvalues typically form a descending series. The most important exception to this is when the number of axes exceeds the number of environmental variables. For example, if there are two environmental variables then there are only two canonical axes. Any subsequent axes are derived simply by CA. Without the environmental constraint, the eigenvalue will almost always increase with the shift from a canonical to a noncanonical axis.

Variance in species data. For a particular axis, the percentage of variance in the community matrix that is explained by that axis is calculated as 100 times the ratio of the eigenvalue to the total variance. As an alternative, you can calculate an after-the-fact correlation between ordination distance and distances in the unreduced species space. Although CCA and CA have an intrinsic basis in the chi-square distance measure, empirically we have found that the correlation between points in a CCA or CA ordination space and the points in the original space can sometimes be higher when distances in the original space are measured as relative Euclidean distances rather than as chi-square distances.

Percent variance in species-environment relation explained. CANOCO reports this statistic, but CCA in PC-ORD does not. It is calculated as the ratio of the eigenvalue for a particular axis to the total amount of community variation represented by the environmental variables. This is statistic is frequently misunderstood. It will always be large if the number of environmental variables is large relative to the number of observations, even if the environmental matrix contains random numbers. The statistic is dangerous because it invites interpretation as the amount of variation in the community that is explained by the environmental variables. This is false in both the statistical sense and the biological sense.

Pearson correlation, species-environment. This is a standard correlation coefficient between sample scores for an axis derived from the species data (the WA scores) and the sample scores that are linear combinations of the environmental variables (the LC scores). This statistic has the same dangers for misinterpretation as the preceding statistic. If the number of environmental variables is large relative to the number of sample units, a large correlation coefficient will result, regardless of any underlying biological meaning.

Kendall (rank) correlation, species-environment. This statistic is a rank correlation coefficient for the same relationship as in the preceding paragraph.

5. Multiple regression results (Table 21.5). As explained above, the key feature of CCA that differentiates it from other ordination techniques is that the ordination of the community matrix is constrained by a multiple regression on the environmental matrix. The regression results report the effectiveness of the environmental variables in structuring the ordination and describe the relationships of the environmental variables to the ordination axes.

Canonical coefficients are conceptually similar to the usual regression coefficients and, as described above, are found by weighted least-squares regression. Canonical coefficients differ from regression coefficients because the regression is made within the iterations of the weighted averaging, rather than an after-the-fact regression of site scores on a set of environmental variables. Although the interpretation is similar to multiple regression, canonical coefficients have a larger variance than regression coefficients (ter Braak 1988). Because of this, canonical coefficients cannot be tested for significance with a Student's t-test.

The canonical coefficients represent the unique contribution of individual variables to the regression solution, as opposed to the simple correlation coefficient between a variable and an ordination axis. As for coefficients from multiple regression, the canonical coefficients are unstable if the environmental variables are moderately to strongly intercorrelated (multicollinear).

The canonical coefficients are presented in two forms. The standardized coefficients are based on the environmental variables centered and standardized to unit variance. The coefficients are also given for the variables in their original units.

The **R-squared** value for a given axis is the coefficient of determination for the multiple regression and expresses how well the environmental variables could be combined to "predict" site scores. Of course as the number of environmental variables increases, the R^2 value will go up, even if the variables contain random numbers. The silly extreme to this problem is that as the number of variables approaches the number of sites, R^2 must go to 1.0.

6. Final scores for sites and species. Ordination scores (coordinates on ordination axes) are given for each site, **x**, and each species, **u** (Tables 21.6, 21.7, 21.8). As in the parent method, CA, these scores are superimposable on the same ordination diagram. The scaling of these scores depends on the options selected for rescaling the ordination axes.

Note that: (1) Each site point falls approximately on the centroid of the species points that occur at that site, so that one may infer which species were most likely to have occurred at a given site. (2) The species points represent their optima, the peaks of their hump-shaped response surfaces to the environmental gradients (assuming unimodal humps and that important environmental variables were measured).

CCA calculates two sets of scores for the rows in the main matrix (usually sites or sample units). Either set can be graphed. The two sets are:

- **LC scores (x)**: Scores for rows in the main matrix are linear combinations of the columns in the second matrix (usually environmental variables). LC scores are in environmental space, each axis formed by linear combinations of environmental variables.

- **WA scores (x*)**: Scores for rows in the main matrix are derived from the columns in the main matrix (usually species). WA scores are in species space.

Early versions of CANOCO and early papers on CCA used the WA scores. More recently ter Braak (1994) recommended the LC scores. LC scores are currently the default in both PC-ORD and CANOCO. These represent the best fit of species abundances to the environmental data. The WA scores best represent the observed species abundances. The differences in the resulting diagrams for the two sets of scores can be quite marked in some cases.

The choice between LC and WA scores has caused considerable confusion (Palmer 1993):

> Since CCA, by any algorithm, produces two sets of site scores, it is unclear which is the most appropriate to use in an ordination diagram. The initial publications on CCA do not advise whether to plot WA scores [derived from species matrix] or LC scores [linear combinations of environmental variables]... Even the manual for CANODRAW (Smilauer 1990) designed to plot CCA results does not state which set of scores is used, although a computer file accompanying the program indicates that LC scores are the default. The most recent version of CANOCO... employs LC scores as the default, whereas previous versions utilized WA scores (ter Braak 1990). I suggest that ecologists use linear combinations in most cases, for reasons to be discussed below.

The choice is crucial, yet many papers showing CCA ordinations have not reported which set of scores was used. An important consequence of this choice is how the ordination reacts to noise in the environmental data. Scores that are linear combinations of environmental variables are highly sensitive to noise. If the environmental data are irrelevant or noisy, then the resulting ordination with LC scores will strongly distort the representation of community similarity (McCune 1997).

Figure 21.2 shows the pronounced influence of noise on the LC scores. In this example, 40 species respond to two underlying environmental gradients.

Chapter 21

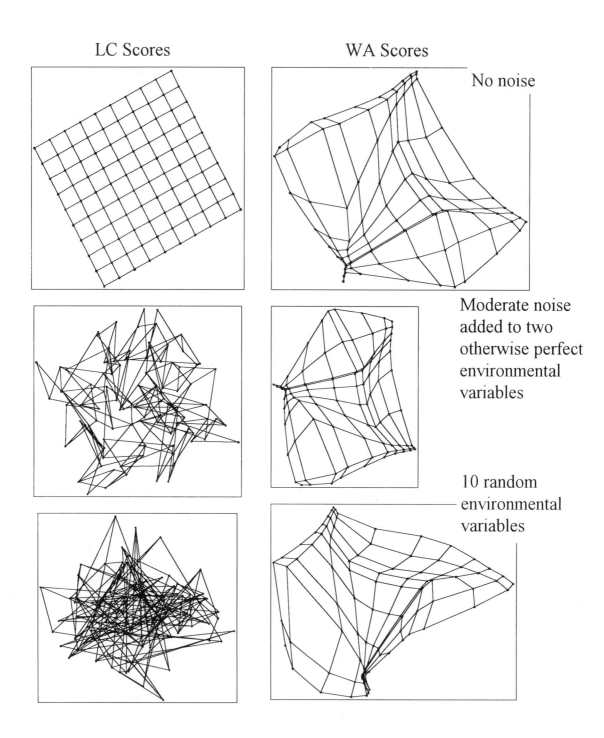

Figure 21.2 Influence of the type and amount of noise in environmental data on LC site scores (left column) and WA site scores (right column) from CCA, based on analysis of simulated responses of 40 species to two independent environmental gradients of approximately equal strength. Top row: noiseless environmental data. Middle row: a small amount of noise added to the two environmental variables. Bottom row: the two underlying environmental variables replaced with ten random variables. See text and McCune (1997) for more details.

The species response curves were sampled with a 10 × 10 grid of 100 points (sample units). In the first row of Figure 21.2 species have smooth unimodal responses to the environmental gradients. The LC scores show a perfect grid, but in this case it is a tautological result, such as $X = X$. Any 2-D plot of LC scores, when there are only two environmental variables, can do nothing but replot a rotation of those two variables. In contrast, the WA scores show some degree of distortion arising from the imperfect representation of the species data by the underlying weighted averaging process.

In the second row of Figure 21.2, a small amount of noise was added to the environmental variables, such that the r^2 between the original variables and the noisy variables was 0.83. This simulates the normal measurement error associated with most environmental variables. Because there are two axes and only two environmental variables, the ordination must perfectly represent those two variables. The small amount of noise has, however, partially scrambled the community patterns in the LC scores, as shown by the folded and contorted grid. In contrast, the WA scores are much more stable, being largely determined by the species data rather than being a direct projection in environmental space.

In the third row of Figure 21.2, the two underlying environmental variables were replaced by ten random environmental variables. This simulates the failure of an investigator to measure the important variables. In this case, the ordination by LC scores completely scrambles the community relationships in its attempt to fit the species to the environment. Still, the WA scores show a reasonably good representation of the grid, and therefore a reasonably good representation of community relationships. Once again this demonstrates the insensitivity of the WA scores to the environmental variables.

In simple linear regression the conceptual equivalent to LC scores is a scatterplot of "predicted" rather than observed data values versus the independent variable — a straight line is formed regardless of the quality of the prediction. Use of LC scores will similarly misrepresent the observed community relationships unless the environmental data are both noiseless and meaningful. Because environmental data in community ecology usually contain measurement error, CCA with LC scores is inappropriate where the objective is to describe community structure (McCune 1997; see also Økland 1996). CCA is insensitive to noise if the final scores are derived from the species scores (WA scores), but in this case, it is not a "direct gradient analysis" because the axes are not combinations of environmental variables.

Based on the simulated grid approach, Palmer (1993) concluded that CCA was insensitive to noise because he introduced noise into the environmental matrix only by augmenting the matrix with random numbers, leaving the noiseless environmental variables intact. In contrast, adding a modest amount of noise to the existing environmental variables destroyed the representation of community structure (McCune 1997).

7. Weights for sites and species. Sites and species are weighted by their totals. These weights can be useful in interpreting the ordination. For example, species at the edges of the ordination often have low weights and therefore have little influence on the analysis (ter Braak 1988).

8. Correlations of environmental variables with ordination axes. Remember that CCA generates two sets of site scores: (1) scores obtained by weighted averaging of the species scores (x^*, the WA scores) and (2) the LC scores, x, obtained by regressing x^* on the environmental variables then calculating fitted values, x, from the regression equation (step 5 under "The basic method" above). Ter Braak (1986) calls correlations of environmental variables with x^* the "**interset correlations**" and correlations with x the "**intraset correlations.**" The intraset correlations will almost always be higher than the interset correlations, because of the dependence of x on the environmental variables that is built into the regression process. The intraset correlations indicate which environmental variables were more influential in structuring the ordination, but intraset correlations cannot be viewed as an independent measure of the strength of relationship between communities and those environmental variables.

The interset correlations have the same problem but in a weaker form: because the ordination axes were constrained by the environmental variables, the interset correlations of the environmental variables with the ordination axes will be higher than for an unconstrained ordination (i.e., indirect gradient analysis).

For consistency in interpretation, it is reasonable to use the intraset correlations if the site scores x were used as the final scores, and to use the interset correlations if the site scores x^* were used as the final scores.

9. Biplot scores for environmental variables (Table 21.9). The environmental variables are often represented as lines radiating from the centroid of the ordination. The biplot scores give the coordinates of the tips of the radiating lines (Fig. 21.3). Given the usual methods of scaling the axes of the community ordination (mean score = 0), the origin of the environmental lines coincides with the origin of the community matrix.

The longer the environmental line, the stronger the relationship of that variable with the community. The position of species relative to the environmental lines can be used to interpret the relationships between species and the environmental variables. Imagine the environmental line extended through the whole ordination space. If perpendiculars are run from each species point to the environmental line, then one can see (1) the approximate ranking of species response curves to that environmental variable, and (2) whether a species has a higher-than-average or lower-than-average optimum on that environmental variable (higher than average if the species falls on the same side of the centroid as the tip of the line).

The coordinates for the environmental points are based on the intraset correlations. These correlations are weighted by a function of the eigenvalue of an axis and the scaling constant (α):

$$v_{jk} = r_{jk}\sqrt{\lambda_k^\alpha}$$

where v_{jk} = the biplot score on axis k of environmental variable j, r_{jk} = intraset correlation of variable j with axis k, and alpha is a scaling constant described above ("Optimizing species or sites").

If Hill's scaling is used, then

$$v_{jk} = r_{jk}\sqrt{\lambda_k^\alpha(1-\lambda_k)}$$

10. Monte Carlo tests of significance. The simplest kinds of Monte Carlo (randomization) tests for CCA test the following hypotheses:

H_o: No linear relationship between matrices. For this hypothesis, the rows in the second matrix are randomly reassigned within the second matrix. This destroys the relationship between the main and second matrices, but it leaves the correlation structure of the variables in the second matrix intact.

H_o: No structure in main matrix and therefore no linear relationship between matrices. For this hypothesis, elements in the main matrix are randomly reassigned *within* columns. This preserves the column totals and the matrix totals, but alters the row totals and destroys the relationships among columns in the main matrix (usually species).

For each hypothesis, you must select the number of runs (the number of randomizations). The larger the number of runs, the more precision you can obtain in the resulting *p*-values. You must also specify how to obtain the random number "seeds" to initialize the random number generator. Your output will contain a table showing how the observed eigenvalues and species-environment correlations compare with those from randomized runs (Table 21.10). For example, to evaluate the significance of the first CCA axis, a *p*-value is computed that is simply the proportion of randomized runs with an eigenvalue greater than or equal to the observed eigenvalue. If:

n = the number of randomizations (permutations) with an eigenvalue greater than or equal to the corresponding observed eigenvalue, and

N = the total number of randomizations (permutations),

then

$$p = (1 + n)/(1 + N)$$

This *p* value is the probability of type I error for the null hypothesis that you selected.

What to report

☐ Contents and dimensions of the two matrices.

☐ Justification or explanation for why you chose to ignore community structure that was unrelated to your environmental variables.

☐ How many axes were interpreted, proportion of variance explained by those axes.

☐ Scaling options (optimizing species or sites? Biplot scaling or Hill's scaling?).

☐ Monte Carlo test methods and results (null hypothesis, number of randomizations, resulting *p*-value).

☐ If an ordination diagram is included, state whether the points are LC or WA scores.

☐ Standard interpretive aids (overlays, correlations of variables with axes).

Example

Consider a hypothetical example where CCA can be applied, providing that you accept the implicit distance measure and its associated assumptions. A lawsuit is filed against a manufacturing company, the suit claiming that the company has altered stream chemistry and therefore damaged the native fauna of a stream. Because replication is impossible, part of the case depends on a clear demonstration of the relationship between the pollutant and the stream communities. The species data set has counts of 38 species in 100 sample units. The environmental data set contains the log-transformed concentration of the pollutant (the variable "LogPoll") from associated water samples, plus concentrations of two other compounds (Var2 and Var3) that possibly interact with the pollutant.

CCA was chosen because we are not interested in community structure in general, only whether some portion of the community structure is more strongly related to the pollutant than expected by chance. We know that stream velocity is also important, but our data show that stream velocity and LogPoll are virtually independent. We therefore omit stream velocity from the model. This focuses the analysis on the specific question of whether LogPoll is related to stream community structure.

The correlation matrix (Table 21.3) among the environmental variables showed each variable to be virtually independent of the others. This simplifies interpretation of the results.

The iteration report showed that a stable solution was quickly found for each of the first three axes. The tolerance level of 0.100000E-12 (= 10^{-10}) was achieved after 7, 14, and 3 iterations for the first three axes, respectively.

Because there are three environmental variables, we can find at most three canonical axes. The total variance ("inertia") in the species data is 4.42, so statistics on variance explained are based on ratios against this number (Table 21.4).

Table 21.3. Correlations among the environmental variables.

	LogPoll	Var2	Var3
LogPoll	1	0.107	-0.119
Var2	0.107	1	-0.039
Var3	-0.119	-0.039	1

Table 21.4. Axis summary statistics

	Axis 1	Axis 2	Axis 3
Eigenvalue	0.636	0.044	0.016
Variance in species data			
% of variance explained	14.4	1.0	0.4
Cumulative % explained	14.4	15.4	15.8
Pearson Correlation, Spp-Envt	0.900	0.307	0.213
Kendall (Rank) Corr., Spp-Envt	0.717	0.167	0.158

Table 21.5. Multiple regression results (regression of sites in species space on environmental variables).

	Canonical Coefficients						
	Standardized			Original Units			
Variable	Axis 1	Axis 2	Axis 3	Axis 1	Axis 2	Axis 3	S.Dev
LogPoll	-0.799	0.014	0.009	-2.385	0.041	0.027	0.335
Var2	0.033	-0.194	0.048	0.11	-0.638	0.159	0.304
Var3	0.003	0.075	0.118	0.01	0.242	0.378	0.312

Table 21.6. Sample unit scores that are derived from the scores of species. These are the WA scores. Raw data totals (weights) are also given.

	WA scores			Raw Data
	Axis 1	Axis 2	Axis 3	Totals
Site1	1.298381	1.555888	-0.98204	1131
Site2	1.17872	1.19812	-1.26412	1000
Site3	0.808255	0.145479	-1.10749	721
Site4	0.335053	-1.16647	-0.34654	635
Site5	0.204182	-1.40531	-0.05847	735
...				
Site99	-1.15441	0.044354	0.100729	580
Site100	-1.31167	-0.45384	-0.23881	748

Table 21.7. Sample unit scores that are linear combinations of environmental variables for 100 sites. These are the LC Scores that are plotted in Fig. 21.3.

	Axis 1	Axis 2	Axis 3
Site1	0.857	0.213	0.012
Site2	0.423	-0.100	-0.103
Site3	0.646	0.103	-0.024
Site4	0.474	-0.238	-0.104
Site5	-0.107	-0.297	0.078
...			
Site99	-0.807	0.159	0.147
Site100	-1.405	0.011	-0.084

Table 21.8. Species scores and raw data totals (weights).

	Axis 1	Axis 2	Axis 3	Raw Data Totals
Sp1	-0.769	4.211	4.643	16
Sp2	-1.608	0.240	-2.377	627
Sp3	-1.051	1.623	1.862	68
Sp4	1.344	1.817	-0.794	3464
...				
Sp37	-1.640	1.072	-1.748	164
Sp38	-1.158	3.689	2.161	7

Table 21.9. Biplot scores and correlations for the environmental variables with the ordination axes. Biplot scores are used to plot the vectors in the ordination diagram. Two kinds of correlations are shown, interset and intraset.

Variable	Axis 1	Axis 2	Axis 3
BIPLOT scores			
LogPoll	-0.797	-0.008	0.002
Var2	-0.028	-0.196	0.045
Var3	0.073	0.081	0.115
INTRASET correlations			
LogPoll	-0.999	-0.038	0.018
Var2	-0.035	-0.933	0.357
Var3	0.092	0.386	0.918
INTERSET correlations			
LogPoll	-0.899	-0.012	0.004
Var2	-0.032	-0.286	0.076
Var3	0.083	0.118	0.195

Table 21.10. Monte Carlo test results for **eigenvalues** and **species-environment correlations** based on 999 runs with randomized data.

		Randomized data			
Axis	Real data	Mean	Minimum	Maximum	p
Eigenvalue					
1	0.636	0.098	0.033	0.217	0.001
2	0.044	0.046	0.009	0.112	0.511
3	0.016	0.020	0.004	0.076	0.620
Spp-Envt Corr.					
1	0.900	0.378	0.224	0.553	0.001
2	0.307	0.287	0.155	0.432	0.333
3	0.213	0.218	0.107	0.396	0.522

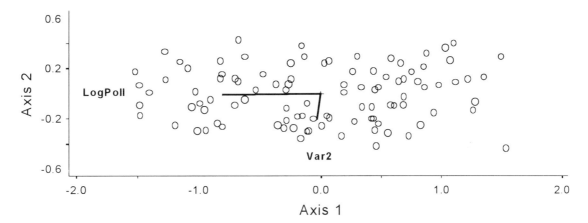

Figure 21.3. Ordination of sites (circles) in environmental space, as defined by CCA, using LC scores. The biplot overlay shows vectors related to the two strongest environmental variables, LogPoll and Var2. Axes are scaled by standard deviates.

Most of the variation (14.4%) explained is in the first axis (Table 21.4). Axes 2 and 3 should be largely ignored — they explain little variance and, as we shall see later, they fail the Monte Carlo test. If variance explained is calculated with the after-the-fact distance-based method, 12.6% of the variance in chi-square distances in the original space is represented by the first axis (13.0% if measured with relative Euclidean distance in the original space).

Following the axis summary statistics, we show the multiple regression results (Table 21.5), the WA scores (Table 21.6), the LC scores (Table 21.7), the species scores (Table 21.8), the biplot scores for environmental variables and correlations between axes and the environmental variables (Table 21.9), and the results of the randomization test (Table 21.10).

What can we conclude from these results? First, we reject the hypothesis of no relationship between the species data and the environmental data. The eigenvalue for the first axis is much higher than the range expected by chance (Table 21.10). This axis is strongly related to the variable LogPoll (Table 21.9; Fig. 21.3). The other two environmental variables have weak relationships to the species data.

Comparison with NMS

An unconstrained ordination of the same data set should result in a lower correlation between the key environmental variables and the ordination scores, but should provide a better representation of the overall community structure. We ran NMS on Sørensen distances and obtained a much stronger description of community structure with only a slightly lower correlation between the variable LogPoll and the first ordination axis. The results of NMS and CCA are compared in Table 21.11. The NMS was based on 100, 2-D runs with random starts, a maximum of 500 iterations per run, and a stability criterion of 0.000001.

Randomization tests for both methods indicated stronger structure in the data than expected by chance alone, but the tests are not directly comparable. In CCA, the randomization tests whether the eigenvalue is larger than for random reassignments of sample units in one of the two matrices. Thus, the test is whether or not the relationship between matrices is stronger than expected by chance. In NMS, the randomization tests whether the final stress is lower with the real data than for the species data shuffled within columns. Thus, the randomization test in NMS is for whether the structure in the species matrix is stronger than expected by chance, and makes no reference to the environmental matrix.

Table 21.11. Comparison of CCA and NMS of the example data set.

	CCA	NMS
Variance represented (%)		
Axis 1	14.4	29.5
Axis 2	1.0	38.9
Cumulative	15.4	68.4
Correlation with LogPoll		
Axis 1	-0.899	0.673
Axis 2	-0.012	-0.038

Variations

Prior transformations

Sample units and species are not usually relativized before CCA, because CCA has a built-in double weighting by sample unit and species totals. This can be considered a form of relativization. Monotonic transformations (such as a log or square root) should, however, be considered. See the guidelines in Chapter 9.

Detrended CCA

The results from CCA can be detrended and rescaled, resulting in detrended canonical correspondence analysis (DCCA) in much the same way that correspondence analysis (CA or RA) can be detrended and rescaled (detrended correspondence analysis or DCA; Ch 20). If detrending and rescaling are not selected, then one often finds the same "arch problem" that occurs in CA: later axes often simply introduce an upwards or downwards bulge into the first axis. If one chooses detrending and rescaling, then the arch is eliminated, but new problems are likely to be introduced by the detrending process, similar to the problems encountered in DCA (Minchin 1987).

Canonical correlation

Canonical correlation is similar to performing PCA simultaneously on two matrices, but subjecting the analyses to an overriding goal of maximizing the correlation between the two sets of axes. That is, the goal is to analyze the linear relationships between the matrices. Before the introduction of CCA, canonical correlation was the most widely used technique for simultaneous analysis of two related matrices in ecology. Examples, nevertheless, were fairly rare (e.g., Austin 1968, Gauch et al. 1976, Kercher & Goldstein 1977).

Tabachnik and Fidell (2001) gave an excellent description of the method, with examples. They call this "the least used and most impoverished of the techniques," despite its apparent generality. Among the theoretical limitations they discuss, difficulty of interpretation and sensitivity to nonlinearities are perhaps the most important to ecologists. Canonical correlation assumes linear relationships within each matrix and between matrices. It is, therefore, fundamentally incompatible with hump-shaped species responses to environmental gradients and with the dust-bunny distribution of sample units in species space (Ch. 5).

Co-inertia analysis

Co-inertia analysis has been described as an alternative to CCA (Dolédec & Chessel 1994, Franquet et al. 1995). In this method, PCA is first performed on the species and environmental matrices separately. Co-inertia Analysis answers questions about the nature and degree of agreement between these two sets of axes. The method calculates new axes to maximize the covariance between the sample scores in the two ordination spaces. Note that co-inertia analysis seeks linear relationships among species, so it will have the same problems with the dust bunny distribution as does PCA (Ch. 14). Furthermore, it faces the same pitfalls as ordinating environmental data directly (Ch. 2).

Redundancy analysis

Redundancy analysis (RDA; ter Braak 1994, Van Wijngaarden et al. 1995) is analogous to CCA but is based on a linear model of species responses, rather than a unimodal model as in CCA. RDA is the canonical counterpart of PCA, just as CCA is the canonical counterpart of CA. What these authors call "redundancy analysis" is also known as "maximum redundancy analysis" in fields outside of ecology (e.g., Rao 1964, Fortier 1966, van den Wollenberg 1977). The term "redundancy analysis" has also been applied to a technique used to interpret canonical correlation results.

Assume we begin with a matrix of response variables (**A**) and a matrix of explanatory variables (**E**). The basic steps of RDA as applied in community ecology are:

1. Center and standardize columns of **A** and **E**.
2. Regress each response variable on **E**.
3. Calculated fitted values for the response variables from the multiple regressions.
4. Perform PCA on the matrix of fitted values
5. Use eigenvectors from that PCA to calculate scores of sample units in the space defined by **E**.

Numerous variants are possible by various ways of initially transforming the data and rescaling the resultant axes (Legendre & Legendre 1998). Use of this method in ecology has increased in recent years, despite it rarely being appropriate for community data because of the assumption of linear relationships between variables in **A** and **E**.

RDA is made much more useful for ecologists by an extension allowing choice of any distance measure (Legendre & Anderson 1999, McArdle & Anderson 2000). This is known as "distance-based RDA" (db-RDA).

Regression with multiple dependent variables

Regression with multiple dependent variables is mathematically essentially the same as regression with a single dependent variable. In the usual case of regressing a single dependent variable (**Y**) on multiple independent variables (**X**), the regression coefficients (**B**) are found by:

$$\mathbf{B} = (\mathbf{X'X})^{-1} \mathbf{X'Y}$$

With multiple dependent variables, **Y** and **B** are matrices rather than vectors. Like canonical correlation, this linear model is usually inappropriate when the dependent variables are species abundances. Austin (1972) found it important to include curvilinear terms and competitive effects of dominant species when using regression. Pursuing this approach led Austin (1979, 1981) to conclude that satisfactory regression models could be obtained only for narrower subsets of the data and only when the dominants were used as predictors.

Simultaneous regression

Simultaneous regression involves simultaneously solving more than one regression equation, where dependent variables from one or more equations are used as independent variables in another equation. Linear simultaneous equation models are a special case of structural equation models (Chapter 30). Nonlinear simultaneous equation models are possible but have apparently not been applied in ecology.

Multivariate regression trees

De'ath (2002) used multivariate regression trees (MRT) to represent the relationships between species composition and environment. (See Chapter 29 for more information on classification and regression trees.) MRT is a method of constrained clustering (Chapter 10). De'ath used MRT to construct a hierarchical classification of communities, selecting division points from environmental variables such that the within-group variation was minimized.

MRT differs from CCA and RDA in that MRT is a divisive hierarchical method for forming groups, while CCA and RDA model continuous structure. De'ath (2002) further pointed out that MRT emphasizes local structure and interactions between environmental effects, while RDA and CCA seek global structure. MRT assumes no particular relationship between species abundance and environment, while RDA and CCA assume linear and unimodal distributions of species along gradients, respectively. Last, one form of MRT allows you to choose any distance measure, similar to db-RDA, while CCA forces an optimization of the chi-square distance measure.

CHAPTER 22

Reliability of Ordination Results

The most common method of evaluating the reliability of ordinations is to look for consistency in interpretation when using various options or ordination methods. When a given pattern emerges over and over, but it varies somewhat in strength and detail, the analyst develops faith that the underlying structure is strong. This subjective approach can be greatly aided by several more quantitative tools, as described below.

When we interpret an ordination, we assume that the configuration of points represents some definite structure, defined by the relationships among the rows and columns (sample units and species for community data). If someone else repeated the study in the same time and place we would hope to see that same configuration emerge. Obviously, it would not be exactly the same, but how different would it be?

Measurement error, **sampling error**, **methodological artifacts**, and **ecological inconsistencies**, and are sources of variation in ordination results. We can quantitatively express the reliability of an ordination with a **proportion of variance represented**, evaluation of **strength of pattern** compared to a null model, **consistency** of a pattern with different subsets of data, and **accuracy** of a pattern with respect to a true underlying pattern. Each of these is discussed below.

Sources of error

Measurement error

Measurement errors include observer effects, estimation error, and instrument error. See Chapter 3 on Community Sampling and Measurements. Although measurement errors can be considerable, they are often dwarfed by the biological inconsistencies in the relationships among species.

Sampling error

Each sample unit represents some area or volume at a point or span of times. In other words, each sample unit is a chunk of space-time that often represents a larger chunk of space-time — but with what reliability? If a row in the matrix represents an average across 40 subsamples from an area, how different would another subsample be? If each bird community is characterized by five listening points, how different would the ordination be if slightly different listening points were used? If a stand of trees is represented by a single large plot, how much do the results depend on the plot location within the stand?

Methodological artifacts

Analytical methods may introduce their own component of apparent community structure. For example, reciprocal averaging compresses the ends of a gradient. This is a repeatable structure, but it is nearly independent of the data and it is purely a methodological artifact. Another example is that the detrending in DCA (Ch. 20) both removes and introduces methodological artifacts. These artifacts intervene between the analyst and the underlying ecological pattern that transcends particular methods.

Ecological inconsistencies

The strength of multivariate analysis of ecological communities comes from the strength of species associations. The complexity of species interactions, numerous influential and interacting environmental factors, persistent effects of historical events, and genetic variation within species all act to prevent simple, strong, general associations. In almost all cases, however, associations among species are sufficiently strong to make ordination a valuable tool. The reliability and importance of the pattern that emerges can be described in several ways, discussed in the following paragraphs.

Measures of error

Proportion of variance represented

If we assume no measurement error and no sampling error, still the ordination may not represent a strong structure. Remember that an ordination attempts a low-dimensional representation of a high-dimensional data matrix. If the variables are largely independent of each other, then the ordination will represent only a small fraction of the variation in the data. In the introduction to ordination (Ch. 13), we introduced the concept of proportion of variance in the data that is represented by the ordination. This is one aspect of pattern strength, but it is not the whole story. For example, we saw that outliers can result in statistics indicating a high variance represented, but this may represent nothing about the main mass of data. Just knowing the proportion of variance represented tells us nothing about sensitivity of the apparent pattern to the inclusion of particular rows or columns of the matrix.

Strength of pattern

If we assume no measurement error, no sampling error, and a reasonably high proportion of the variance represented, still the ordination may not represent a stable, strong structure. Is the pattern stronger than expected by chance? We can answer this question by comparing our ordination results to a null model.

Note that there are other kinds of null models in community ecology (e.g., Bell 2001). One is concerned with comparing observed distributions of relative abundances of species with that expected by random allocation of abundance of species. See, for example, Hopf and Brown (1986) and numerous papers cited there. A wide variety of approaches have been designed for an equally wide array of specific problems (e.g., Connor & Simberloff 1979, Manly 1995, Pielou 1984, Wilson 1987). Sanderson et al. (1998) reviewed the literature on null models for presence-absence data.

The null hypothesis of "no correlation structure" among the variables in a community matrix is often trivially easy to reject and hardly needs testing, except in particular circumstances when the structure of the data calls into question the strength of the correlation structure. Nevertheless, it is reassuring to readers if you provide results of this hypothesis test.

Many different parametric methods have been used to test for significant correlation structure in eigenanalysis of multivariate data. One common example is Bartlett's test of sphericity (this and other techniques reviewed by Grossman et al. 1991, Jackson 1993, and Franklin et al. 1995). Heuristic methods are also available (see Chapter 14).

Nonparametric methods, such as randomization or Monte Carlo tests, can also be used to compare the degree of structure extracted by multivariate techniques with that from repeatedly shuffled data sets or distance matrices. Randomization tests can be applied to any ordination method.

For ordinations of a single matrix, the values within variables (columns) can be randomly shuffled, the shuffling performed independently for each column. The result is a data set having the same column totals, the same matrix total, and the same number of elements containing zeros. This is a more realistic null model than, for example, using a table of random numbers for comparison, because a random numbers table is unlike typical community data. For example, most community data sets contain a large proportion of zeros and have a very skewed distribution of species abundances (Ch. 5). Ideally, the randomization should retain as many of the properties of the matrix as possible while destroying the relationships among species. A sample unit × species matrix shuffled within species destroys the resemblances among species and the sample unit totals. Other properties of the matrix are preserved.

For presence-absence data, Sanderson et al. (1998) and Sanderson (2000) applied the knight's tour algorithm to the problem of creating null matrices. This provides an efficient solution for generating a series of null matrices whose row and column totals are fixed, a crucial improvement over previous attempts at generating null matrices. Unfortunately a comparable approach to quantitative data is not available.

For comparisons of two matrices (e.g., CCA; Ch. 21) or a matrix and a grouping variable (e.g., MRPP; Ch. 24), one can shuffle one matrix relative to the other. For example, consider a sample unit × species matrix and a sample unit × environmental matrix. We preserve all aspects of the matrices while destroying their relationship by randomly permuting the order of sample units in one of the matrices.

Consistency

Knowing whether an observed pattern is stronger than expected by chance is not a fully satisfying description of the reliability of an ordination. Wilson (1981) in comparing independent rankings of species along gradients, based on multiple partitions of the same data set, stated "the rejection of this null hypothesis (independence of rankings) is not a strong statement. Non-independence does not indicate the nature or the degree of dependence, and the establishment of non-independence holds little encouragement for the investigator seeking to interpret the patterns of nature. A more appropriate null model is one in which there exists a true underlying species ranking."

Wilson (1981) distinguished between consistency and accuracy and provided operational definitions of these. **Consistency** is the ability of an ordination to produce concordant results from different subsets of the data. Wilson (1981) used nonoverlapping subsets of the data to evaluate consistency, but you can also use other strategies, such as bootstrapping. **Accuracy** is the degree to which an ordination correctly portrays the true community structure. We discuss this below.

Accuracy

Pattern accuracy is the aspect of reliability most difficult to assess. The concept implies knowledge of the true underlying pattern. That is completely known only with simulated data sets that are generated from a known underlying structure. Wilson (1981) approximated the true underlying pattern by obtaining a maximum likelihood estimate of species ranks based on multiple independent subsets of large data sets (see "Wilson's method" below).

Bootstrapped ordination

A "bootstrap" refers to using the original data set to generate new data sets. This is one application of a class of computationally intensive methods for testing the significance of almost any statistic with a minimum of assumptions (Efron and Tibsharani 1991). Typically the original data set is sampled with replacement. From an $n \times p$ matrix, n items are randomly drawn for a bootstrap sample, usually with some items being drawn more than once. Based on a large number of bootstrap samples, the standard error of the mean is then calculated for whatever parameter is being evaluated. The individual data points can be single numbers, vectors, or matrices. If the parameter relates to positions on individual ordination axes, it is essential to rotate or otherwise adjust the scores to eliminate trivial differences such as one axis in one ordination being a reflection of an axis in another ordination. Procrustes rotation accomplishes this by rotating and dilating or expanding an ordination to maximize the correspondence with a second ordination (see Chapter 15).

Knox and Peet (1989) and Knox (1989) used bootstrapped ordination methods to assess the consistency of ordinations for both simulated and field data. They used the variance in rank of species scores across bootstrap replicates. These variances were averaged across species. The average variance was then rescaled to range from 0 to 1:

$$\text{scaled rank variance} = \frac{\text{observed variance in rank}}{\text{expected variance in rank}}$$

The expected variance in rank = $(n^2 - 1)/12$ where n = the number of items. See Knox and Peet (1989) for suggestions on how to interpret the scaled rank variance.

Pillar (1999b) proposed a general method for bootstrapped ordinations, comparing a bootstrapped real data set with a parallel bootstrapping of randomized data sets. He demonstrated its application to metric ordinations, but its applicability to nonmetric multidimensional scaling (Ch. 16) has not been investigated.

The steps in Pillar's (1999b) method are:

1. Save the usual ordination scores for k axes from the complete data set ($n \times p$). Call the $n \times k$ scores the original ordination.
2. Draw a bootstrapped sample of size n.
3. Ordinate the sample.
4. Perform Procrustes rotation of the k axes from the bootstrapped ordination, maximizing its alignment with the original ordination.
5. Calculate the correlation coefficient between the original and bootstrapped ordination scores, saving a separate coefficient for each axis. The higher the correlation, the better the agreement between the scores for the full data set and the bootstrap.
6. Repeat all of the steps above for a randomization of the original data set. The elements of the complete data set are randomly permuted within columns.
7. For each axis, if the correlation coefficient from step 5 for the randomized data set is greater than or equal to the correlation coefficient from the nonrandomized data set, then increment a frequency counter, $F = F + 1$.
8. Repeat the steps above a large number of times (B = 40 or more, depending on computation time).
9. For the null hypothesis that the ordination structure of the data set is no stronger than expected by chance, calculate a probability of type I error:

$$p = F/B$$

The interpretation of p is fundamentally the same as for other statistical tests. The probability indicates the strength of the structure in an ordination, as compared to a null data set containing variables with the same observed distribution but random association. A p that is smaller than a previously stated threshold (alpha) indicates an ordination dimension that is more stable than expected for the same dimension in the ordination of a random data set. In this case, we reject the null hypothesis of structure no stronger than expected by chance. Alternatively, one may view the p as a continuous value indicating the consistency ("reliability" in Pillar's words) of the ordination.

Like other statistical tests, a small sample size means low power and high probability of type II error.

Failure to reject null hypothesis in this case may simply mean that a larger sample size was needed for a confident conclusion.

Wilson's method

Wilson (1981) performed ordinations on replicate subsamples, tested for independence of species rankings, and described accuracy and consistency, based on the methods of Feigen and Cohen (1978). This technique is data hungry because it requires nonoverlapping partitions of the data.

Definitions

w_0 = the true underlying species ranking

$\tilde{w}_i$ = an estimate of the true ranking, based on species scores on an ordination axis

$X(w_0, w)$ = the number of discordant pairs between two rankings, w_0 and w (the basic measure of disagreement between rankings; see Kendall 1975 and Feigen and Cohen 1978). Two species constitute a discordant pair if their rank order in one ranking is reversed within a second ranking.

τ = Kendall's tau, a commonly used rank correlation coefficient, which is a linear function of X.

q = the number of rankings (subsets)

k = the number of objects (species)

$\hat{w}_0$ = the value of w to minimize $\sum_{i=1}^{q} X(w, \tilde{w}_i)$

According to Wilson (1981), "Average ranks are an excellent basis for a first approximation of $\hat{w}_0$, with only a few permutations of ranks necessary to identify the maximum likelihood estimate." The measure of overall disagreement between the observed rankings based on subsets of the data and the maximum likelihood estimated ranking is

$$\overline{X} = (1/q) \sum_{i=1}^{q} X(\hat{w}_0, \tilde{w}_i)$$

The expected value of Kendall's rank correlation (τ) between the true underlying species ranking and the ordination species ranking is estimated by

$$\hat{E}[\tau] = 1 - 2 \binom{k}{2}^{-1} \overline{X}$$

Kendall's τ ranges from -1 (complete disagreement) to 1 (complete agreement), and it can be used as a measure of **accuracy** of the ordination.

The **consistency** of the ordination is measured as the ratio of the observed variation to the expected variation:

$$C = s_x / \sqrt{var(X)}$$

where $var(X)$ is the expected variance of X (see tables in Feigin and Cohen, 1978). The lower the value of C, the higher the consistency.

Procedure (The procedure is written for sample units to be partitioned, then the rankings of species compared. In principle, one could also partition species, then compare the ranking of samples, but this would address a question about statistical redundancy of species rather than reliability of ordination.)

1. Randomly partition the sample into q subsets. The number q needs to be a compromise between a small enough number to yield many subsets and a large enough number to allow satisfactory ordination results (Wilson suggested 20-25 items in each subset).
2. By ordination, produce q rankings of the p species.
3. Test for overall independence of the rankings (e.g., using Kendall's coefficient of concordance, W). If the hypothesis of independence is not rejected, stop.
4. Calculate the maximum likelihood estimate ($\overline{X}$) of the true species ranking.
5. Measure the accuracy (τ) of the ordination rankings.
6. Measure the consistency (C) of the rankings.
7. Wilson (1981) also recommended testing the fit of the observations to the model, by comparing observed and expected frequencies of X with a Kolmogorov-Smirnov or chi-square test. If the model is inappropriate, reject the analysis and stop.

Part 5. Comparing Groups

CHAPTER 23

Multivariate Experiments

The ecological literature shows an unnecessary separation between the statistical tools used by experimental and observational studies. Experimentalists tend to use univariate statistics, even when they pose multivariate questions, while pattern seekers range freely between univariate and multivariate statistics. Analysts of experimental community studies can often improve statistical power and the directness of the answer to their questions by using multivariate tools.

The standard parametric multivariate tools are usually inappropriate for experimental studies of community structure, because they often encounter the same problems describe in Chapters 5 and 6. Here we introduce some alternative methods for comparing groups, such as permutation tests (MRPP), indicator species analysis, and Mantel tests. We then develop these more fully in the subsequent chapters.

What is an experiment?

Although the word "experiment" is often used loosely, in a true experiment, the investigator manipulates one or more variables in a consistent way and contrasts responses with controls or other treatments. The key word is **manipulates**, with the implicit consequence that the scientist can completely control the allocation and uniformity of treatments.

"Natural experiments" take advantage of events in nature that are similar to manipulative treatments. For example, if part of a slope burned, one may contrast paired burned and unburned sample units across the edges of the burn. Natural experiments are often simpler in design than planned, manipulative experiments. Natural experiments can be criticized as having less control over extraneous variables than is possible with laboratory experiments. This same criticism can, however, also be applied to virtually any field experiment in a natural setting.

Nonreplicated experiments present special problems (Hurlbert 1984, Eberhardt and Thomas 1991), but they have been viewed as unavoidable in many situations. See the recent literature on what we can extract from these experiments (Carpenter et al. 1989, Carpenter 1990, Eberhardt and Thomas 1991). In another examples, Peterson et al. (1993) repeatedly fertilized a tundra river and examined changes in diatom community composition (univariate approach to multivariate problem; refer to path 3 in Fig. 23.1), among other ecosystem responses.

Univariate vs. multivariate analyses

One can use univariate and multivariate analyses on the same design. Univariate analyses allow you to see responses of individual variables clearly. Multivariate analyses can be criticized as obfuscating the interpretation, while the results of univariate analyses are supposed to be easy to communicate. But what about a question like: "Did channelizing the stream affect the insect communities?" Can univariate analyses provide a satisfying answer to this multivariate question?

Multivariate analyses integrate responses of individual variables. One can have significant multivariate differences in the absence of significant univariate differences. A series of univariate analyses of intercorrelated response variables does not express the strength and nature of the relationships among the response variables.

Frequently, investigators pose a general multivariate question but then set up an experiment to test one or two species responses. This common solution conveniently avoids addressing the question directly.

How do ecologists approach the problem of univariate vs. multivariate analyses of community experiments? The paths we take can be summarized by Figure 23.1. Each path represents an alternative approach. Examples of these approaches are given below, with numbers corresponding to their position in Figure 23.1.

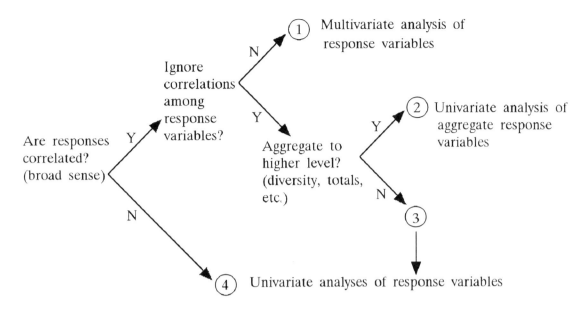

Figure 23.1. Four approaches to analysis of multivariate experiments in community ecology.

1. Multivariate responses

Gilbert et al. (1993) and Van Wijngaarden et al. (1995) used multivariate methods to answer multivariate questions. Gilbert studied the effects of an introduced bacterium on bacterial communities on roots. Van Wijngaarden uses RDA (Redundancy Analysis, a regression-constrained version of PCA, analogous to CCA; Ch. 21) to evaluate insecticide effects on communities of aquatic macroinvertebrates.

Gough and Grace (1999) used structural equation modeling (Ch. 30) to analyze the simultaneous response of biomass and species density to two experimental factors and a suite of abiotic variables. This approach allowed the separation of direct and indirect effects on the response variables. In some cases direct and indirect effects canceled each other out — this could not have been detected with univariate analyses.

2. Univariate analysis of aggregate response variables

Tilman (1993) analyzed aggregate univariate measures including diversity, proportion of species gained or lost, and biomass. Harte and Shaw (1995) analyzed biomass of growth forms, each growth form being an aggregate of species. They used an in-situ climate manipulation experiment to study the responses of montane meadow vegetation to climate warming. Although these meadows contained about 100 species of flowering plants, they simplified the concept of "vegetation" by using as response variables the aggregate biomass of three growth forms: graminoids, forbs, and shrubs. They did not report a direct analysis of species composition

3. Univariate analyses of individual response variables

Arnott and Vanni (1993), de Swart et al. (1994), Schopp-Guth et al. (1994), and Turkington et al. (1993), among many others, analyzed individual species separately. In most cases, they analyzed only the dominant species, even though it is probable that species responses were correlated. Ignoring the correlations among species sacrifices statistical power and biological insight.

4. Univariate analyses of uncorrelated responses

It is possible, though improbable, that a series of experimental treatments would result in species responding singly and without correlation to other species responses. We have found no examples of species being demonstrated to be uncorrelated, as would be needed to justify restricting the analysis to a series of univariate comparisons for individual species.

What to report

The exact nature of the analysis dictates what you should report, but some general guidelines are possible.

- Effect size and direction (e.g., difference between means, slopes of lines, etc.). This is a description of the treatment effect.
- Partitioning of variance (often applicable in one form or another, be it proportion of variance on various discriminant functions or the r^2 for a covariate).
- sample size.
- p(type I error).
- p(type II error), if possible. See Cohen (1988).

Tools and examples

MANOVA

Multivariate analysis of variance (MANOVA) and variants of the general linear model (GLM) apply to multiple response variables. The object is to represent the covariation in the response variables as a function of the experimental groups and covariates. MANOVA tests whether the mean differences among groups are likely to have occurred by chance. The differences are evaluated for a combination of response (dependent) variables,. New, synthetic dependent variables are constructed to maximize the group differences. The synthetic variables are linear combinations of the original variables.

Stroup and Stubbendieck (1983) prescribed MANOVA of experimental data with species composition as dependent variables. They point out that "correlation among the various species counts is inevitable," and that a multivariate approach increases the power of the statistical tests.

Gilbert et al. (1993) used MANOVA to determine whether bacterial communities on roots differed among habitats in an experiment. Some of the habitats were treated with an introduced bacterium; others were not. If they found significant differences among habitats, then they used stepwise discriminant analysis (Ch. 26) to explore the differences among communities from the various habitats.

If relationships among the response variables can be reasonably approximated by straight lines, then MANOVA is a good tool. This is not, however, the case for typical community data. Relationships among species are usually strongly nonlinear (Ch. 5). The dust bunny distribution of sample units in species space cannot be simply transformed into a multivariate normal distribution and linear relationships among species. In community analysis MANOVA should, therefore, be reserved for the few special cases where the assumption of linear relationships is reasonable and demonstrable.

Discriminant analysis (DA)

Experimental groups can define the a priori groups required for discriminant analysis (DA; Ch. 26). Mathematically, DA and MANOVA are the same, but DA differs from MANOVA in interpretation. To quote Tabachnik and Fidell (1996), "MANOVA asks if mean differences among groups on the combined dependent variable are larger than expected by chance; discriminant function analysis asks if there is some combination of variables that reliably separates groups." One can think of the groups as predictors (independent variables) in MANOVA and the community matrix as a response (dependent variables), while the reverse is true in DA. In DA, one attempts to combine the independent variables in such a way as to maximize the group separation. See Tabachnik & Fidell (2001) for an excellent discussion of the differences between DA and MANOVA.

We have not seen DA applied to more complex group structures (like factorial designs) in community ecology. It is, however, possible to do this (Tabachnik & Fidell 2001).

Mantel tests

Mantel tests provide tremendous flexibility for nonparametric analysis of multivariate responses to experimental treatments, based on constructing distance matrices to match the design. Mantel tests compare more than one distance matrix from the same set of sample units. Note the contrast (Fig. 23.2) with a multi-response permutation procedure (MRPP), which compares different groups of sample units. For more information, see Chapters 24 and 27.

MRPP

Multi-response permutation procedures (MRPP) provide a nonparametric multivariate test of differences between two or more groups, based on analysis of a distance matrix. In experimental studies, the treatments define the groups. The method works very well with community data, but is restricted to simple designs (one-way classification with two or more groups or one-way classification with blocking or matched pairs).

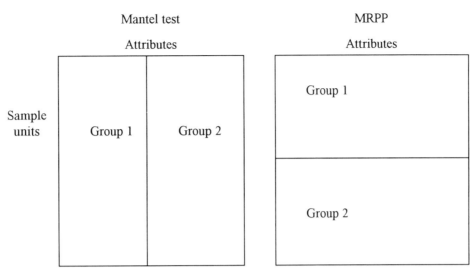

Figure 23.2. Comparison of data grouping for Mantel tests and MRPP.

Biondini et al. (1988) gave results from two case studies as part of a methodological study of MRPP and blocked MRPP. In one case, successional responses of vegetation were compared between four different topsoil treatments. In the second case, they compared biomass of shrub species on reclamation sites receiving six different treatments.

Zimmerman et al. (1985) compared biomass of various plant species on burned and unburned sites in North Dakota. Biondini et al. (1985) compared plant species composition, rate of successional change, and soil characteristics after four levels of disturbance.

The Q_b method

The Q_b Method (Pillar and Orloci 1996) is a nonparametric randomization method similar in use to MRPP except that the method allows multifactor experiments and inclusion of interation terms. Unfortunately for community ecology, a Euclidean metric is required to accomplish the partitioning of the sums of squares of the distance matrix. We describe the method further in the MRPP chapter (Ch. 24). The method is not currently available in the major statistical packages.

NPMANOVA

Nonparametric MANOVA (Anderson 2001) allows multivariate comparisons based on non-Euclidean distances, not just for one-way classifications (as with Mantel tests and MRPP), but also for factorial and nested designs. We describe this method in Chapter 24 (MRPP). Anderson (2001) gave an example for an experiment involving three factors, one of them nested.

BATC and BACI designs

BATC (before-after-treatment-control) or **BACI** (before-after-control-impact) designs present special problems. The problems derive from the use of repeated measures and the nonreplicated treatments that are inherent in their designs.

These are difficult and important problems. They are important from the standpoint of environmental impact assessment from single sources. They are inherently difficult from the impossibility of replicating treatments. Thus, these designs have received a fair amount of attention in the recent literature (Eberhardt and Thomas 1991; Osenberg et al. 1994; Stewart-Oaten et al. 1986, 1992; Thrush et al. 1994; Philippi et al. 1998). Nonparametric multivariate analyses of changes in community structure under these designs need more study.

Matched pairs

Matched pair designs, also known as paired-sample or matched case-control studies, seek statistical

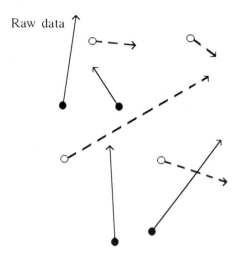

Figure 23.3. Vectors representing trajectories of sample units in species space. Solid vectors represent treated sample units; dashed vectors represent controls.

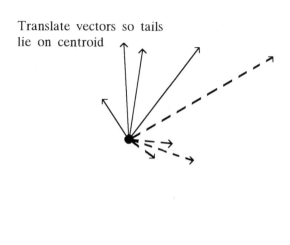

Figure 23.4. Treatment and control vectors translated to put the tails on the centroid. Vectors represent trajectories of sample units in species space.

power by comparing two treatments (or treatment vs. control) in paired sample units that are otherwise as close to identical as possible. Conceptually blocked MRPP is to MRPP as a paired-sample t-test is to an independent-sample t-test. But a multivariate response demands different analytical tools. Many methods can accommodate matched-pair designs, including MANOVA and matched-case logistic regression. For ecological community data, blocked MRPP (Ch. 24) is often appropriate.

Compositional vectors

Community changes can be compared between experimental treatments by analyzing the differences in length and direction of compositional vectors (McCune 1992, Phillipi 1998). These vectors describe the trajectory of sample units in species space.

Permanent plot data, models of community dynamics, and long-term field experiments are all proliferating. The growth of data and theory is creating a growing need for a logical and precise language for comparing and describing trajectories in species space. The concept of communities as points moving in species space provides an intuitive basis for further development of that language (McCune 1992). The vector manipulations suggested in this section are so far not available in commercial software. The calculations can, however, be readily performed in spreadsheets or scripting languages.

We can examine model predictions by comparing predicted trajectories with actual change. The total error in those predictions can be partitioned into differences in rate and direction. Those components of error can be further partitioned into components contributed by individual species.

For example, assume four treatment and four control sample units, each measured before and after treatment. Further assume a simple system of two species to aid visualization, although in practice the calculations can be made in as many dimensions as desired.

Sample units varied in pretreatment condition (this would be expected in actual field experiments; Fig. 23.3). For each sample unit, a vector extends from the position in species space before treatment to the position in species space after treatment. The vectors show movements of the sample units in species space. We wish to test whether treated sample units (Fig. 23.3, solid vectors) moved in a different direction and or rate than control (dashed vectors) sample units. At least three kinds of hypotheses can be tested:

Overall test of differences (combining rate and direction) of treatment and control vectors. First translate the vectors to a common origin (Fig. 23.4) by subtracting the score of the tail from the scores of both the head and tail. Then pose the hypothesis of no difference between the two (or more) groups of vector heads. This is equivalent to a test of whether the heads of the vectors occupy the same region in n-dimensional species space. One could use MRPP, MANOVA, or discriminant analysis to perform the test.

Differences in direction only. One can isolate and test for differences in direction between treatment and control vectors by translating the vectors to a common origin, standardizing the vectors to unit length (Fig. 23.5), then applying one of the techniques listed above.

Differences in rate only. Differences in mean length of treatment and control vectors can be compared with a t-test, one-way ANOVA, or corresponding nonparametric techniques.

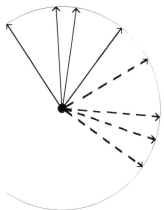

Figure 23.5. Compositional vectors first translated to put the tails on the centroid, then standardized to unit length.

CHAPTER 24

MRPP (Multi-response Permutation Procedures)
and Related Techniques

Background

MRPP is a nonparametric procedure for testing the hypothesis of no difference between two or more groups of entities. For example, one could compare species composition between burned and unburned plots to test the hypothesis of no treatment effect.

MRPP is just one of numerous independently developed nonparametric methods for testing for group differences. Two variants (ANOSIM and the Q_b method) can be found at the end of this chapter. A recent breakthrough (NPMANOVA, Anderson 2001) allowing the use of non-Euclidean distance measures in multifactorial designs is also described at the end of this chapter.

A good introduction to MRPP is the appendix in Biondini et al. (1985). More details can be found in Berry et al. (1983), Mielke (1984), and Mielke and Berry (2001).

The method requires pre-existing groups of entities or sample units. Commonly these groups are inherent in the sampling design or experimental design. For exploratory analysis they can be defined in any way you want. For example, two groups of sample units could be based on the presence-absence of a species of particular interest. You can also define groups based on a categorical environmental variable, such as whether the bedrock is limestone, granitic rock, or mafic rock.

MRPP provides little more than the test statistic, a measure of "effect size," and a p-value. Describing the differences among groups is usually done with one or more of the following methods:

1. **Overlay** the variable that defines group membership on an ordination of sample units. This gives a graphical representation of the relationships among groups. It is particularly useful when there are only two or three strong underlying patterns of variation.

2. For community data, describe the indicator value of individual species for separating the groups with Dufrêne and Legendre's (1997) **indicator species analysis** (Ch. 26).

3. Use **discriminant analysis** (Ch. 26) in a descriptive mode to identify variables most effective in predicting group membership. The size of the standardized discriminant function coefficient associated with each variable indicates the contribution of that variable to the discriminant function.

4. Use a **mean similarity dendrogram** (Van Sickle 1997) to graphically show relationships among groups. This is essentially a cluster analysis of within-group averages (or medians). Conceptually it is closely linked to MRPP, being based on mean similarities of replicate objects within the same class. The results are plotted on a scale of the original distance units rather than by Wishart's Objective Function (see Chapter 10). The end of each branch of the tree is clipped at the average within-group distance for the group represented by that branch.

When to use it

Discriminant analysis and multivariate analysis of variance (MANOVA) are parametric procedures that can be used on the same general class of questions as MRPP. However, MRPP has the advantage of not requiring distributional assumptions (such as multivariate normality and homogeneity of variances) that are seldom met with ecological community data. The default choice for community data should, therefore, be MRPP or a similar randomization technique.

With multivariate normal data, MRPP can be more powerful or less powerful than MANOVA (and a test based on Hotelling's T^2), depending on the nature of the data (Smith 1998). But with the nonlinear relationships and extremely skewed frequency distributions typical of community data, MANOVA is seldom a reasonable choice.

Complex experimental designs are difficult to analyze with MRPP and related techniques. If you must use a complex design, you can either use MANOVA, an MRPP-related technique based on Euclidean distance (e.g., see the Q_b method below), or analyze your data piecewise with MRPP, slicing it various ways to answer different questions. The last approach sacrifices your ability to analyze the interaction terms. You can, however, visualize those interactions using your experimental factors in a joint plot overlay on an ordination diagram (Ch. 13). Independent, noninteracting factors will show up as perpendicular vectors.

Assumptions

It is sometimes said, though it is untrue, that nonparametric statistics require no assumptions. Nonparametric statistics such as MRPP do, however, avoid *distributional* assumptions. In some cases this comes with a loss in power. The following assumptions should be considered when applying MRPP and related techniques:

1. The distance measure chosen adequately represents the variation of interest in the data.
2. The sample units are independent. The usual problems with pseudoreplication, subsampling, and repeated measures are conceptually the same as with ANOVA.
3. The relative weighting of the variables has been controlled prior to calculating the distance measure, such that the weighting of variables is appropriate for assessing the ecological question at hand.

How it works

1. Calculate distance matrix, **D**. Various distance measures can be used. Euclidean distance and squared Euclidean distance were used most often in the early literature of MRPP. Zimmerman et al. (1985) recommended Euclidean distance over squared Euclidean distance. Proportional city-block distance measures (e.g., Sørensen distance) are increasingly used in published studies with MRPP and community data. If MRPP is used in conjunction with ordination, then it is often desirable to choose the same distance measure for both.

2. Calculate the average distance x_i within each group i.

3. Calculate delta (the weighted mean within-group distance)

$$delta = \delta = \sum_{i=1}^{g} C_i x_i$$

for g groups, where C is a weight that depends on the number of items in the groups (normally $C_i = n_i / N$, where n_i is the number of items in group i and N is the total number of items). Note that all $n_i \geq 2$. For a given mean overall distance, smaller values of δ indicate tighter clustering within groups.

Various other weightings have been used and are generally based on group size. Those available in PC-ORD are in Table 24.1. In each case, n_i is the number of items in group i, g is the number of groups, and C_i is the weight applied to each item in group i.

Table 24.1. Methods for weighting groups in MRPP.

Formula	Comments
$C_i = \dfrac{n_i}{\sum n_i}$	A natural weighting recommended by Mielke (1984) and used in most recent applications of MRPP.
$C_i = \dfrac{n_i - 1}{\sum (n_i - 1)}$	With squared Euclidean distance this weighting results in an MRPP statistic that is equivalent to a 2-sample t-test or one-way ANOVA F-test (Mielke et al. 1982; Zimmerman et al. 1985). While this option accounts for degrees of freedom, this is a foreign concept to permutation procedures.
$C_i = \dfrac{1}{g}$	Not recommended but available for experimentation.
$C_i = \dfrac{n_i * (n_i - 1)}{\sum (n_i * (n_i - 1))}$	Not recommended but available for experimentation. Used in some early applications of MRPP.

4. Determine probability of a δ this small or smaller. The brute force or exact method (applicable only to tiny data sets) is to calculate δ for all possible partitions of the same sizes. This is the same as the number of possible permutations of the values assigning cases (sample units) to groups (Fig. 24.1). The number of possible partitions (M) for two groups is

$$M = N!/(n_1! * n_2!)$$

For example, for $n_1 = n_2 = 15$, $M = 1.55 \times 10^8$

Assume all partitions could have occurred with equal chance. Then calculate the proportion of these that have δ smaller than the observed δ (Fig. 24.2).

$$p = \frac{1 + \text{no. smaller deltas}}{\text{total no. possible partitions}}$$

A more reasonable method that is also applicable to medium or large data sets is to approximate the distribution of δ from a continuous distribution (Pearson type III). This distribution accommodates the fact that the underlying permutation distribution is often substantially skewed. The Pearson type III distribution incorporates three parameters, the mean m, standard deviation s, and gamma g (skewness of δ under the null hypothesis).

The test statistic, T is

$$T = (\delta - m_\delta) / s_\delta$$

where m_δ and s_δ are the mean and standard deviation of δ under the null hypothesis. In words, the test statistic is the difference between the observed and expected deltas divided by the square root of the variance in delta. (Do you see the resemblance to the Student's t-test?)

$$T = \frac{\text{observed } \delta - \text{expected } \delta}{\text{s. dev. of expected } \delta}$$

The test statistic, T, describes the separation between the groups. The more negative is T, the stronger the separation. The observed delta (the average within-group distance) is compared to an expected delta, the latter calculated to represent the mean delta for all possible partitions of the data. The variance and skewness of delta are descriptors of the distribution of all possible deltas corresponding to the possible partitions of the items. The question then becomes, is the observed delta at all unusual considering the distribution of possible deltas? The probability value expresses the likelihood of getting a delta as extreme or more extreme than the observed delta, given the distribution

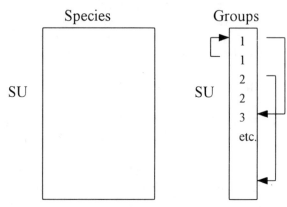

Figure 24.1. Schematic showing random reassignment of sample units to groups. The assignments of SUs to groups are shuffled, while the species matrix is held constant.

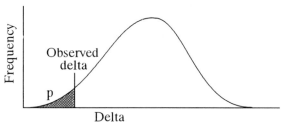

Figure 24.2. Frequency distribution of delta under the null hypothesis, compared to the observed delta. The area under the curve less than the observed delta is the probability of type I error under the null hypothesis of no difference between groups.

of possible deltas. The p-value associated with T is determined by numerical integration of the Pearson type III distribution.

5. The p value is useful for evaluating how likely it is that an observed difference is due to chance, but we also need a description of the *effect size* that is independent of the sample size. This is provided by the **chance-corrected within-group agreement** (A):

$$A = 1 - \frac{\delta}{m_\delta} = 1 - \frac{\text{observed } \delta}{\text{expected } \delta}$$

The agreement statistic A describes within-group homogeneity, compared to the random expectation. (Originally shown as script R in Mielke's papers and R in older versions of PC-ORD, we have since adopted the symbol A to avoid confusion with regression statistics.) When all items are identical within groups, then the observed delta = 0 and $A = 1$, the highest possible value for A. If heterogeneity within groups equals expectation by chance, then $A = 0$. On the other hand, if there is LESS agreement within groups than expected by chance, then $A < 0$.

In community ecology, values for A are commonly below 0.1, even when the observed delta differs significantly from the expected. An $A > 0.3$ is fairly high. Remember that for $A = 1$, all items must be identical within groups.

Statistical significance (small p) may result even when the "effect size" (A) is small, if the sample size is large. For example, you may find that $A = 0.01$ is associated with a statistically significant delta with $N = 200$. In such cases, you need to carefully consider the ecological significance of the result, not just the statistical significance. On the other hand, with a small sample size, a large effect size is needed to achieve statistical significance.

What to report

Because many readers will not be familiar with this method, you may wish to give a brief statement of the **purpose and general approach** of the method. Likewise, you should cite a paper that gives details of the method. Mielke (1984) and Mielke and Berry (2001) are good references for the method. Biondini et al. (1985) and Zimmerman et al. (1985) were some of the earliest applications to ecological problems.

You should also report:

☐ The software (because MRPP is not available in most statistical packages).

☐ Distance measure (consider including a justification for this).

☐ How groups were defined (this should directly or indirectly include a statement of the size of each group).

☐ Chance-corrected within-group agreement, A.

☐ p-value.

Examples

Simple example

A simple hypothetical example illustrates the basic concepts of MRPP. Assume we have three a priori groups of five plots each. In each plot we measured the abundance of two species (Table 24.2). The bivariate scatterplot (Fig. 24.3) shows a dust-bunny distribution typical of community data.

We first calculate a distance matrix (Table 24.2), but only a portion of these values are actually used. The between-group distances (shaded cells of Table 24.2) are ignored. Omitting the redundant cells above the diagonal leaves just the within-group distances, in this case ten distance values for each group. From these, we calculate the average within-group distance for each group (Table 24.3). For comparison we redid the calculation for Euclidean distance and the Sørensen distances converted to ranks. The latter yields a nonmetric MRPP, analogous to treatment of the distance matrix in nonmetric multidimensional scaling.

The average within-group distances (Table 24.3) show that Groups 1 and 3 have relatively high dispersions, while Group 2 is relatively tight. This accords with our intuition from Figure 24.3.

Clearly, we can reject the null hypothesis of no difference among groups (Table 24.4). The three groups occupy different regions of species space, as shown by the strong chance-corrected within-group agreement (A) and test statistic (T). In this case, we would have reached the same conclusion with any of the distance matrices. Selection of a distance measure should, however, be based on properties of the distance measures rather than comparison of results.

A particular pair of groups can be compared using the same methods as the comparison across all groups.

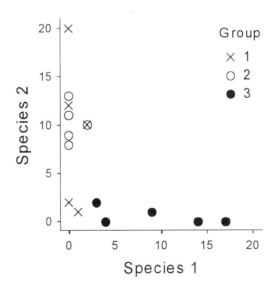

Figure 24.3. Fifteen sample units in species space, each sample unit assigned to one of three groups.

Table 24.2. A species data matrix of 15 plots by 2 species, their assignments to three groups, and Sørensen distances among plots. Shaded cells are between-group distances, ignored by MRPP.

Plot	Sp1	Sp2	Group	1	2	3	4	5	6	7	8	9	10	11	12	13	14	15
1	0	20	1	0.000	0.250	0.375	0.818	0.909	0.212	0.290	0.375	0.379	0.429	0.840	1.000	0.933	1.000	1.000
2	0	12	1	0.250	0.000	0.167	0.714	0.857	0.040	0.043	0.167	0.143	0.200	0.765	1.000	0.909	1.000	1.000
3	2	10	1	0.375	0.167	0.000	0.714	0.714	0.200	0.130	0.000	0.143	0.200	0.529	0.750	0.727	0.846	0.862
4	0	2	1	0.818	0.714	0.714	0.000	0.500	0.733	0.692	0.714	0.636	0.600	0.429	1.000	0.833	1.000	1.000
5	1	1	1	0.909	0.857	0.714	0.500	0.000	0.867	0.846	0.714	0.818	0.800	0.429	0.667	0.667	0.875	0.895
6	0	13	2	0.212	0.040	0.200	0.733	0.867	0.000	0.083	0.200	0.182	0.238	0.778	1.000	0.913	1.000	1.000
7	0	11	2	0.290	0.043	0.130	0.692	0.846	0.083	0.000	0.130	0.100	0.158	0.750	1.000	0.905	1.000	1.000
8	2	10	2	0.375	0.167	0.000	0.714	0.714	0.200	0.130	0.000	0.143	0.200	0.529	0.750	0.727	0.846	0.862
9	0	9	2	0.379	0.143	0.143	0.636	0.818	0.182	0.100	0.143	0.000	0.059	0.714	1.000	0.895	1.000	1.000
10	0	8	2	0.429	0.200	0.200	0.600	0.800	0.238	0.158	0.200	0.059	0.000	0.692	1.000	0.889	1.000	1.000
11	3	2	3	0.840	0.765	0.529	0.429	0.429	0.778	0.750	0.529	0.714	0.692	0.000	0.333	0.467	0.684	0.727
12	4	0	3	1.000	1.000	0.750	1.000	0.667	1.000	1.000	0.750	1.000	1.000	0.333	0.000	0.429	0.556	0.619
13	9	1	3	0.933	0.909	0.727	0.833	0.667	0.913	0.905	0.727	0.895	0.889	0.467	0.429	0.000	0.250	0.333
14	14	0	3	1.000	1.000	0.846	1.000	0.875	1.000	1.000	0.846	1.000	1.000	0.684	0.556	0.250	0.000	0.097
15	17	0	3	1.000	1.000	0.862	1.000	0.895	1.000	1.000	0.862	1.000	1.000	0.727	0.619	0.333	0.097	0.000

Table 24.3. Average within-group distance calculated from three different distance matrices. The average within-group distance is used as the test statistic.

	Average within-group distance		
Group	Sørensen	Ranked Sørensen	Euclidean
1	0.602	0.453	9.78
2	0.149	0.159	2.79
3	0.449	0.337	7.79
Average	0.400	0.316	6.79

Table 24.4. Summary statistics for MRPP of simple example. Results are given for three different distance matrices, comparing across all groups, as well as for multiple pairwise comparisons for the Sørensen distances. The pairwise comparisons were also made with MRPP.

		δ under null hypothesis					
	Observed δ	Expected	Variance	Skewness	T	p	A
Sørensen distances	0.400	0.625	0.0019	-1.24	-5.14	0.0007	0.359
Ranked Sørensen	0.316	0.495	0.0012	-1.26	-5.18	0.0007	0.361
Euclidean	6.79	9.97	0.5177	-1.18	-4.43	0.0017	0.320
Multiple comparisons (Sørensen)							
1 vs 2	0.376	0.397	0.0005	-0.31	-0.93	0.1730	0.055
1 vs 3	0.526	0.699	0.0019	-1.26	-3.89	0.0039	0.248
2 vs 3	0.299	0.627	0.0036	-2.11	-5.49	0.0017	0.523

The pairwise comparisons show that Groups 1 and 2 are broadly overlapping, and we cannot reject the null hypothesis (Table 24.4). The comparisons of Groups 1 vs. 3 and 2 vs. 3 yielded statistics comparable to the overall comparisons. The A values are similar in size, but the p-values are somewhat smaller than for the overall test, reflecting the smaller sample size. Note that the hazards of multiple comparisons apply here. A conservative approach would be to apply a Bonferroni procedure to control the "experimentwise" error rate.

Examples in the literature

Lesica et al. (1991) tested the null hypothesis of no difference in species composition between old-growth and second-growth forests. The test was separately applied to four different groups of species. The average within-group distance allowed the reader to assess how tightly the old-growth and second-growth sample units grouped in species space for each of the species groups.

That study and most of the other early applications of MRPP used Euclidean distance between sample units in species space. Informal comparisons of MRPP based on Sørensen distances vs. Euclidean distances have often shown similar results. Sørensen distances are, however, less prone to exaggerate the influence of outliers. Use of MRPP with Sørensen distances or ranked Sørensen distances is increasingly frequent (McCune et al. 2000).

Variations

Rank-transformed MRPP

One can rank transform the distance matrix, then conduct an otherwise normal MRPP. Any of the distance measures can be rank transformed. The rank transformation can help to correct the loss of sensitivity of distance measures as community heterogeneity increases. It also subtly changes the null hypothesis being answered from "average within-group distance no smaller than expected by chance" to "no difference in average within-group ranked distances." It also makes the MRPP results more analogous in theory to from those from nonmetric multidimensional scaling. In theory, MRPP on ranked distances is more similar to ANOSIM (Clarke & Green 1988, Clarke 1993) than is MRPP with raw distances.

With community data, the test statistic, skewness of the test statistic under the null hypothesis, and the resulting p-value are often similar, whether the data are ranked or not. The chance-corrected within-group agreement, however, is often considerably higher after the distance measure is converted to ranks.

The ranking procedure operates on nonredundant distances (i.e., one triangle of the square, symmetrical distance matrix). Ties are assigned the average rank of the tied elements. For example, the values 1, 3, 3, 9, 10 would receive ranks 1, 2.5, 2.5, 4, 5. After elements are assigned initial ranks, they are adjusted by subtracting the rank given to zero distances. This results in all raw distances of zero being assigned a rank distance of zero. For example, if there were five zero distances in the matrix, they would each initially be assigned a rank of 3, taking into account the five-way tie. Then 3 would be subtracted from each element in the matrix.

McCune et al. (2000) used nonmetric MRPP (based on a rank-transformed Sørensen distance matrix) to compare species composition among groups of sample units in a data set with high beta diversity. This choice of distance matrix enhanced the correspondence of the MRPP results with the nonmetric multidimensional scaling that they used to illustrate the relationships among sample units.

Blocked MRPP (MRBP)

Randomized block experiments, paired-sample data, and simple repeated-measures designs are very common in ecology. These designs can be analyzed with a variant of MRPP called MRBP or blocked MRPP. The method is explained by Mielke (1984, 1991), Mielke and Berry (1982), Mielke and Iyer (1982), and Biondini et al. (1988). It has been applied to ecological data by Biondini et al. (1988), Peterson & McCune (2001), Ponzetti & McCune (2001), and Zimmerman et al. (1985).

Given b blocks and g groups (treatments), the MRPP statistic is modified to:

$$\delta = \left[g \binom{b}{2} \right]^{-1} \sum_{i=1}^{g} \sum_{j<k} \Delta(x_{ij}, x_{ik})$$

where $\Delta(x,y)$ is the distance between points x and y in the p-dimensional space. Each of the p dimensions should be measured on the same scale or standardized to a common scale. In community ecology, each of the p dimensions would normally correspond to a different species or a different environmental variable. Note that for paired-sample data, b is the number of pairs of observations and $g = 2$. The combinatoric term is simply the number of items represented in the double summation. Thus δ is the average distance between blocks within treatments, including only nonredundant off-diagonal elements of the distance matrix. Normally δ is calculated after alignment of the blocks.

The null hypothesis assigns equal probabilities to each of the $M = (g!)^b$ possible allocations of the g, r-dimensional measurements to the g treatments within each of the b blocks. In other words, the observed values are randomly reassigned to different treatments in each block.

Like MRPP, small values of δ imply a concentration of treatments in the p dimensional space. The added twists of MRBP are that these distances are summed with respect to the blocks, and that the user has the option of aligning blocks so that all treatments in a given block have a median of zero.

Equations for estimating the standard deviation and skewness of δ are relatively complex (see Mielke 1984) and are not reproduced here. The test statistic, T, is based on these values as in MRPP and the p-value is again approximated from a Pearson type III distribution.

Note that this analysis requires a balanced design: there must be one sample unit for each combination of block and treatment. The number of treatments must be the same among blocks and each treatment must be present in each block.

Consider the following application to a paired-sample design. Imagine a study of predation effects on rocky intertidal communities. We paired treatment and control plots, the treatment being exclosure of a predator. The purpose of the pairing was to control for other variables, such as wave exposure, that influenced the communities. Each pair of plots was considered a block, and groups were defined by the treatments. Using blocked MRPP focuses the analysis on within-block differences, presumably due to the treatment alone. In contrast, randomly placed plots without blocking would have added the variation due to wave exposure, reducing our ability to detect the treatment effect.

Several options provide flexibility with MRBP:

Average distance function commensuration. This option equalizes the contribution of each variable to the distance function. For each variable m the sum of deviations (Dev_m) is calculated:

$$Dev_m = \sum_{i=1}^{g}\sum_{j=1}^{b}\sum_{k=1}^{g}\sum_{l=1}^{b} |x_{mij} - x_{mkl}|^V$$

V is set to 2 for squared Euclidean distance or 1 for Euclidean distance. Then each element x of the data matrix is divided by the sum of the deviations for the corresponding variable to produce the transformed value y:

$$y_{mij} = x_{mij} / Dev_m$$

Distance function commensuration relativizes your variables. If you have already relativized your data then it is unnecessary. If you have not already relativized your data, you may or may not want to at this point. The same criteria apply as discussed under relativization in Chapter 9.

Median alignment within blocks. If the median for each variable in each block is subtracted from the raw data for each block, then the medians are said to be aligned to zero for all blocks (Table 24.5). Usually, alignment in a randomized block design is desirable, focusing the analysis on within-block differences among treatments. But if the problem is conceptualized as paired agreement, say between model predictions and observed data, then alignment is not used. In this case, each event or observation represents a group, one block contains the observed data, and one block contains the predicted data. Because the goal is an exact match of predicted and observed, rather than the two just being correlated, then the medians should not be aligned. If, however, the goal is to determine whether or not the two sets of numbers are correlated apart from any exact agreement in value, then the blocks should be aligned.

Distance measure. Euclidean distance has been used most often with MRBP. If squared Euclidean distance is chosen, then the resulting statistics are comparable to permutation tests of Pearson correlation. Methods dependent on the normal distribution require sample statistics based on squared Euclidean distances. Because MRPP, MRBP, and related statistics are based on a permutation distribution, there is no necessity for choosing squared Euclidean distance. Mielke (1991) and Mielke & Berry (2001) argued convincingly against using squared Euclidean distances. Mielke (1991, pp. 56-57) stated that "absolute deviations" (meaning Euclidean distances) are more logical and intuitive than squared deviations (squared Euclidean distance). Although Mielke's papers did not consider proportional city-block distance measures, Sørensen distance is incompatible with median alignment. Alignment results in both positive and negative values, but Sørensen distances require nonnegative data. A nonproportional city-block distance could, however, be used in conjunction with median alignment.

Table 24.5. Example comparing results from raw data versus data aligned within blocks to zero as input to Blocked MRPP.

	Raw Data		Aligned Data	
	Block 1	Block 2	Block 1	Block 2
Group 1	4	9	1.5	1.5
Group 2	2	7	-0.5	-0.5
Group 3	3	8	0.5	0.5
Group 4	1	2	-1.5	-5.5
Median	2.5	7.5	0	0
Observed δ	5 = (5+5+5+1)/4		1 = (0+0+0+4)/4	
Expected δ	4.375		2.225	
Agreement (A)	0.086		0.556	
p	0.184		0.016	

Permutation test for bivariate Pearson correlation coefficient. This method offers a nonparametric test of the null hypothesis that a bivariate correlation coefficient is equal to zero. Assume we are looking at the relationship between N measurements of variables X and Y. If there is only one response variable, squared Euclidean distance is selected, each variable is selected as a block, and there are N groups, then MRBP yields a permutation test for the Pearson correlation coefficient, r. Under the null hypothesis, δ is equivalent to r (Mielke 1984, p. 824). See that same paper for a description of the relationship between δ and the standard F statistic.

Analysis of similarity (ANOSIM)

ANOSIM (Clarke & Green 1988; Clarke 1993) is similar to MRPP in concept but uses a different test statistic. Elements of a similarity matrix among all sample units are ranked. The highest similarity is given a rank of 1.

$$R = (\bar{r}_B - \bar{r}_W) / (M/2)$$

where:

r_B = rank similarity for each between-group similarity

r_W = rank similarity for each within-group similarity

$M = n(n-1)/2$

n = the total number of sample units

The denominator constrains R to the range -1 to 1. Positive values indicate differences among groups. The interpretation is thus similar to the A statistic of MRPP. We know of no formal comparisons of the results of ANOSIM and MRPP.

The Q_b method

Randomization tests can be devised for almost any problem. One difficult group of problems is testing for interactions in multivariate experiments (multiple response variables and complex experimental designs). Unfortunately, this is one of the most common classes of experiments in ecology, although ecologists usually dodge the multivariate nature of their experiments. A method has been proposed, however, that allows randomization tests from such experiments (Pillar and Orloci 1996).

Pillar and Orloci (1996) present a method for partitioning variance in a distance matrix into sums of squares for multiple factors, including interaction terms and multiple contrasts. This promising method has at least one serious shortcoming in that it is limited to Euclidean distance measures (they use the "chord distance;" see Chapter 6). The method is apparently not available in the large statistical packages.

The test criterion is the sum of the squared distances between groups:

$$Q_b = Q_t - Q_w$$

Note that the additivity of these terms depends on Euclidean metric properties of the distance matrix.

The component terms are calculated as follows. The total sum of squares (Q_t) is based on one triangle of the distance matrix, the triangle having $n(n-1)/2$ terms, each term being a squared distance between two entities j and k:

$$Q_t = \frac{1}{n} \sum_{j=1}^{n-1} \sum_{k=j+1}^{n} d_{kj}^2$$

The within-group sum of squares Q_{wg} is summed across all g groups:

$$Q_w = \sum_{i=1}^{g} Q_{wg}$$

where

$$Q_{wg} = \frac{1}{n_g} \sum_{j=1}^{n-1} \sum_{k=j+1}^{n} d_{jk}^2 \, \delta(j,k,g)$$

The last term, $\delta(j,k,g)$, is an indicator variable that takes the value of one if the entities belong to the same group g; otherwise the value is zero.

The problem is more complex if there are two or more factors. Calculation of the interaction terms is based on a joint classification by the two factors. For example, in a 2 × 2 factorial design, there are four groups defined by the joint classification of the two factors X and Y. The sum of squares is partitioned as follows:

$$Q_{b|X+Y} = Q_{b|X} + Q_{b|Y} + Q_{b|XY}$$

$Q_{b|X}$ and $Q_{b|Y}$ are the sums of squares between groups specific to factors X and Y alone. The last term is the interaction term. The calculations are given in detail in Pillar and Orloci (1996) but are too extensive to repeat here. The results can be summarized in a table much like a traditional ANOVA table.

NPMANOVA

Anderson (2001) solved the problem of using non-Euclidean distance measures (such as Sørensen distance) in multifactor designs. The key to the solution is that "the sum of squared distances between points and their centroid is equal to (and can be calculated directly from) the sum of squared interpoint distances divided by the number of points [Fig. 24.4]... The

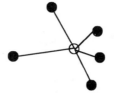 

Figure 24.4. The sum of squared distances from points to the centroid (left) can be calculated from the average squared interpoint distance (right).

relationship between distances to centroids and interpoint distances for the Euclidean measure has been known for a long time... What is important is the implication this has for analyses based on non-Euclidean distances. Namely, an additive partitioning of sums of squares can be obtained for any distance measure directly from the distance matrix, without calculating the central locations of groups." Avoiding calculating the position of the centroid is important, because the centroid does not represent the central tendency in a non-Euclidean space.

The total sum of squares of a distance matrix **D** with N rows and N columns is

$$SS_T = \frac{1}{N} \sum_{i=1}^{N-1} \sum_{j=i+1}^{N} d_{ij}^2$$

The residual (within-group) sum of squares for a one-way classification is

$$SS_R = \frac{1}{n} \sum_{i=1}^{N-1} \sum_{j=i+1}^{N} d_{ij}^2 \varepsilon_{ij}$$

where n is the number of observations per group, N is the number of sample units, and $\varepsilon_{ij} = 1$ if i and j are in the same group, but $\varepsilon_{ij} = 0$ if in different groups.

The sum of squares between groups is then $SS_A = SS_T - SS_R$ so we can calculate a pseudo-F-ratio:

$$F = \frac{SS_A / (a-1)}{SS_R / (N-a)}$$

where a is the number of groups. If the distance matrix contains Euclidean distances, then this gives the traditional parametric univariate F ratio.

Evaluating the significance of this F requires a permutation test, even if the distance measure is Euclidean and the variables are normally distributed (Anderson 2001). As with MRPP, the number of possible permutations becomes enormous, even with small samples sizes. Anderson (2001) recommended at least 1000 permutations for a test with $\alpha = 0.05$ and 5000 permutations for a test with $\alpha = 0.01$.

The permutation is simple with a one-way design: randomly reassign group labels. Nested and factorial designs are more complicated. Equations for these sums of squares are given in Anderson (2001). For a two-factor design (say factors A and B), one calculates the following terms:

SS_A = within-group sum of squares for A, ignoring any influence of B

SS_B = within-group sum of squares for B, ignoring any influence of A

SS_R = residual sum of squares, pooling the sum of squares within groups defined by each of the combinations of factors A and B

SS_{AB} = interaction sum of squares for AB, by subtraction:

$$SS_{AB} = SS_T - SS_A - SS_B - SS_R$$

If factor B is nested within A, then

$$SS_{B(A)} = SS_T - SS_A - SS_R$$

and there is no interaction term. Calculation of the mean-squared errors and pseudo-F-ratios for the nested and factorial designs are analogous to parametric ANOVA and the formula for the one-way NPANOVA above.

Permutation tests for the nested and factorial designs are, unfortunately, not straightforward. Decisions must be made for each study, based on sample sizes and the nature of the design. Anderson (2001) gave two examples. A factorial design with small sample sizes was handled by unrestricted permutation of the raw data. This weakens the test, but there is little choice if the sample sizes are so small that the number of possible permutations is small. Exact permutation tests were used for a design with three factors, one of them nested. Variability in a higher level factor was tested against the nested factor by permuting the subsamples but not mixing replicates within subsamples.

Like the other permutation tests discussed above, one of the appeals of NPMANOVA for community data is that we can avoid assumptions of linear species responses and normally distributed errors. Anderson (2001) stated that "the only assumption of the test is that the observations (rows of the original data matrix) are exchangeable under a true null hypothesis." Exchangeability requires that the observations (rows) be independent and that they have similar dispersions. ANOSIM, MRPP, and NPMANOVA are all sensitive to different dispersions. Imagine two clouds of points, one much larger than another, but with the same centroid. If you consider these two clouds different, then this sensitivity is an asset; if you consider them the same, then homogeneity of dispersions is another assumption you must meet.

CHAPTER 25

Indicator Species Analysis

Background

A very common goal in community analysis is to detect and describe the value of different species for indicating environmental conditions. If environmental differences are conceptualized as groups of sample units, then Dufrêne and Legendre's (1997) method of calculating species indicator values provides a simple, intuitive solution. Groups are commonly defined by categorical environmental variables, levels of disturbance, experimental treatments, presence-absence of a target species, or habitat types.

This method combines information on the concentration of species abundance in a particular group and the faithfulness of occurrence of a species in a particular group. A perfect indicator of a particular group should be *faithful* to that group (always present). It should also be *exclusive* to that group, never occurring in other groups. Dufrêne and Legendre's indicator species analysis produces indicator values for each species in each group, based on these standards of a perfect indicator. Indicator values are tested for statistical significance using a randomization (Monte Carlo) technique.

When to use it

Indicator species analysis can be used anytime we wish to contrast performance of individual species across two or more groups of sample units. The method is applicable only to species data, because it is based on concepts of both abundance and frequency (concentration of abundance in particular groups and relative frequency within a group). For most other kinds of data, one or the other of these concepts does not make sense. Some common applications are:

- Describe species relationships to environmental categories or experimental groups.
- Describe hierarchical structure of communities.
- Choose optimum number of clusters.
- Describe community types.
- Ordinate species on axes defined by categorical variables.

We give examples of these below.

Indicator species analysis is a natural companion to multi-response permutation procedures (MRPP; Ch. 24) of community data, supplementing the test of no multivariate difference between groups with a description of how well each species separates among groups.

How it works

Dufrêne and Legendre's method requires two or more a priori groups of sample units, and data on species abundance or species presence in each of the sample units.

For each species, the following steps are taken in the implementation in PC-ORD:

1. Calculate the proportional abundance of a particular species in a particular group relative to the abundance of that species in all groups. Also express as a percentage and display these intermediate results. You can think of this proportion as the concentration of abundance into a particular group.

Let $\mathbf{A}$ = sample unit × species matrix

a_{ijk} = abundance of species j in sample unit i of group k

n_k = number of sample units in group k

g = total number of groups

First calculate the mean abundance x_{kj} of species j in group k:

$$x_{kj} = \frac{\sum_{i=1}^{n_k} a_{ijk}}{n_k}$$

Then calculate the relative abundance RA_{kj} of species j in group k (this measures exclusiveness, the concentration of abundance into a particular group):

$$RA_{jk} = \frac{x_{kj}}{\sum_{k=1}^{g} x_{kj}}$$

2. Calculate the proportional frequency of the species in each group (the proportion of sample units in each group that contain that species). Also express this as a percentage and display these intermediate results. You can think of these percentages as the

faithfulness or constancy of presence in a particular group.

First transform **A** to a matrix of presence-absence, **B**:

$$b_{ij} = a_{ij}^0$$

Then calculate relative frequency RF_{kj} of species j in group k:

$$RF_{kj} = \frac{\sum_{i=1}^{n_k} b_{ijk}}{n_k}$$

3. Combine the two proportions calculated in steps 1 and 2 by multiplying them. Express the result as a percentage, yielding an indicator value (IV_{kj}) for each species j in each group k.

$$IV_{kj} = 100 \, (RA_{kj} \times RF_{kj})$$

Because the component terms are multiplied, both indicator criteria must be high for the overall indicator value to be high. Conversely, if either term is low, then the species is considered a poor indicator.

4. The highest indicator value (IV_{max}) for a given species across groups is saved as a summary of the overall indicator value of that species.

5. Evaluate statistical significance of IV_{max} by a Monte Carlo method. Randomly reassign SUs to groups 1000 times. Each time, calculate IV_{max}. The probability of type I error is the proportion of times that the IV_{max} from the randomized data set equals or exceeds the IV_{max} from the actual data set. The null hypothesis is that IV_{max} is no larger than would be expected by chance (i.e., that the species has no indicator value).

Properties of the indicator values

The indicator values range from zero (no indication) to 100 (perfect indication). Perfect indication means that presence of a species points to a particular group without error, at least with the data set in hand.

When more than two groups are defined, the IV for a particular species in a particular group depends on the set of sample units belonging to other groups. In the first example below, the indicator values in group 1 would be different if group 3 were omitted from the analysis.

Species with only one or two occurrences never yield an indicator value stronger than expected by chance. This is true because the observed indicator value will be very frequent (or constant in the case of one occurrence) in the randomizations.

What to report

☐ Software used, and a brief description of the method, because it is not available in most statistical packages.

☐ Number of randomizations used in the Monte Carlo test.

☐ Often a table of indicator values for each species in each group, sometimes including only the statistically significant indicators.

☐ *p*-values for each species or a list of species that meet a specified criterion of statistical and/or biological significance.

Examples

1. Species relationships to environmental categories.

Three groups of 18 plots each were defined by their topographic position in a recently burned area (upper bench, slope, and lower bench). Although the main focus of the study was temporal change following fire, Indicator Species Analysis can be used to contrast the species present in the three topographic positions. Excerpts from the two components of the indicator value, as well as the indicator value itself, and the results of the randomization test, are shown in Tables 25.1 to 25.4.

The first species, "Larocc" occurred only in one group, though it occurred in only 6% of the plots in that group (one plot; Table 25.2). Singleton species such as this, as well as very infrequent species, have no possibility of being a statistically significant indicator species, because the result of all of its occurrences falling in one group is quite likely (Table 25.4).

"Acergl," however, was a significant indicator of group 1, with an indicator value of 33. All of its occurrences were in the first group, and it occurred in one-third of those plots. The randomization test showed that the probability of an indicator value of 33 or higher, given this species' distribution of abundances, was 0.005.

Table 25.1. **Relative abundance** (%) of each species in each group defined by topographic position. The data matrix contains 54 plots and 85 species. Each group contains 18 items. "Sequence" is the sequence of occurrence of the group in the data; "Max" is the maximum relative abundance of the species.

			Group		
	Sequence		1	2	3
	Identifier:		1	2	3
	N of items:		18	18	18
Species	Avg	Max			
1 Larocc	33	100	0	0	100
2 Piceng	33	63	13	63	25
3 Pincon	33	50	50	25	25
4 Pinpon	33	100	0	100	0
5 Acergl	33	100	100	0	0
6 Adebic	33	86	0	14	86
7 Amealn	33	75	25	75	0
8 Antluz	33	44	44	19	37
9 Antmic	33	100	0	0	100
10 Apoand	33	100	0	100	0
11 Aranud	33	54	30	17	54
12 Arncor	33	100	0	100	0
etc.					
83 Phleum	33	67	67	6	28
84 Poapra	33	53	53	5	42
85 UnkGr1	33	100	0	100	0
Averages	33	73	31	33	36

Table 25.2. **Relative frequency** (%) of each species in each group defined by topographic position. The data matrix contains 54 plots and 85 species. "Max" is the maximum relative frequency of each species.

			Group		
	Sequence		1	2	3
	Identifier:		1	2	3
	N of items:		18	18	18
Species	Avg	Max			
1 Larocc	2	6	0	0	6
2 Piceng	15	28	6	28	11
3 Pincon	6	6	6	6	6
4 Pinpon	6	17	0	17	0
5 Acergl	11	33	33	0	0
6 Adebic	13	33	0	6	33
7 Amealn	37	72	39	72	0
8 Antluz	26	39	39	17	22
9 Antmic	2	6	0	0	6
10 Apoand	15	44	0	44	0
11 Aranud	56	83	56	28	83
12 Arncor	2	6	0	6	0
etc.					
83 Phleum	17	28	28	6	17
84 Poapra	17	22	22	6	22
85 UnkGr1	2	6	0	6	0
Averages	21	32	20	21	21

Table 25.3. **Indicator values** (% of perfect indication) of each species for each group, rounded to the nearest whole percentage. These values were obtained by combining the relative abundances and relative frequencies in Tables 25.1 and 25.2. The data matrix contains 54 plots and 85 species. "Max" is the maximum indicator value of the species across the three groups.

	Sequence		Group		
			1	2	3
	Identifier:		1	2	3
	N of items:		18	18	18
Species	Avg	Max			
1 Larocc	2	6	0	0	6
2 Piceng	7	17	1	17	3
3 Pincon	2	3	3	1	1
4 Pinpon	6	17	0	17	0
5 Acergl	11	33	33	0	0
6 Adebic	10	29	0	1	29
7 Amealn	21	54	10	54	0
8 Antluz	9	17	17	3	8
9 Antmic	2	6	0	0	6
10 Apoand	15	44	0	44	0
11 Aranud	22	45	16	5	45
12 Arncor	2	6	0	6	0
etc.					
83 Phleum	8	19	19	0	5
84 Poapra	7	12	12	0	9
85 UnkGr1	2	6	0	6	0
Averages	9	20	8	10	10

Table 25.4. **Monte Carlo test** of significance of observed maximum indicator value (IV) for each species, based on 1000 randomizations. The means and standard deviations of the IV from the randomizations are given along with p-values for the hypothesis of no difference between groups. The p-value is based on the proportion of randomized trials with indicator value equal to or exceeding the observed indicator value.

Species	Observed Indicator Value (IV)	IV from randomized groups		
		Mean	S.Dev	p
1 Larocc	5.6	5.6	0.17	0.999
2 Piceng	17.4	11.6	5.08	0.237
3 Pincon	2.8	7.4	3.84	0.999
4 Pinpon	16.7	7.2	4.03	0.110
5 Acergl	33.3	10.2	4.85	0.005
6 Adebic	28.6	10.9	4.69	0.019
7 Amealn	54.2	20.5	5.13	0
8 Antluz	17	15.9	5.27	0.299
9 Antmic	5.6	5.6	0.17	0.999
10 Apoand	44.4	11.5	4.68	0
11 Aranud	44.8	26.5	5.08	0.005
12 Arncor	5.6	5.6	0.17	0.999
etc.				
83 Phleum	18.5	12.6	4.86	0.133
84 Poapra	11.7	12.7	5.1	0.550
85 UnkGr1	5.6	5.6	0.17	0.999

2. Describe hierarchical structure of communities

Indicator species analysis can also be used in combination with a hierarchical grouping of sample units. The groups may be formed by any method including cluster analysis. Indicator species corresponding with each division into subgroups can be shown diagrammatically (Fig. 25.1).

3. Choose optimum number of clusters

Dufrêne and Legendre (1997) used indicator species analysis to choose a stopping point in cluster analysis. If groups are too finely divided then indicator values will be low. On the other hand, if groups are too large, then their internal heterogeneity will reduce the indicator values. Dufrêne and Legendre found that indicator values peak at some intermediate level of clustering, and that the position of this peak will vary by species. Taken collectively the method can be used to decide on an appropriate level of clustering for species data, illustrated in the following example.

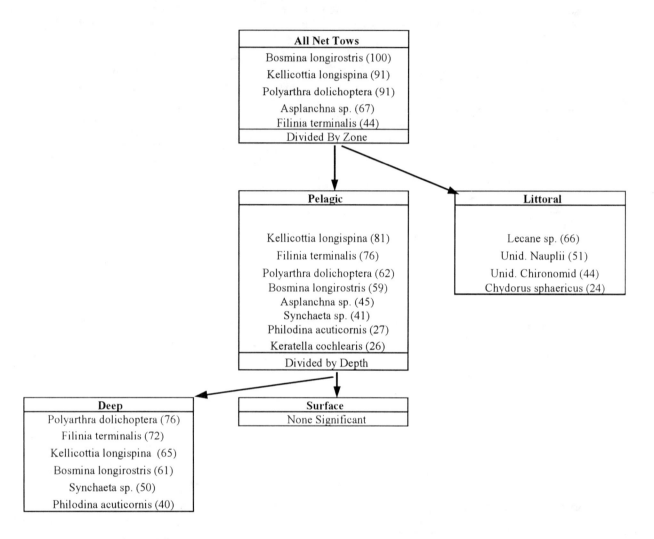

Figure 25.1. Portion of an indicator species hierarchy for freshwater zooplankton, based on Warncke (1998). Only statistically significant indicator species are shown. Groups and subgroups were based on a hierarchical division of sample units. The numbers for each species represent the percentage of perfect indication (*IV*) of that species in that subgroup.

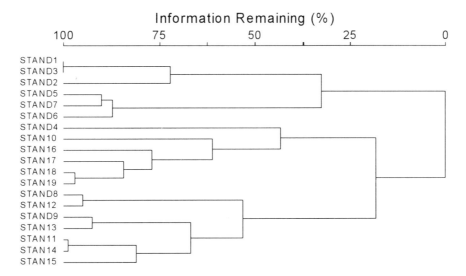

Figure 25.2. Dendrogram from cluster analysis of the Bison Range data set. Is there an optimum number of clusters?

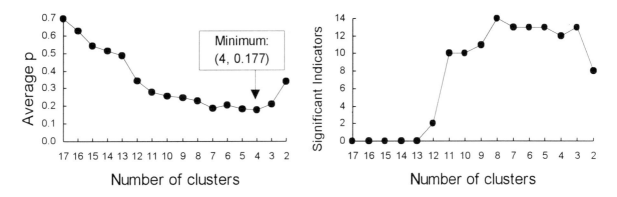

Figure 25.3. Use of indicator species analysis as an objective criterion for pruning a dendrogram. Left: change in p-value from the randomization tests, averaged across species at each step in the clustering. Right: number of species with $p \leq 0.05$ for each step of clustering.

Indicator Species Analysis can provide a quantitative, objective criterion for picking the most ecologically meaningful point to prune a dendrogram from cluster analysis. We illustrate this application with a data set of 19 plots × 50 plant species, the data collected at the National Bison Range in western Montana, USA. We first clustered the plots using flexible beta linkage (Ch. 11), with $\beta = -0.25$ and Sørensen distances (Fig. 25.2). Group membership at each step of cluster formation was written into a file. Indicator values were then calculated for each species at each level of grouping, including only species occurring in three or more plots. We averaged the resulting p-values across all species, repeating this for each step of clustering. The cluster step yielding the smallest average p-value can be taken as the most informative level in the dendrogram (Fig. 25.3). In this case, four groups (clusters) provided the lowest average p-value.

We also tallied the number of species shown to be significant indicators, then plotted this number against cluster step. The number of significant indicators was high and fairly even at 12-14 species (about 40% of the total number of species) across the 8-group to 3-group levels. Either the highest average *p*-value or the highest number of significant indicators, or both, could be the basis for choosing the optimum number of groups.

4. Describe community types

A classification of communities, regardless of how it was derived, should be communicated in part by describing differential species: those that can be used to differentiate among classes. Indicator species analysis is an ideal tool for this. It provides not only a measure of the indicator value of each species, but also an evaluation of the likelihood of obtaining a given result by chance. For example, Qian et al. (1999) sought indicator species from cryptograms growing on rotting logs, differentiating among 12 vegetation types based on the vascular plants. They tabulated indicator values for 46 indicator species by 12 community types.

5. Ordinate species on categorical variables

Peterson & McCune (2001) used indicator species analysis to ordinate species with respect to categorical factors. For example, they contrasted two mountain ranges, ordinating species on an indicator value axis ranging from species associated with one mountain range to species associated with the other mountain range (Fig. 25.4). Each six-letter acronym denotes a species and its position on a gradient from a strong Cascade Range indicator to a strong Coast Range indicator. Levels of statistical significance are indicated by * and **. Species were included only if they had an indicator value larger than expected by chance ($p < 0.05$).

Each category of the grouping variable can be the basis of a separate axis, but because the categories are not independent, the traditional two-dimensional ordination diagram is inappropriate for this application. Nevertheless, species scores for each category can be viewed as separate one-dimensional ordinations, or, if two categories are used, the axes for each category can be placed end-to-end, with the origins touching (Fig. 25.4).

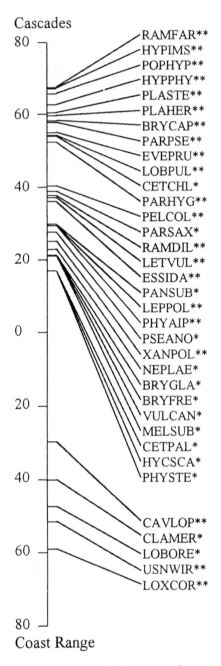

Figure 25.4. Ordination of species based on indicator species analysis, contrasting association with two mountain ranges, the Cascades and the Coast Range (from Peterson & McCune 2001). See text for details.

CHAPTER 26

Discriminant Analysis

Background

Discriminant analysis (DA) is an eigenanalysis technique that maximally separates a fixed number of groups. The object is to find axes ("discriminant functions," "canonical variates," or "canonical axes") that best discriminate among groups. DA maximizes the among-group variation relative to the within-group variation. Discriminant analysis is sometimes called "canonical analysis," "canonical variates analysis," or "multiple discriminant analysis."

DA differs from all the other ordination methods in that it requires predefined groups. In the literature, groups have been defined by species presence, animal behavior, season, sex, geographic area, soil type, etc. (see list of examples in Williams 1983). While DA is a very important method, it has limited application to community data.

Mathematically, DA is the same as multivariate analysis of variance (MANOVA), but the emphasis is different and the results are summarized differently. In MANOVA, the independent variables are groups. For example, you might have treatment and control groups and use MANOVA to evaluate a multivariate response to the groups. MANOVA seeks differences in the dependent variables among the groups. In DA, the independent variables are used as predictors of group membership. With this emphasis, the groups may be considered the dependent variables.

DA is available in all of the major statistical packages. The representation differs among them. For example, all phases of DA are combined into one function in SPSS, while SAS splits different kinds of DA into different procedures (DISCRIM, CANDISC, and STEPDISC).

When to use it

Discriminant analysis is closely allied to logistic regression in the kinds of problems to which it is applicable. The emphasis in logistic regression is, however, on constructing a predictive model rather than describing multivariate correlation structure. See Tabachnik and Fidell (1996, chapter 12) for a discussion of the differences. Note particularly that logistic regression requires fewer assumptions.

DA has been used in ecology for many purposes, including:

1. Summarizing the differences between groups (often used as a follow-up to clustering, to help describe the groups); "descriptive discriminant analysis." With community data, you could use indicator species analysis as a nonparametric alternative.
2. Multivariate testing of whether or not two or more groups differ significantly from each other. For ecological community data this is better done with MRPP, thus avoiding the assumptions listed below.
3. Determining the dimensionality of group differences.
4. Checking for misclassified items.
5. Predicting group membership or classifying new cases ("predictive discriminant analysis").
6. Comparing occupied vs. unoccupied habitat to determine the habitat characteristics that allow or prevent a species' existence. DA has been widely used for this purpose in wildlife studies and rare plant studies.

Assumptions

1. **Homogeneous within-group variances** (generally not too serious if not testing a formal hypothesis). Violations are most severe when sample sizes are unequal and small. We can statistically evaluate whether we met this assumption (Legendre & Legendre 1998, p. 623), but for a quick evaluation, look for approximate equality of the within-group scatters in ordinations (either based on DA or another method).

2. **Multivariate normality** within groups. Meeting this is a matter of judgement, usually based on evaluating univariate normality. Formal tests of multivariate normality are often overly conservative (see Ch. 14). DA is relatively robust with respect to skew but is sensitive to outliers.

3. **Linearity** among all pairs of variables. We rarely meet this important assumption with community data.

4. **Prior probabilities**. "Prior probabilities" are the sampling probabilities for each group (the probability that a sample unit will belong to a certain group). Prior probabilities are important in the classification phase of discriminant analysis because they control the relative weights assigned to elements of the cross-products matrix. In most cases, we have poor information on these probabilities so we make one of two assumptions: (1) that the priors are equal or (2) that the prior probabilities are proportional to the number of items in each group.

Williams (1983) commented on the applicability of these assumptions in ecology: "Unfortunately these assumptions are almost never met by ecological data. For example, frequency distributions are almost always nonnormal, usually highly skewed, often bimodal, and in many cases discrete, and dispersions are almost universally heterogeneous. The statistical robustness of discrimination procedures under such conditions is poorly known at best."

He further showed that when priors are replaced by relative sample sizes that bear no relationship to the true prior probabilities, an uncontrolled and unknown amount of arbitrariness is introduced into the discriminant analysis.

Clearly, the potential applications of DA in community analysis are strongly limited by the required assumptions. Because relationships among environmental variables or species traits are more readily linearized than are relationships among species abundance, discrimination of groups in environmental space or in trait space will usually be more appropriate than discrimination of groups in species space.

How it works

The "direct" procedure is described below. All variables are entered into the discriminant functions simultaneously. This is the usual case in community ecology. One can also use a stepwise procedure to use a series of selected variables, analogous to stepwise multiple regression.

1. Calculate variance/covariance matrix for each group.

2. Calculate pooled variance/covariance matrix (S_p) from the above matrices.

3. Calculate between group variance (S_g) for each variable.

4. Maximize the F-ratio:

$$F = \frac{\mathbf{y}' \mathbf{S_g} \mathbf{y}}{\mathbf{y}' \mathbf{S_p} \mathbf{y}}$$

where the $\mathbf{y}$ is an the eigenvector associated with a particular discriminant function. We seek $\mathbf{y}$ to maximize F.

5. Maximize this ratio by finding the partial derivatives with a characteristic equation:

$$|\mathbf{S_p^{-1} S_g} - \lambda \mathbf{I}| = 0$$

The number of roots is $g-1$, where g is number of groups. In other words, the number of functions (axes) derived is one less than the number of groups. The eigenvalues thus express the percent of variance among groups explained by those axes. If there are only two groups then the single function will express all of the variance between groups.

6. Solve for each eigenvector $\mathbf{y}$ (also known as the "canonical variates" or "discriminant functions").

$$[\mathbf{S_p^{-1} S_g} - \lambda \mathbf{I}]\mathbf{y} = 0$$

7. Locate points (sample units) on each axis.

$$\mathbf{X} = \mathbf{AY}$$

$\mathbf{X}$ = scores (coordinates) for n rows (sample units) on m dimensions, where $m = g-1$.

$\mathbf{A}$ = original data matrix of n rows by p columns

$\mathbf{Y}$ = matrix of m eigenvectors with loadings for p variables. Each eigenvector is known as a discriminant function. These unstandardized discriminant functions can be used as (linear) prediction equations, assigning scores to unclassified items. Most statistical packages will also produce "standardized discriminant function coefficients." By standardizing to unit variance, the absolute value of these coefficients can be interpreted directly as an indication of the relative importance of the individual variables in contributing to the discriminant function.

8. Classification phase. A classification equation is derived for each group, one term in the equation for each variable, plus a constant. By inserting data values for a given SU, one calculates a classification score for each group for that SU. The SU is assigned to the group in which it had the highest score. The coefficients in the equation are derived from the $p \times p$ within-group variance-covariance matrix (S_p) and a $p \times$

1 vector of the means for each variable in group k, $\mathbf{M}_k$. First, calculate $\mathbf{W}$ by dividing each term of $\mathbf{S}_p$ by the within-group degrees of freedom. Then:

$$\mathbf{C}_k = \mathbf{W}^{-1} \mathbf{M}_k$$

The constant is derived as:

$$constant_k = -\frac{1}{2} \mathbf{C}_k \mathbf{M}_k$$

The constant and the coefficients in $\mathbf{C}_k$ define a linear equation of the usual form, one equation for each group k. Substituting the data for each variable into this linear equation yields a classification score for each sample unit (or case) in group k. See Tabachnik and Fidell (2001, pp. 467-469) for a complete example. The steps given above describe the simplest case, assuming equal prior probabilities of group membership.

Interpretation

Interpretation of discriminant analysis is based primarily on the following statistics:

Summary statistics

- Wilk's lambda (λ). Wilk's λ is the error sum of squares divided by the sum of the effect sum of squares and the error sum of squares. Thus, it is the variance among the objects *not* explained by the discriminant functions. It ranges from zero (perfect separation of groups) to one (no separation of groups). By itself, it is not a measure of statistical significance. That is tested with a chi-square approximation.
- Chi-square (derived from Wilk's lambda).
- Variance explained.

Standardized discriminant function coefficients.

The unstandardized coefficients are useful if you wish to use DA in a predictive mode, but the standardized coefficients are more useful for assessing the relative importance of various predictors.

Classification table

A matrix of predicted group membership versus actual group membership. A successful DA is correct in most of its predictions. Titus et al. (1984) showed how Cohen's kappa statistic can be used to compare the prediction success with that expected by chance, a particularly useful procedure if group sizes are strongly unequal.

One can also use resampling procedures to assess the stability and prediction bias of the discriminant functions (Verbyla 1986). Prediction bias is likely to occur when a model contains many independent variables relative to sample size, or when many different sets of independent variables are tested with stepwise DA.

What to report

- ☐ Basis for defining the groups and the number of items in each group.
- ☐ Number and nature of the predictor variables.
- ☐ If applied to community data, a justification of why you consider suitable a linear model of relationships among species.
- ☐ Statement of whether you used the direct or stepwise method.
- ☐ How many axes (functions) you interpreted and proportion of variance represented by those axes.
- ☐ Tests of significance for each discriminant function (not necessary for many applications).
- ☐ Description of most influential variables based on discriminant function coefficients, means in each group, and univariate F-ratios.
- ☐ Percent of cases misclassified.
- ☐ Unstandardized coefficients for the discriminant functions if the model is to be used to assign scores to new data points.
- ☐ Standard interpretive aids (overlays and correlations of variables with axes).

Examples

Discriminant analysis is occasionally useful in community analysis for helping to describe the differences between a priori groups of sample units, including those defined by dominance types (McCune and Allen 1985), experimental groups, or other community types (Matthews 1979, Norris & Barkham 1970). DA can be used to detect whether community types can be "predicted" from environmental variables (e.g., on the basis of soil characteristics; Gerdol et al. 1985). DA is a useful supplement to cluster analysis in helping to identify misclassified items (McCune 1988, Böhm et al. 1991), or to classify items into existing groups (Böhm et al. 1991).

The assumptions of DA are more likely to be met in other applications that are only tangentially related to community analysis. For example, McCune (1988) used cluster analysis to define groups of North American pines based on life history and morphological

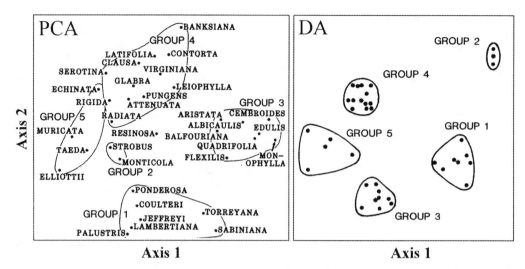

Figure 26.1. Comparison of DA and PCA. Groups are tighter in DA than in PCA because DA maximizes group separation while PCA maximizes the representation of variance among individual points. Groups were superimposed on an ordination of pine species in ecological trait space (after McCune 1988). *Pinus resinosa* was not assigned to a group, so it does not appear in the DA ordination.

characteristics. DA was first used to find misclassified species, then to describe group differences. These ecological groups were then compared to phylogenetic groups

DA can be useful for summarizing variation in environmental variables. Böhm et al. (1991, 1995a, b) used cluster analysis to classify 24-hour patterns of ambient ozone near the ground in western United States. Each group had a characteristic diurnal curve. DA identified misclassified days. After reclassifying those days, equations for the final DA were used to classify another 10,000 days from various sites. The pattern of centroids in the DA ordination was related to theory of ozone formation.

Refining and summarizing a classification

A typical procedure for refining and summarizing a classification is as follows:

1. Run cluster analysis to define groups.
2. Check for misclassified items with DA.
3. Refine the groups.
4. Describe groups based on the loadings of individual variables on standardized discriminant functions.
5. Examine group means, standard deviations, and univariate F-ratios for individual variables.
6. Describe groups with respect to another set of variables (e.g., classify the communities, then determine whether and how the community types differ with respect to environmental variables).

Contrast between PCA and DA

Principal components analysis (PCA) and discriminant analysis (DA) are both eigenanalysis techniques, but they have fundamentally different goals. DA maximizes the separation of prior groups using linear combinations of variables, while PCA represents as much variation among the individual points as possible in a low dimensional space, also using a linear combination of variables. Prior groups have no role in PCA but can be superimposed on the ordination space (Fig. 26.1).

We illustrate this contrast with a data matrix of 18 ecological traits for 34 species of North American pines (McCune 1988). Ecological groups of pines were found by cluster analysis. DA and PCA graphically described the relationships among groups and individual species, respectively (Fig. 26.1). Because DA maximizes group separation, while PCA knows of no grouping structure, the groups are much more sharply separated by DA than by PCA. The 2-D DA represented 72% of the variation among the 5 groups, while the 2-D PCA represented 44% of the variation among the 34 species.

Table 26.1. Predictions of goshawk nesting sites from DA compared to actual results, in one case using equal prior probabilities, in the other case using prior probabilities based on the occupancy rate of landscape cells. The first value of 0.83 means that 83% of the sites that were predicted by DA to be nesting sites actually were nesting sites.

		Predicted with EQUAL PRIORS		Predicted with UNEQUAL PRIORS	
		Nest	Not nest	Nest	Not nest
Actual	Nest	0.83	0.17	0.48	0.52
	Not nest	0.17	0.83	0.02	0.98

Use of prior probabilities

A common application of DA is in studies of habitat of one or more focal species (e.g., Mann & Shugart 1983, Seagle et al. 1987). We give an extended example below to show the importance of prior probabilities in classification based on discriminant functions. Twenty habitat variables were measured at each of 45 goshawk nest sites and 45 randomly located sites. A discriminant function was derived that used the habitat variables to best separate the nest sites from the non-nest sites. Two runs were compared. The first used equal prior probabilities: p(occupied) = 0.5 and p(not occupied) = 0.5. The second used the actual fraction of landscape cells that are occupied by goshawks: p(occupied) = 0.07, p(not occupied) = 0.93. Table 26.1 shows the results of the classifications.

With equal priors, the likelihood of making an error turned out to be about equal in both directions; we are no more likely to be wrong when predicting a nest site than when predicting an unoccupied site. With unequal priors, this changes dramatically. We are almost always right when we predict that a site is unoccupied, but we are right only about half the time when we predict that the site is occupied.

This may seem to suggest that using equal priors was overall a better approach, but a closer look at what happens when we apply the classification function reveals the opposite. Assume we selected 100 potential sites from the landscape at random, then predicted whether or not those sites would be occupied by goshawks. Given the error rates above, we can calculate the total number of errors that we will make, comparing the two different assumptions about prior probabilities.

We know that the total proportion of the sites that are occupied will be about 0.07 and the proportion unoccupied will be 0.93. The total error rate with EQUAL priors is calculated as follows, summing the number of errors with each kinds of misprediction:

No. non-nests predicted nests = p(predicted nest but not nest) × number of non-nests
= 0.17 × 93
= 15.8

No. nests predicted non-nests = p(predicted not nest but nest) × number of nests
= 0.17 × 7
= 1.2

Total number of errors = 15.8 + 1.2
= 17

Repeating the same calculations for the UNEQUAL prior probabilities:

No. non-nests predicted nests = p(predicted nest but not nest) × number of non-nests
= 0.02 × 93
= 1.9

No. nests predicted non-nests = p(predicted not nest but nest) × number of nests
= 0.52 × 7
= 3.6

Total number of errors = 1.9 + 3.6
= 5.5

In this example, using the unequal prior probabilities substantially reduced the total number of errors. The advantage in using the unequal prior probabilities is that we can take advantage of prior information about the likelihood of a site being a nest site. If we were to apply the goshawk discriminant function to all of the cells in a landscape, then map the results, the total number of errors on the map can be lowered by using appropriate prior probabilities. The result, however, is that the discriminant functions are very cautious about predicting nest sites, such that about half the actual nest sites will be missed. With different goals for the prediction, it might be desirable to use equal priors. For example, if we wish to lower the probability of one particular kind of error, that of failing to predict actual nest sites, the equal priors will give better results in this case.

Variations

Because DA is closely allied to MANOVA, a similarly large number of variants are possible. In practice, however, the main variation one sees is the choice between stepwise and direct entry of the variables. With community data (species as variables), the usual choice is direct entry. A discriminant function coefficient is calculated for each species. With a goal of constructing a parsimonious model, particularly when it is to be used for predictive purposes, stepwise entry of variables is the logical choice. The criteria for variable selection can be set, similar to multiple regression.

One variant of canonical correspondence analysis (CCA; Ch. 21) is similar to DA. If the environmental matrix consists of a single binary variable or a series of binary variables representing alternative states of a multistate categorical variable, then the results from CCA will be analogous to DA. Note, however, that a CCA of species against a binary variable will always produce two perfect piles of sample units in an ordination, if sample units are plotted as linear combinations of the predictors. See Legendre and Legendre (1998, p. 626) for more information on the commonalities between CCA and DA.

CHAPTER 27

Mantel Test

Background

The Mantel Test is used to test the null hypothesis of no relationship between two square symmetrical matrices. The Mantel test was originally developed to evaluate spatial and temporal clusterings of diseases, such as leukemia (Mantel 1967). Sokal (1979) introduced the technique to systematics and quantitative biogeography. Douglas and Endler (1982) gave the mathematical details in a palatable form.

The Mantel test evaluates correlation between distance (or similarity or correlation or dissimilarity) matrices. Normally each matrix is calculated from a different set of variables, measured for the same sample units. The Mantel test is an alternative to regressing one matrix against the other, circumventing the problem of partial dependence within each matrix. Because the cells of distance matrices are not independent of each other, we cannot accept the *p*-values from standard techniques that assume independence of the observations (for example, Pearson correlation). Nevertheless, the Pearson correlation (r) can be used as a measure of the strength of relationship between two distance matrices. In this context r is called the **standardized Mantel statistic** (Sokal & Rohlf 1995), and r ranges from -1 to 1.

When to use it

Use a Mantel test to evaluate the congruence between two distance (or similarity) matrices of the same dimensions. The two matrices must refer to the same set of entities in the same order. For example, we may wish to evaluate the correspondence between:
- two groups of organisms from the same set of sample units (e.g., plants and invertebrates),
- dissimilarity matrices representing community structure before and after a disturbance,
- geographic distance and ecological distance, or
- genetic distance and geographic distance.

In contrast to MRPP, which compares multiple groups of various sizes, each group consisting of separate sample units, Mantel tests are appropriate when two distance matrices are to be compared from the same set of sample units. While Mantel tests seek linear relationships between two matrices, the ability to construct those matrices from any distance measure, similarity measure, or design variable lends great power and flexibility to the approach. Mantel tests are, for example, flexible enough to test almost the same hypothesis as with MRPP (see example below).

How it works

Mantel's method tests the significance of the correlation between matrices by evaluating results from repeated randomization. The basic question is, "How often does a randomization of one matrix result in a correlation as strong or stronger than the observed correlation?" If randomizations frequently result in correlations between matrices that are as strong as the correlation with the original data, then we have little confidence that the correlation meaningfully differs from zero. On the other hand, a strong correlation structure between matrices will rarely be preserved or enhanced if one of the matrices has been shuffled.

The exact nature of the randomization is important. For the Mantel test, we shuffle the order of the rows *and* columns of one of the two matrices (it doesn't matter which matrix). The rows and columns are permuted simultaneously, such that for each i, the ith row and ith column will correspond to the same case. The diagonal elements always remain in the diagonal (but in different positions), but nondiagonal elements appear with equal probability in each of the $n(n-1)$ nondiagonal positions (Box 27.1). After each permutation, Mantel's Z statistic is calculated and the resulting values provide an empirical distribution that is used for the significance test. The Z statistic from the nonrandomized data is compared with the distribution of the Z statistic from the shuffled matrix. Note that Z is simply the sum of the product of corresponding nonredundant elements of the two matrices, excluding the diagonal.

$$Z = \sum_{i=1}^{n-1} \sum_{j=i+1}^{n} x_{ij} \, y_{ij}$$

The standardized Mantel statistic (r) is calculated as the usual Pearson correlation coefficient between the two matrices. This is Z standardized by the variances in the two matrices.

Chapter 27

Box 27.1. Method of permuting a matrix for the Mantel test.

Schematic showing the simultaneous permutation of rows and columns

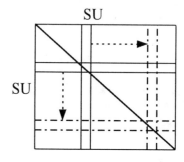

Matrix 1   Matrix 2

The starting matrix for the second of two symmetrical matrices is shown below. In this example, each matrix has four sample units. Only one matrix is permuted. For clarity, the contents of the matrix have been replaced with two digits indicating the original row and column in the matrix. For example, "23" originated in row 2, column 3.

	SU1	SU2	SU3	SU4
SU1	11	12	13	14
SU2	21	22	23	24
SU3	31	32	33	34
SU4	41	42	43	44

Select SU3 at random to swap with first item. Swap rows 1 and 3

	SU1	SU2	SU3	SU4
SU3	31	32	33	34
SU2	21	22	23	24
SU1	11	12	13	14
SU4	41	42	43	44

Swap columns 1 and 3

	SU3	SU2	SU1	SU4
SU3	33	32	31	34
SU2	23	22	21	24
SU1	13	12	11	14
SU4	43	42	41	44

This completes one step in the permutation. The step is repeated for items 2, 3, and 4.

You can choose **Mantel's asymptotic approximation** method (much faster with large data sets) or perform a **randomization (Monte Carlo) test** by conducting the actual permutations of one of the distance matrices as described above. Like other randomization tests, a p-value is calculated from the number of randomizations that yielded a test statistic equal to or more extreme than the observed value. Alternatively, Mantel's asymptotic approximation transforms the Z statistic into t, which is asymptotically normal (Mantel 1967, Douglas & Endler 1972). The asymptotic approximation is most useful with a large sample size, because the calculation is much faster than performing the permutations and the approximation is closer than with a small sample size.

Just like a correlation coefficient, the association between matrices can be either negative or positive, as indicated by the sign of the standardized Mantel statistic. If r is negative, then the randomization test counts the number of randomized runs with Z less than or equal to the observed value.

What to report

- Content and size of the two distance (or similarity) matrices.
- Method for evaluating test statistic (Mantel's asymptotic approximation or randomization).
- Standardized Mantel statistic (r) as a measure of "effect size."
- p-value.

Examples

Literature

All of the uses of Mantel tests in community ecology that we know have used nonexperimental data, but there is no reason why Mantel tests could not be used for experimental research. See Sokal and Rohlf (1995) for applications to other areas in biology.

McCune and Allen (1985) used Mantel tests to compare site difference matrices with compositional dissimilarity matrices. Although the site variables could be combined into a single composite site difference matrix, they did not do this, because site variables vary greatly in their importance to vegetation, and the inclusion of weak factors with strong on an equal basis can only weaken the site difference matrix.

Burgman (1987) compared four community matrices (total data, common plants only, rare plants only, and guilds) against each of three environmental matrices.

Tuomisto et al. (1995) compared similarity matrices based on different families of plants and trees in the Amazonian rainforest. They used the "R package" software (Legendre and Vaudor 1991).

Ritchie et al. (2000) compared microbe community profiles with two types of molecular markers, fatty acid methyl esters (FAME) and length-heterogeneity PCR (LH-PCR). Both methods showed differences in soil microbe communities related to agricultural practices. They used a Mantel test on Sørensen distance matrices to demonstrate the fundamental similarity of community structure indicated by FAME vs. LH-PCR ($r = 0.79$, $p = 0.001$). This evidence helped to convince them that the newer method, LH-PCR, gave reliable results.

The Mantel tests offers a very useful solution to the perplexing problem of incorporating *spatial proximity* into models of ecological communities (Urban et al. 2002). By combining geographic coordinates (such as UTM's) into a distance matrix, we can use Mantel tests to evaluate the relationship between distances in species space and distances in geographic space. Urban et al. distinguished between two kinds of spatial models: "distance apart" and "location." A Mantel test using a geographic distance matrix evaluates distance apart, consistent with an interpretation of local spatial structure due to autocorrelated environmental variables or a contiguous spatial process. On the other hand, by incorporating geographic coordinates in CCA and RDA (redundancy analysis), we represent location. Urban et al. (2002) gave the example:

> ...consider a case where vegetation pattern is controlled by a strong directional gradient (say, elevation) in addition to a local spatial process (say, seed dispersal). As an explanatory variable, *distance apart* would capture the dispersal process, while *location* might not. In fact, for the residuals from elevation (i.e., variation in species composition *not* predicted by elevation) to be correlated with *location*, the residuals would have to either increase or decreases monotonically with location (say, longitude or UTM easting). But we would expect seed dispersal to act similarly over the elevation gradient, an expectation consistent with *distance apart* as an explanatory variable.

Comparison of asymptotic and randomization methods

1. We have a data set of vegetation (**A**) and associated site characteristics (**E**) for 19 sample units.

In the first example, we test the hypothesis of no relationship between vegetation and site characteristics (H_0: **A** is unrelated to **E**).

The problem is approached in two ways, first with Mantel's asymptotic approximation, then with a randomization procedure (see example output below). The standardized Mantel statistic r is, of course, the same, because the choice of method only pertains to evaluating the statistical significance of r. In both cases we can soundly reject the null hypothesis. Note that the p-value is considerably higher with the randomization test. Yet it is the smallest possible p-value given the number of randomizations selected. If n is the number of randomized runs with $Z \geq Z_{obs} = 28.3756$, and N is the number of randomized runs, then

$$p = \frac{1+n}{1+N} = \frac{1+0}{1+1000} = 0.001$$

As in this example, the smallest possible p-value is always $1/(1 + N)$.

Example Output (from PC-ORD, McCune & Mefford 1999):

With **Mantel's asymptotic approximation** the results are:
```
DATA MATRICES
Main matrix:
   19 STANDS    (rows)
   50 SPECIES   (columns)
Distance matrix calculated from main matrix.
Distance measure = SORENSEN

Second matrix:
   19   STANDS    (rows)
    6   ENVIRON   (columns)
Distance matrix calculated from second matrix.
Distance measure = SORENSEN

TEST STATISTIC: t-distribution with infinite degrees of freedom
using asymptotic approximation of Mantel (1967).
If t < 0, then negative association is indicated.
If t > 0, then positive association is indicated.

STANDARDIZED MANTEL STATISTIC:      .481371 = r
         OBSERVED Z =    .2838E+02
         EXPECTED Z =    .2645E+02
       VARIANCE OF Z =   .1222E+00
  STANDARD ERROR OF Z =  .3496E+00
                  t =       5.4969
                  p =     .00000005
```

With the **randomization method**, the form of the results is somewhat different:
```
MANTEL TEST RESULTS: Randomization (Monte Carlo test) method
    .481371 = r = Standardized Mantel statistic
  .283756E+02 = Observed Z (sum of cross products)
  .264450E+02 = Average Z from randomized runs
  .120814E+00 = Variance of Z from randomized runs
  .255765E+02 = Minimum Z from randomized runs
  .278301E+02 = Maximum Z from randomized runs
         1000 = Number of randomized runs
            0 = Number of randomized runs with Z > observed Z
     .001000 = p (type I error)
-------------------------------------------------------------
p = proportion of randomized runs with Z ≥ observed Z; i.e.,
p = (1 + number of runs >= observed)/(1 + number of randomized runs)
Positive association between matrices is indicated by observed
Z greater than average Z from randomized runs.
```

Comparison of two species groups

Using the same data set, we now partition the species matrix (**A**) into two parts, grasses (**A**$_g$) and other components of the vegetation (**A**$_v$). Our new hypothesis is that the distribution of grasses is independent of the other components of the vegetation. (H_0: **A**$_g$ is unrelated to **A**$_v$. Test statistic: Z.)

Dissimilarity matrices are calculated separately for **A**$_g$ and **A**$_v$ using the quantitative version of the Sørensen coefficient. The relationship between the two matrices is shown in a scatterplot (Fig. 27.1). There is considerable scatter in the relationship, though a weak positive relationship is apparent. This appearance is supported by a smallish standardized Mantel statistic ($r = 0.24$) that is, nevertheless, statistically significant ($p = 0.032$) based on the randomization test. We can conclude that dissimilarity in grasses is positively, but weakly related to dissimilarity in other components of the vegetation.

The randomization test yielded the following results: Of the 1000 randomizations, 31 had a Z as large or larger than the observed Z (Fig. 27.2).

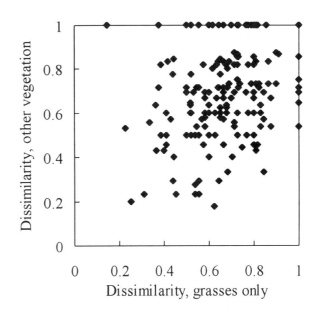

Figure 27.1. Scatterplot of dissimilarity of plots based on grasses against dissimilarity of plots based on other species.

Example output, comparison of two species groups:

```
Method chosen is a randomization (Monte Carlo) test.
    Monte Carlo test: null hypothesis is no relationship between matrices
    No. of randomized runs:      1000
    Random number seeds:         1217

MANTEL TEST RESULTS: 'Randomization (Monte Carlo test) method
------------------------------------------------------------
    0.238500 = r = Standardized Mantel statistic
    0.158331E+03 = Observed Z (sum of cross products)
    0.155539E+03 = Average Z from randomized runs
    0.171850E+01 = Variance of Z from randomized runs
    0.152315E+03 = Minimum Z from randomized runs
    0.160216E+03 = Maximum Z from randomized runs
        1000 = Number of randomized runs
          31 = Number of runs with Z > observed Z
           0 = Number of runs with Z = observed Z
         969 = Number of runs with Z < observed Z
    0.032000 = p (type I error)
------------------------------------------------------------
```

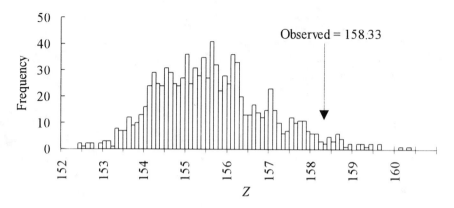

Figure 27.2. Frequency distribution of Z, the Mantel test statistic, based on 1000 randomizations for the same example as in Figure 27.1.

Comparison of Mantel test and MRPP

Refer to Chapter 24 on MRPP (p. 190) for a small example comparing three groups of five plots. MRPP tested the hypothesis of no difference between groups in species space. A similar comparison can be made with a Mantel test (though less conveniently), by constructing a design matrix (Table 27.1), then comparing the design matrix to a matrix of Sørensen distances in species space (Fig. 27.3).

The design matrix in this case is a matrix of 15 × 15 plots, each element indicating whether the comparison is a between-group comparison or a within-group comparison (Table 27.1). The design matrix is treated as a distance matrix, with zero "distance" for a within-group comparison and ones indicating a between-group comparison.

The standardized Mantel statistic equals 0.44. The Z statistic in this case is 107.2, which is compared to an expected value of 93.7 with a variance of 6.9. Mantel's asymptotic approximation yields $t = 5.1$ and $p \cong 10^{-7}$. Using the randomization method with 9999 trials yielded p = 0.0003, which is quite similar to the result from MRPP ($p = 0.0007$) using the same distance measure.

Although in this case the results are quite similar, MRPP should be the preferred approach for this problem. For one-way classifications, it is easier to explain and more convenient to run than the Mantel test.

Variations

Rank-transformed distance matrices (Dietz 1983, Lefkovitch 1984) should be explored as a useful transformation of species data prior to a Mantel test, judging by the success of ranked distance matrices in nonmetric multidimensional scaling of community data (Ch. 16). This would be analogous to using a Spearman rank correlation coefficient instead of a Pearson correlation coefficient.

Smouse et al. (1986) extended the method to deal with more than two matrices at once (analogous to multiple regression). They calculated a partial Mantel statistic, expressing the relationship between two matrices while controlling for a third matrix. The calculation of the partial Mantel statistic is essentially the same as a partial correlation coefficient (see Ch. 30). The significance of the partial Mantel statistic is tested by permutation. Two permutation methods are given by Smouse (1986) and Legendre and Legendre (1998).

Table 27.1. Design matrix for Mantel test of same hypothesis as MRPP: no multivariate difference among three groups. Within-group comparisons are assigned a zero, between group comparisons a one.

	Plot 1	2	3	4	5	6	7	8	9	10	11	12	13	14	15
plot1	0	0	0	0	0	1	1	1	1	1	1	1	1	1	1
plot2	0	0	0	0	0	1	1	1	1	1	1	1	1	1	1
plot3	0	0	0	0	0	1	1	1	1	1	1	1	1	1	1
plot4	0	0	0	0	0	1	1	1	1	1	1	1	1	1	1
plot5	0	0	0	0	0	1	1	1	1	1	1	1	1	1	1
plot6	1	1	1	1	1	0	0	0	0	0	1	1	1	1	1
plot7	1	1	1	1	1	0	0	0	0	0	1	1	1	1	1
plot8	1	1	1	1	1	0	0	0	0	0	1	1	1	1	1
plot9	1	1	1	1	1	0	0	0	0	0	1	1	1	1	1
plot10	1	1	1	1	1	0	0	0	0	0	1	1	1	1	1
plot11	1	1	1	1	1	1	1	1	1	1	0	0	0	0	0
plot12	1	1	1	1	1	1	1	1	1	1	0	0	0	0	0
plot13	1	1	1	1	1	1	1	1	1	1	0	0	0	0	0
plot14	1	1	1	1	1	1	1	1	1	1	0	0	0	0	0
plot15	1	1	1	1	1	1	1	1	1	1	0	0	0	0	0

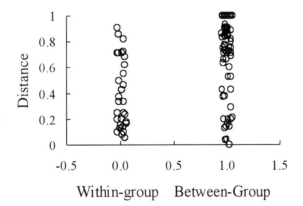

Figure 27.3. Scatterplot of Sørensen distances in species space against values in the design matrix (0 = within group, 1 = between group). The 0/1 values in the design matrix have been jittered by adding a small random number.

CHAPTER 28

Nested Designs

Left to their own devices, most biologists will sample with nested, hierarchical designs. As a sampling strategy they are logical, efficient, and provide a multilayered view of a complex world. But our tools for analyzing nested designs are poorly developed at best. The most common approach is to collapse the design at some level by averaging or summing observations at lower levels and analyzing the averages or sums. A few analysts will take the time to use separate analyses for the same question at different levels of the design.

We need multivariate tools that partition variation among levels in the design. Hierarchical data need a hierarchical viewer, not the flat view of a single level in the design. For example, consider a spatially nested design in stream ecology. Assume we have collected stream bottom community data from 6 Surber samples (microplots) randomly located in each of 20 sites in each of 5 reaches of stream, for a total of 600 Surber samples. Some variables may show large microplot to microplot variation (e.g., because of local substrate and current characteristics), yet they may be fairly uniform at larger scales. On the other hand, microplots may be quite consistent within sites, even though sites may differ greatly from each other, perhaps because of local disturbance or the nature of the surrounding vegetation. At the reach level, large-scaled factors (e.g., geomorphic differences among reaches) may be strongly expressed in the community data.

If we are analyzing *single* "well-behaved" response variables, nested ANOVA is an appropriate tool for giving us a good multilayered view of the variation. But when studying ecological *communities*, our response variables are typically numerous, nonlinearly interrelated, and far from normally distributed.

Before 2001, we had no good tools for analyzing these kinds of problems. But Anderson (2001) devised a method of nonparametric MANOVA (NPMANOVA) that can accommodate nested designs, even with non-Euclidean distance measures. We have not yet tried Anderson's method, but we give an example nested ANOVA to show how nesting is handled with traditional parametric statistics. Last, we give an example of multivariate nested NPMANOVA from Anderson (2001).

Examples

We begin with some univariate examples, to illustrate how situations in which doing the correct nested ANOVA does and does not make a difference in the results. Assume we measured the variable RICHNESS (for species richness) in three plots (PLOT) on each of four streams (STREAM) in each of two watersheds (WSHED). We ask:

- Does species richness differ among streams?
- Does species richness differ among watersheds?

Three examples using the software SPSS are given below, each example comparing a nested analysis with the incorrect factorial analysis. In all three examples, the correct design is plots nested within streams within watersheds. The examples differ in the relative importance of different sources of variation.

Example 1

In this example (Tables 28.1 and 28.2) most of the variation is concentrated in the difference between streams, with essentially no differences between watersheds. Because there is little interaction between STREAM and WSHED, the results of the two analyses are not qualitatively different (compare Tables 28.1 and 28.2). Notice that in the correct analysis, the F-ratio for WSHED is the ratio of mean squares MSwshed/MSstream and the F-ratio for STREAM is MSstream/MSwithin. Also notice how the sums of squares differ between the two analyses. The SS for the interaction term in the incorrect analysis is included in the SS for STREAM in the nested design. Otherwise, the sums of squares are identical.

Correct analysis (nested):
> MANOVA RICHNESS by WSHED(1,2) STREAM(1,4)
> /DESIGN WSHED VS 1, STREAM WITHIN WSHED = 1 VS WITHIN.

Table 28.1. Nested sampling design analyzed with a univariate nested ANOVA.

Source of Variation	SS	DF	MS	F	p
WSHED	0.67	1	0.67	0.01	0.933
STREAM W/IN WSHED (err 1)	513.83	6	85.64	17.72	0.000
WITHIN CELLS	77.33	16	4.83		

Incorrect analysis (factorial):
> MANOVA RICHNESS by WSHED(1,2) STREAM(1,4) /DESIGN.

Table 28.2. Nested sampling design **incorrectly** analyzed with a factorial ANOVA.

Source of Variation	SS	DF	MS	F	p
WSHED	0.67	1	0.67	0.14	0.715
STREAM	508.83	3	169.61	35.09	0.000
WSHED BY STREAM	5.00	3	1.67	0.34	0.793
WITHIN CELLS	77.33	16	4.83		

Example 2

In this example, the watersheds differ greatly, and there is little variation among streams within watershed. The results are again qualitatively similar between the correct and incorrect analysis (Tables 28.3 and 28.4).

Correct analysis (nested):
> MANOVA RICHNESS by WSHED(1,2) STREAM(1,4)
> /DESIGN WSHED VS 1, STREAM WITHIN WSHED = 1 VS WITHIN.

Table 28.3. Nested sampling design analyzed with a univariate nested ANOVA.

Source of Variation	SS	DF	MS	F	p
WSHED	198.37	1	198.37	124.20	0.000
STREAM W/IN WSHED (err 1)	9.58	6	1.60	0.44	0.841
WITHIN CELLS	58.00	16	3.63		

Incorrect analysis (factorial):
> MANOVA RICHNESS by WSHED(1,2) STREAM(1,4) /DESIGN.

Table 28.4. Nested sampling design **incorrectly** analyzed with a factorial ANOVA.

Source of Variation	SS	DF	MS	F	p
WSHED	198.37	1	198.37	54.72	0.000
STREAM	3.13	3	1.04	0.29	0.834
WSHED BY STREAM	6.46	3	2.15	0.59	0.628
WITHIN CELLS	58.00	16	4.83		

Example 3

In this case, streams differ greatly (Table 28.5), but the factorial analysis detects an interaction between stream and watershed (Tables 28.6 and 28.7). The results from the two analyses differ greatly. We include the data so you can try the same analyses on your software.

Table 28.5. Data set for Example 3.

WSHED	STREAM	PLOT	RICHNESS	WSHED	STREAM	PLOT	RICHNESS
1	1	1	14	2	1	1	13
1	1	2	12	2	1	2	12
1	1	3	8	2	1	3	9
1	2	1	3	2	2	1	22
1	2	2	5	2	2	2	16
1	2	3	1	2	2	3	21
1	3	1	8	2	3	1	14
1	3	2	9	2	3	2	13
1	3	3	12	2	3	3	11
1	4	1	3	2	4	1	10
1	4	2	7	2	4	2	10
1	4	3	10	2	4	3	12

Correct analysis (nested):

MANOVA RICHNESS by WSHED(1,2) STREAM(1,4)
/DESIGN WSHED VS 1, STREAM WITHIN WSHED = 1 VS WITHIN.

Table 28.6. Nested sampling design analyzed with a univariate nested ANOVA.

Source of Variation	SS	DF	MS	F	p
WSHED	315.37	1	315.37	7.71	0.032
STREAM W/IN WSHED (err 1)	245.58	6	40.93	6.42	0.001
WITHIN CELLS	102.00	16	6.38		

Incorrect analysis (factorial):

MANOVA RICHNESS by WSHED(1,2) STREAM(1,4) /DESIGN.

Table 28.7. Nested sampling design **incorrectly** analyzed with a factorial ANOVA.

Source of Variation	SS	DF	MS	F	p
WSHED	315.37	1	315.37	49.47	0.000
STREAM	0.13	3	0.04	0.01	0.999
WSHED BY STREAM	245.46	3	81.82	12.83	0.000
WITHIN CELLS	102.00	16	6.38		

Example 4

The method for adding more levels to the design is not immediately obvious in SPSS. In this case, we added to the correct, nested design another level "VALLEY," representing three different valleys within each watershed. Streams are nested within valleys. The syntax for requesting the analysis is:

MANOVA RICHNESS by WSHED(1,2) VALLEY(1,3) STREAM(1,4)
/DESIGN
WSHED VS 1,
VALLEY WITHIN WSHED = 1 VS 2
STREAM WITHIN VALLEY WITHIN WSHED = 2 VS WITHIN.

As in the other examples, variation at each level in design is compared with the variation in the next lower level.

Example 5

This example from Anderson (2001) incorporates nesting, temporal change, and the size of artificial patches. The objective was to determine whether colonization of patches was influenced by patch size. Species were recorded on five dates, but individual patches were sampled only once, rather than repeatedly. Patches of a given size were set up in pairs. Colonization was recorded in each patch. Thus, individual patches were nested within a pair of patches, both attached to the same spot. The experimental unit was a pair of patches, while the sample unit was an individual patch. Upper level factors (time, patch size) were tested against pair-to-pair variation, rather than patch-to-patch variation.

The analysis (NPMANOVA) was based on Sørensen (Bray-Curtis) distances calculated from "double-root" transformed data. The distance matrix was calculated among all sample units. An exact permutation test for the nested factor (patch) was achieved by permuting the observations among pairs, but always staying within the same combination of date and patch size. For the upper level terms, pairs were permuted as a unit. For each test, 4999 permutations were used.

The communities clearly differed among dates and among patch sizes (Table 28.8). The date × patch size interaction was statistically significant, meaning that the influence of patch size varied among dates. Interestingly, pairwise comparisons of differences between patch sizes, separated by date, showed no consistent temporal trend of increasing or decreasing differences. The nested factor (patch pair) turned out to be unimportant — differences among pairs within date × patch size combinations were not statistically significant.

Anderson's NPMANOVA can be extended to more deeply nested designs (e.g., Example 4 above). This approach should allow ecologists to better see hierarchical structures in their data. Nested designs afford an opportunity to study scaling in community ecology, avoiding the two commonly used but unappealing alternatives: pseudoreplicating by using an error term too low in the design, or just averaging away and ignoring fine-scale variation.

Table 28.8. Nonparametric MANOVA of a Sørensen distance matrix, adapted from the example from Anderson (2001).

Source	Error term	d.f.	SS	MS	F	p
1. Date	4	4	30,306	7,576	20.50	0.0002
2. Patch Size	4	2	6,415	3,207	8.68	0.0002
3. Date × patch size	4	8	6,224	778	2.10	0.0062
4. Patch pair	5	15	5,544	370	1.28	0.3384
5. Residual		30	8,697	290		
Total		59	57,186			

Part 6. Structural Models

CHAPTER 29

Classification and Regression Trees

Dean L. Urban
Nicholas School of the Environment and Earth Sciences
Duke University

Introduction

Community ecology has a long tradition in which discrete habitats or community types are distinguished in terms of species-compositional or environmental variables. Familiar applications include classification of wildlife habitat and typal plant communities. In each case, we address the companion questions, (a) *Do these groups differ significantly from one another?* and (b) *Which variables best account for these differences?* Specifically, we might ask (1a) Do groups defined on species composition differ in terms of environmental variables? (1b) If so, which environmental variables best differentiate the groups? Reciprocally, (2a) Do groups defined on environmental or other variables differ in species composition? (2b) If so, which species might serve as diagnostic indicators for these groups? Similarly, in habitat classification we define groups based on the presence or abundance of a species of interest, and ask (3a) Do occupied "habitats" differ from other sites? and (3b) Which variables (environmental or species-compositional) best distinguish habitat? Note that in the first two cases we might also ask the obvious question, Which variable(s) — species or environmental factors — best distinguish groups defined on these same variables? In this case the first question of *whether* the groups differ is trivial (they differ by definition), but exploring the second question can provide for easier interpretation or parsimonious predictive models.

We have already considered a number of common approaches for these applications: multivariate analysis of variance (MANOVA) and discriminant functions analysis (DA) are the classical approaches, but we have also considered nonparametric MANOVA (Anderson 2001), multi-response permutation procedures (MRPP, Mielke et al. 1981), Indicator Species Analysis (ISA, Dufrene & Legendre 1997), the Mantel equivalent of MANOVA (a group-contrast or design matrix Mantel test, Legendre & Legendre 1998), and the program ANOSIM (Clark et al. 1993). Of these techniques, MANOVA, NPMANOVA, MRPP, the group-contrast Mantel test, and ANOSIM focus on whether the groups differ statistically without indicating which variables are responsible for the group differences. Indicator species analysis, DA, and partial Mantel tests identify the variables that best distinguish the groups. Note that any technique that explicitly identifies which variables distinguish the groups also implicitly tests for overall group differences, in that failure to find distinguishing variables implies that the groups are not different. Only DA provides an explicit vehicle for the classification of unknown samples, although the indicators in ISA could be used to develop such a classification. Organizing these analyses by question provides for a crude matrix that matches each application to the techniques best suited to address it (Table 29.1).

Classification and regression tree (CART) models offer a rather new and compelling alternative for exploring differences among groups (Breiman et al. 1984, Moore et al. 1991, Vayassieres et al. 2000). CART partitions the dataset recursively into subsets that are increasingly homogeneous with respect to the defined groups, providing a tree-like classification and an associated dichotomous key to classify unknown samples into the groups.

As a nonparametric method, CART is robust to many of the data issues that sometimes plague parametric models. As a recursive model, the approach is also able to capture some relationships that make sense ecologically but that are difficult to reconcile with conventional linear models in statistics.

Table 29.1. A matrix matching statistical techniques to various applications that require group classification or discrimination. Applications are discussed in the Introduction, coded here as groups defined on species composition (SPP) or environmental variables (ENV). Techniques are discriminant analysis (DA), group-contrast Mantel test (GC-Mantel), multivariate analysis of variance (MANOVA), nonparametric MANOVA (NPMANOVA), multi-response permutation procedures (MRPP), classification and regression trees (CART), generalized linear models (GLM), and generalized additive models (GAM).

Application	Appropriate Techniques
Exploratory data analysis:	
1a. Do SPP groups differ?	CART, DA, GC-Mantel, MANOVA, NPMANOVA, MRPP
1b. On which ENV variable(s)?	CART, DA, partial GC-Mantel
2a. Do ENV groups differ?	ISA, CART, GC-Mantel, MRPP
2b. On which SPP?	ISA, CART, partial GC-Mantel
3a. Do habitats differ?	DA, CART, MANOVA, NPMANOVA, MRPP, logistic regression, GLM, GAM, etc.
3b. On which variable(s)?	CART, DA, partial GC-Mantel, logistic regression, etc.
Predict group membership:	
1c. on SPP	ISA (with some modification)
2c. on ENV	CART, DA, (multinomial) logistic regression
3c. habitat variables	CART, DA, logistic regression

We will focus on *classification* trees, that is, models in which the dependent variable is categorical (e.g., the white fir forest type). By contrast, *regression* trees predict continuous response variables (e.g., basal area of white fir). The same algorithm works in each case, although terminology, presentation, and interpretation differ somewhat for discrete as compared to continuous cases (Venables & Ripley 1999, and see **Variations** below). *Multivariate regression trees* (De'ath 2002; Chapter 10) are similar to univariate regression trees, but replace the single continuous response with multiple continuous responses.

A heuristic example

Consider the ecologically reasonable but hypothetical pattern in which the occurrence of tree species X is governed by soil moisture along an elevation gradient complex. In this case, we wish to classify a discrete, binary response variable: each sample will be classified as "habitat" or "not habitat" for species X. In mountain systems, as elevation increases temperature declines while precipitation generally increases, resulting in an increase in available soil moisture (Stephenson 1990, 1998; Urban et al. 2000). This pattern is modified locally by topographic exposure and other variables. North-facing slopes are cooler and moister, while south-facing slopes are warmer and drier due to differences in radiation loading. Likewise, high ridges are drier than topographic convergence zones due to local drainage, and deeper or finer-textured soils are wetter than shallower or coarser-textured soils due to soil water-holding capacity. Thus, we can enumerate a variety of environmental settings that might provide equivalently mesic conditions for species X: average exposures at the optimal elevation, slightly higher elevations on south-facing slopes, slightly lower elevations on north-facing slopes, deeper or finer soils at lower elevations (or on south slopes), atypically shallow or sandy soils at suitable higher or protected (north-facing) sites, convergence zones at lower elevations or south-facing slopes, and so on. Accounting for this range of environmental settings that qualify as habitat for species X is difficult in a linear model, because the accounting must be simultaneous in a statistical sense. A CART model solves this problem by identifying the several cases recursively: the model first identifies the average or most typical case, and then iteratively revisits the cases that emerge as exceptions to the rule. The result is a hierarchical tree model in which alternative settings that qualify as habitat are identified as different branches of the tree. The solution identifies the variables associated with each branch (strictly, each split of each branch) while also providing a dichotomous key that can classify new samples accordingly.

When to use it

In ecology, CART models have been applied most often to classify habitats (i.e., "habitat" versus "not habitat") or vegetation types in terms of measured environmental variables. The results of the analysis are an overall assessment of how different the groups are, which variables distinguish the groups, and a predictive model that can classify new samples into the groups. Because of the recursive nature of the analysis, CART models are especially amenable to applications where there might be multiple alternative environmental settings (combinations of variables) associated with the same group. Statistically, CART models are compelling in applications where the effects of the predictor variables are nonadditive (e.g., in cases of substitutable resources), and where interactions among variables are not simply multiplicative. CART models can readily admit a mix of categorical, rank, and continuous variables, and indeed are insensitive to monotonic transformations of the variables. CART models also are reasonably accommodating of missing values, especially in new data that are to be classified into groups using a fitted model. Finally, CART models are very easy to interpret visually, and so are especially useful for exploratory data analysis.

Increasingly, CART applications are used to generate maps of communities, wildlife habitats, or land cover types based on environmental variables held in a geographic information system (GIS), an application that takes advantage of the efficient coding of the predictive tree model as a series of "IF ... THEN ..." conditional statements (Moore et al. 1991, Urban et al. 2002). These applications also take advantage of the increasing availability of geospatial environmental data derived from digital elevation models and remotely sensed imagery.

In theory, there is no reason why CART models could not be used to distinguish groups in terms of species composition. The nonparametric nature of the analysis and the method of variable selection would seem appropriate for species data; in this, CART models would avoid some of the issues that often plague linear models when applied to species abundance data.

How it works

Performing a CART analysis for classification entails identifying the groups, choosing the predictor variables, running the analysis, and (often) "pruning" the tree to optimize its classification efficacy (usually based on optimizing misclassification errors).

Identifying the groups

In most applications the choice of groups to classify is straightforward, but there may be subtle ramifications that are not always appreciated in practice. In particular, choice of groups often has implications for how certain misclassifications are interpreted. In most habitat classification models the approach is a two-group model, contrasting "habitat" and "not habitat." Habitat samples are those on which the species was observed to occur, which explicitly denies the possibility that the species might be observed in marginal or inappropriate habitat. In most cases this is presumed to be a rare exception. The "not habitat" samples can be defined in either of two ways. In the first approach, the occupied samples are contrasted with a second group comprised of samples where the species was not observed, and these latter samples are presumed to be inappropriate habitat for that species. The analysis in this case contrasts the two groups in a way analogous to a two-sample t-test. This approach ignores the possibility that significant amounts of actual habitat might be unoccupied (or that the species might not be encountered during sampling). Alternatively, the occupied samples are contrasted with a second group comprising a random sample of the study area or landscape, representing where the species *might* have been observed. The analysis in this case is analogous to a one-sample t-test, in effect asking whether the samples occupied by the species are different from a random grab of the landscape. This admits that some actual habitat will be included in the "not habitat" sample.

In the multiple-group case, some opportunity for misclassification also arises due to the way the analysis is performed. The critical issue here is the extent to which the groups are exclusive. For example, if the groups are habitats for various bird species, the question is to what extent the same vegetation sample might qualify as "habitat" for different bird species. If habitats overlap substantially, then a multi-group classification model will under-represent available habitat for some species because each sample will be assigned to the most likely group (species).

The solution in this case is to build a series of two-group habitat models, one for each bird species, so that each vegetation sample can be classified as habitat for any bird species independently. Alternatively, if the goal is to emphasize relative differences among species in terms of habitat selection (sensu Green 1971), then the multi-group approach is proper but the misclassifications should be interpreted accordingly. Finally, if the groups are presumed to be rather exclusive, then a multi-group model is reasonable and misclassifications are also readily interpretable. In the case of vegetation types or communities identified via clustering, where the goal is to test group differences and highlight the environmental variables that define group differences, a multi-group classification model is appropriate.

Thus, the first and most straightforward step in the analysis often has no truly correct answer, in that there may be ecologically reasonable and expected exceptions to the definition of groups. The key to this is to ensure that these assumptions are weighed in advance, and subsequently evaluated in terms of misclassification rates in the fitted model (see below). In fairness, it should also be noted that this issue of how to define groups is by no means restricted to CART models; any classification application faces the same issue.

Choosing the variables

CART models are refreshingly indulgent in terms of variables that can be used as predictors. The models are nonparametric, working typically with ranked continuous values, but also can include categorical predictors such as soil classes or topographic positions. Most implementations limit the number of levels for categorical variables to a computationally reasonable number that is still likely to be larger than the number of ecologically relevant distinctions that can be made.

A distinctive feature of CART is that the model identifies a single variable as the indicator variable for each branch of the tree. This is in contrast to discriminant functions or logistic regression, in which the groups are distinguished along multivariate axes.

The CART algorithm

Recall our hypothetical example of tree species X and its occurrence in a variety of environmental settings along an elevation/topographic moisture gradient. This example provides an intuitive illustration of the CART classification algorithm. The data in this case will involve two groups, termed "habitat" for species X or vegetation type X (coded 1 in the data) and "not X" samples where the species or vegetation type was not encountered (coded 0). There are several environmental variables including elevation, aspect (coded to 8 octants), soil depth, and an index of topographic convergence.

First, the continuous variables (elevation, soil depth, convergence) are rank-transformed. The analysis then begins with the first variable, and iteratively partitions the data between each of two values in the ranked data series. At each partition, the model computes an estimate of within-partition heterogeneity or "impurity" of the partitions. Impurity is estimated in terms of *deviance*, a log-likelihood estimator based on the empirical distribution of samples into groups along each branch of the tree. Alternative estimates of impurity may be based on information theory (analogous to the familiar Shannon function) or other indices (Breiman et al. 1984, Venables & Ripley 1999).

The goal is to find the split of the data that maximizes the decrease in deviance with the partition. This split will optimize the separation of groups (0, 1) on either side of the partition. Thus, if we had 100 different elevation values, 99 potential partitions would be considered. For a categorical factor with L levels, there are $2^{L-1}-1$ candidate splits. This partition- ing is then repeated in similar fashion for the other variables (aspect, soil depth, convergence). The best split of the best variable is then selected to define the first branch of the tree. One branch is associated with values of the chosen variable less than the threshold value; the other branch, values equal to or greater than this threshold.

After this first partition, there are two branches of the growing tree. In this example, suppose that the first split occurs on elevation: group X tends to occur at elevations above 1000 m. At each branch (or node) of the tree, there will be some samples that have been misclassified: higher-elevations that do not support X, and lower-elevation samples that do. Within each branch, the process of variable selection is repeated and two new branches are created for a total of four. On each side, different variables can be used as predictors, or the same variable can be re-used. For example, a second split at lower elevations might isolate northern exposures as supporting X, perhaps by separating the octants N and NE from the other six aspects. Likewise, in the higher-elevation branch of the tree, the task is to eliminate "not X" recursively from a branch that is largely habitat type X. Perhaps the very highest elevations are not habitat, in which case elevation would be re-used as a variable within the same main branch.

This recursive splitting continues until no samples are misclassified (rarely), the end nodes (leaves) reach a threshold homogeneity, or until some minimum number of samples per node is met. In the end, the CART model would be expected to generate a tree in which each of the suitable environmental settings enumerated above appears as a terminal leaf of the tree. The branch to each leaf specifies the combination of conditions (sometimes called an "interaction pathway") that defines the leaf.

This approach to variable selection in classification trees thus attends a number of issues that can confound parametric linear models in habitat classification. Monotonic transformations of the data are handled implicitly, in that the rank values are insensitive to such transformations. Nonlinear species responses are also allowable; for example, re-using elevation in a "higher than" split followed by a "lower than" split effectively delimits a unimodal species response to elevation. Likewise, interactions (correlations) among variables are also handled because only the single best predictor is selected at each branch, while different predictors are still free to be selected at other branches of the tree. (Some implementations of CART models also identify a set of "next best" variables at each split, thus acknowledging correlations among predictors.)

Model refinement via tree pruning

This tree-growing algorithm bears some reflection, as it has two important implications for ecological applications. First, while every partition is optimized at each step of the algorithm, this process cannot guarantee an optimal overall partition of the data. This is because CART uses a "one-step look-ahead" algorithm that cannot "see" the entire solution at once. (This is similar to the issue of local versus global minima encountered in other iterative solutions such as non-metric multidimensional scaling; Ch. 16). In practice, this issue of global optimality is probably not a serious drawback. Second, and more importantly, the recursive nature of the solution tends to produce models that are statistically over-fitted to the data used to generate the model, often due to spurious or coincidental relationships. For example, if an atypical specimen of species X happened to occur on a high-elevation rock outcrop, this odd case might ultimately be isolated in its own terminal leaf of a CART tree. This *is* a problem, because it can lead to over-generalized interpretations of the solution ("high-elevation rock outcrops are important habitat for X"). Over-fitting also influences the extent to which the model can be applied to other datasets in predictive mode (other rock outcrops would be predicted as "habitat"). For this reason, CART models are often simplified (the tree is "pruned") to yield a final model that provides a balance of accuracy with the training dataset and robustness to novel data.

Tree pruning uses a "cost-complexity" approach that attempts to balance the cost of pruning the tree (and thus misclassifying some known samples) against the implicit cost of a more complex model (which tends to misclassify novel samples). This logic is crudely analogous to the selection criterion for including a variable in a multiple regression, which balances the improved accuracy of the more complex model against the cost of estimating parameters for the additional variable. In a classification tree, the criterion for pruning can be either deviance or the number of misclassifications. Plotting the error rate as a function of the number of terminal leaves in the tree provides a decreasing curve that typically reaches a minimum or "flattens out" for a larger tree. The tree is pruned by choosing a tree size near the minimum, i.e., the smallest tree for an acceptably low error rate (see **Example**, below).

Model refinement via tree-pruning can occur in two ways depending on available data. In the best case, there is a training set of data used to build the tree and a second, independent data set reserved for testing the model. In this case, the model tests are true validations of the fitted model (sensu Mankin et al 1975). More typically, there is no such luxury of data and the model is assessed via cross-validation by subsetting the data. In this, for example, a tree might be constructed using 90% of the available data and then tested against the reserved 10%. This process is repeated for multiple subsettings of the data, in Monte Carlo fashion, to generate bootstrapped estimates of the classification accuracy of the model.

Classification accuracy is assessed in terms of a matrix of classification success, also known as the "confusion matrix." In a binary classification scheme (e.g., habitat/not habitat), this matrix includes four elements: positive matches, negative matches, and mismatches of two types. In a multiple-group case, the matrix is simply larger. In the simplest sense, classification successes can be quantified as the number of matches divided by the total number of cases. It is at this point that assumptions about the definition of groups come into play, as different assumptions about group membership will admit varying levels of misclassifications as being ecologically reasonable. Thus, while the statistical accounting of misclassification is straightforward, the ecological interpretation of these misclassifications may be less clear-cut.

In tree pruning, a branch of the tree is removed if the change in misclassification rate upon removal is small relative to the overall complexity of the tree. For example, a tree with eight nodes might classify samples only marginally less accurately than a tree with twice as many nodes; in this case, the simpler tree would be more easily interpretable at little cost in accuracy. Pruning is at the discretion of the user. A nice feature of tree models is that they are hierarchical, so the classification can be stopped at any level in the tree, for whatever reason. Thus, the models adapt to missing values, which stop the classification at the level before the missing value is needed. Trees also can be simplified arbitrarily in applications that require varying levels of precision.

What to report

- ☐ Software and algorithm used in the application. Because CART models are relatively new and software implementations vary, it is important to specify the software.

- ☐ As in any classification model, full disclosure should also itemize the variables used and why they were included, as well as the basis for group definition (for groups defined by clustering or similar methods, this is rather straightforward).

- ☐ Stopping rules for the tree (minimum group size that will be split) and the cross-validation procedure used to test the tree (size of subsample, number of trials).

- ☐ A figure of the (pruned) tree. These are typically very intuitive, but should include, at each branch, the variable and level defining the split (Fig. 29.1). For each node, the dendrogram should include the number (or proportion) of cases in that group and their classification success (e.g., the S-plus tree diagram reports the number of samples in each node and the number that are misclassified). The dendrogram thus illustrates alternative settings that define the groups, which variable(s) define these settings, and how "impure" the groups are on each branch.

- ☐ A misclassification table (confusion matrix) for the final tree. Note that because each group in the CART model may occur in more than one branch of the tree, the confusion matrix is a concise summary of the classification success (while also perhaps missing some nuances evident in the tree itself).

- ☐ If definition of groups in the analysis is ecologically ambiguous, some discussion of misclassifications for these groups is warranted. Again, while all misclassifications are statistically "bad," these errors may not be interpreted in the same way ecologically.

- ☐ Some interpretation of the CART tree to guide the reader through it. For example, it might be pointed out that variables that appear near the top of the tree are more important (explain more of the variation) than variables that are invoked lower in the tree. Likewise, a variable that is invoked repeatedly along a branch might be a manifestation of nonlinear responses to that variable—especially if the splits are defined by alternating "greater than" and "less than" decisions. Finally, the degree to which very different variables are invoked in major branches of a tree might suggest that alternative but dissimilar definitions of "habitat" are being discovered. This last instance would suggest that the CART model might be a more appropriate solution to the classification than an alternative model that relies on linear, additive relationships such as discriminant analysis.

Example

Urban et al. (2002) collected vegetation and environmental data on 99, 0.04-ha quadrats in the mixed-conifer forests of Sequoia-Kings Canyon National Park in the southern Sierra Nevada of California, USA. Vegetation data consisted of trees recorded by species and diameter (dbh); these measurements were summarized as basal area (m^2/ha) for 21 tree species. Of these, 13 were retained for analysis while eight uncommon species were discarded (Table 29.2). Species data were relativized by species maximum (Ch. 9) because raw basal area differed by orders of magnitude among species (these range in stature from dogwood, *Cornus nuttallii*, to giant sequoia, *Sequoiadendron giganteum*). Environmental variables included: elevation; slope; aspect; transformed aspect (after Beers et al. 1966); and soil depth (mean and standard deviations of 30 measurements per plot). chemistry (pH, cations, carbon); and texture (% sand/silt/clay). A total of 21 environmental variables were measured; six were eliminated based on redundancies due to correlations among variables.

Vegetation plots were assembled into community types via hierarchical clustering using average linkage and Bray-Curtis dissimilarities (Ch. 11) computed from relative basal area in PC-ORD (McCune & Mefford 1999). Eight clusters were selected as the most interpretable level of classification, based on the A statistic provided by MRPP (Ch. 24). One singleton group was discarded, leaving seven community types for further analysis. These community types were readily identifiable in terms of species composition, based on Indicator Species Analysis (Table 29.3).

Table 29.2. Species names and codes used in the example analyses (from Urban et al. 2000, 2002).

Code	Scientific name	Common name
ABco	*Abies concolor*	white fir
ABma	*Abies magnifica*	red fir
CAde	*Calocedrus decurrens*	incense-cedar
CEin	*Ceanothus integerrimus*	deer brush
COnu	*Cornus nutallii*	Pacific dogwood
PIco	*Pinus contorta*	lodgepole pine
PIje	*Pinus jeffreyii*	Jeffrey pine
PIla	*Pinus lambertiana*	sugar pine
PImo	*Pinus monticola*	western white pine
PIpo	*Pinus ponderosa*	ponderosa pine
QUch	*Quercus chrysolepis*	canyon live oak
QUke	*Quercus kellogii*	California black oak
SEgi	*Sequoiadendron giganteum*	giant sequoia

Table 29.3. Indicator Species Analysis for the seven forest types identified via hierarchical clustering. Indicator values (*IV*) are percentage of perfect fidelity. Indicator values were tested for statistical significance based on 1000 permutations (**, $p < 0.001$; *, $p < 0.005$). Sequence = order of groups in data, Identifier = group identifier, Avg = Average *IV*, Max = Maximum *IV*, MaxGrp = Group with highest *IV*.

				Group						
			Sequence:	1	2	3	4	5	6	7
			Identifier:	1	5	29	38	45	67	76
			Species Label:	ABco	CAde	PIpo	ABma	PImo	PIje	PIco
			Number of items:	37	10	10	14	8	14	5
Species	Avg	Max	Max IV: Grp							
ABco**	13	48	1	48	31	5	1	0	8	0
ABma**	13	60	38	0	0	0	60	12	2	15
CAde**	14	70	5	2	70	23	0	0	0	0
CEin	3	20	29	0	0	20	0	0	0	0
COnu**	8	59	5	0	59	0	0	0	0	0
PIco**	14	95	76	0	0	0	0	1	0	95
PIje**	14	99	67	0	0	0	0	0	99	0
PIla*	11	45	1	45	26	3	0	0	1	0
PImo**	13	86	45	0	0	0	1	86	0	4
PIpo*	9	60	29	1	2	60	0	0	0	0
QUch	3	19	5	0	19	1	0	0	0	0
QUke**	13	77	29	1	11	77	0	0	0	0
SEgi	2	16	1	16	0	0	0	0	0	0
Averages	10	58		9	17	14	5	8	8	9

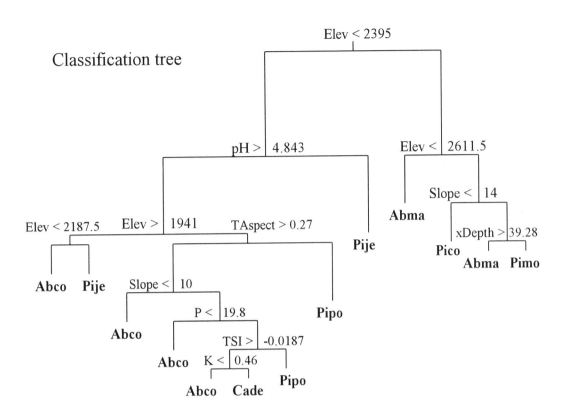
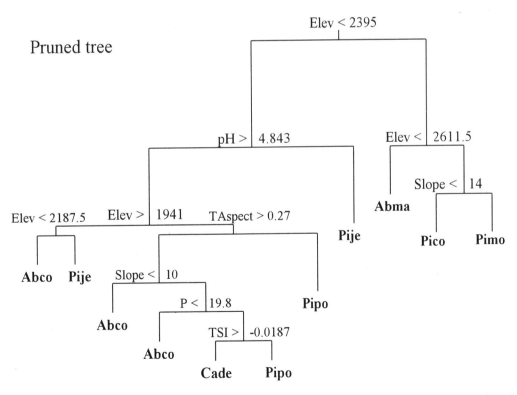

Figure 29.1. *Upper*: Classification tree for 7 forest types on 15 environmental variables (function rpart, complexity parameter (cp) = 0.000001, minsplit = 10, split = information). *Lower*: Pruned classification tree, simplified by stopping the tree at the number of nodes corresponding to the point where the pruning curve crosses the minimum (1 S.E.) line (Fig. 29.2).

Table 29.4. Misclassification table for the 7 forest types, based on a pruned CART model with 11 nodes (Fig. 29.3). Rows are actual forest types, columns are predicted forest types. Row totals are indexed as number correct/number misclassified. Total misclassification rate based on jack-knifing is 39/98 (39.8%).

	ABco	ABma	CAde	PIco	PIje	PImo	PIpo	Totals
ABco	25	1	6	0	1	0	4	25/12
ABma	1	11	0	0	0	2	0	11/3
CAde	8	0	0	0	0	0	2	0/10
PIco	0	0	0	3	0	2	0	3/2
PIje	3	0	0	0	11	0	0	11/3
PImo	0	2	0	1	0	5	0	5/3
PIpo	3	0	3	0	0	0	4	4/6

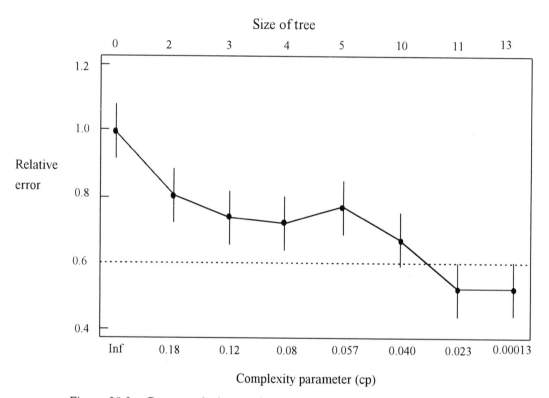

Figure 29.2. Cost-complexity pruning curve for the classification tree in Figure 29.1. Error bars are estimated from 10 cross-validation subsets of the samples. The horizontal line is one standard error above the minimum error rate. "Inf" = infinite. Relative error is calculated by cross-validation.

Further analysis of the community types was conducted in the statistical package S-PLUS (SP2000 for Windows, Insightful, Inc.). S-PLUS provides a native CART model implementation (function tree and its support functions). A somewhat more versatile contributed library for recursive partitioning is also available (rpart, Therneau & Atkinson 1997). Examples shown here use the rpart functions.

A default classification tree for the seven forest types has 13 terminal nodes (Fig. 29.1, upper). This analysis also generates a cost-complexity pruning table based on trees cross-validated using 10 subsets of the data. This table provides a sequence of alternative trees of increasing complexity (number of nodes) and decreasing error rate (Fig. 29.2). The horizontal line in Figure 29.2 is a minimum associated with one standard error in the overall error rates; by convention, the tree can be pruned to the left-most tree for which the error rate is below this minimum. In this case, the pruning results in a tree with 11 nodes (Fig. 29.1, lower). Interestingly, the final tree uses only six environmental variables to classify the six groups; one of these (elevation), it reuses several times.

The CART tree is readily interpretable visually. In this, each branch of the tree can be traced from root to leaf to yield the set of environmental conditions associated with each node. This aspect of the tree is what facilitates its encoding into "IF ... THEN" rules to classify new samples. Because the solution is entire and exhaustive, the tree also has the rather tidy feature that it can be "decoded" in any direction (left/right, up/down) at the user's whim. Indeed, while the CART solution includes a tabular summary of all the splits in the model, the tree itself is typically more easily interpreted.

A more stringent test of the model's classification power can be generated by repeating the analysis using a full jack-knife cross-validation (i.e., by withholding each observation in turn for cross-validation, yielding 98 separate trees in this case). The confusion matrix for this jack-knifed model shows a rather high error rate (Table 29.4). In particular, the lower-elevation pine/oak and incense-cedar types are not well distinguished environmentally. Overall, the CART model suggests that the even though the vegetation types are compositionally distinct, they are not very distinct in terms of these environmental variables.

As a word of caution, it should be mentioned that a default CART model using S-PLUS's native **tree** function results in a slightly different classification tree using the same data (not shown). Back-classification suggests an error rate of only 11% misclassifications — apparently an impressive improvement over the results from the rpart analysis. But because this tree is heavily conditioned on the data and involves no cross-validation, it is highly over-fitted. (Unfortunately, it is not easy to perform cross-validations using the **tree** function in S-PLUS.) This discrepancy underscores the importance of cross-validation. Again, this is true for all classification models, which by definition run a high risk of overfitting.

Variations

As noted at the outset, this discussion has focused on classification trees, in which the dependent variable is categorical. The same CART algorithm can be used to predict a continuous response variable, yielding a regression tree instead. While the algorithm is essentially the same, some terms are defined differently and the model prediction is somewhat different as well. In a regression tree, the prediction for each node (terminal leaf) is a single value of the response variable, which is the same for all samples within that node. Different levels of the response variable are provided by different branches of the tree. Deviance is defined more simply in a regression tree, in terms of the familiar corrected sum of squared deviations from the mean value; the best tree minimizes this residual variance in the response variable.

While regression trees are not yet common in ecological applications, they deserve more attention because of the potential interpretative power of the tree model. For example, consider the hypothetical example discussed above and the real examples from the Sierra Nevada. In a classification tree, the response variable might be "white fir community type" and the solution is free to identify alternative environmental settings that qualify as white fir habitat. In a regression tree, the response variable might be "white fir basal area" and the branches similarly might illustrate alternative settings that provide for high white fir abundance. In a GIS-based rendering of these trees, either approach is capable of mapping the locations where white fir abundance is predicted by soil depth, by topographic exposure, and so on: an immensely informative analysis. In a regression tree, we might also map the explanatory power of each predictor variable, accounted in terms of model deviance at each node. This capability to assess the relative importance of predictor variables — and *where* they matter geographically — is a capability that has not yet been exploited in community and landscape ecology.

Conclusions

CART models are not a panacea for ecologists, nor are they appropriate for every application aimed at interpreting or classifying groups. They do, however, offer a number of compelling advantages that can be motivated on ecological as well as statistical grounds. Ecologically, CART models are amenable to cases where alternative environmental settings can support similar community types or habitats, a condition not well accounted with linear or unimodal models. CART models also admit a variety of data types, including continuous, rank, and categorical, and they are also amenable to data-dictated or user-defined limits on data precision. Finally, they can be tuned via pruning to provide classification models of optimized or user-selected accuracy and bias. Thus, while not appropriate for every application, they certainly warrant inclusion in the ecologist's toolbox.

CHAPTER 30

Structural Equation Modeling

Background

General introduction

Several major kinds of questions can be addressed when analyzing communities. Previous chapters have dealt with a number of issues, but have especially focused on defining and evaluating groups and on ordination of response measurements. Figure 30.1 shows the kind of results that are often found when analyzing communities.

In this chapter, we address in more detail some of the methods that are available for examining how community characteristics, such as patterns in ordination space, relate to environmental variables.

One of the most basic and intuitive ways of examining how environmental conditions relate to community patterns is by overlaying the simple correlations between environmental variables and ordination axes, as shown by the vector lines in Figure 30.1. Often, dominant environmental variables are used as the basis for rotating the ordination so as to maximize correlations between ordination axes and environmental factors, thereby making the ordination axes easier to interpret (Ch. 15).

Environmental overlays (Fig. 30.1) use vectors to show bivariate correlations between a variable and the two axes being shown. For example, the vector for soil calcium is plotted at such an angle so as to represent a positive correlation with Axis 1 and a simultaneous negative correlation with Axis 2.

Environmental correlations with community ordinations are presented as descriptive information, with strong relationships suggesting explanations about the environmental factors affecting community structure. As will be illustrated later in this chapter, when the interrelationships among several environmental variables and the net relationships to ordination space are considered, the complexity of the system typically becomes apparent. A full understanding of the whole suite of relationships among environmental variables and their relationships to ordination space requires analytical techniques capable of dealing with complex systems. A number of approaches will be examined in this chapter, though the emphasis will be placed on one of the more comprehensive approaches, structural equation modeling (SEM).

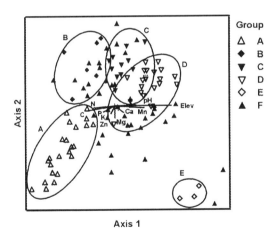

Figure 30.1. Example of an ordination biplot showing results of nonmetric multidimensional scaling, group identities for individual plots, and vectors indicating environmental correlations (modified from Grace et al. 2000). Elev is elevation; C, Ca, K, Mg, Mn, N, P, and Zn are elements in soil samples. Ellipses represent ordination space envelopes for vegetation groups A, B, C, D, and E while an envelope is not given for the heterogeneous group F.

Philosophy, history, and current status of SEM

Mathematical approaches to dealing with complex interrelationships have expanded greatly in the past two decades. Much of this development has taken place out of view from most biologists and ecologists. One of the most comprehensive statistical approaches to dealing with complex data is known as structural equation modeling (SEM).

Structural equation modeling evolved out of a merger of factor analysis and path analysis (Wright 1918) during the early 1970s (Keesling 1972, Jöreskog 1973, Wiley 1973). Arising from the traditions of econometrics, psychology, and sociology, SEM has come to be used in a very wide range of disciplines, but has only recently begun to be used for ecological problems. From an application standpoint, SEM is a method of evaluating complex hypotheses involving multiple causal pathways among variables. From a statistician's standpoint, SEM is a generalized method for the analysis of covariance relationships.

In this chapter we introduce SEM and illustrate its use. To facilitate the discussion, we present a brief glossary of terms related to SEM in Box 30.1.

Due to its comprehensive mathematical framework, multiple regression, factor analysis, analysis of covariance, and path analysis can all be viewed as special cases within SEM. It is important to realize that the "modeling" part of SEM refers to statistical models and is not comparable to dynamic simulation modeling. SEM could just as easily be referred to as structural equation analysis, but the choice of SEM has become the convention.

SEM is designed to accomplish several things. At the most general level, it allows correlational and non-experimental as well as experimental data to be used to evaluate the validity of multivariate hypotheses.

Of importance to ecologists, the application of SEM results in (1) statistical evaluation of the correspondence between hypothesized multivariate path models and data, (2) estimation of unobserved conceptual variables from specific measured variables, (3) estimation of the strength of direct and indirect pathways between variables, and (4) direct estimation of the reliability of predictions.

One of the especially useful features of SEM is its ability to permit statistical evaluation of complex hypotheses involving multiple causal pathways. In addition, SEM explicitly recognizes that we are interested in theoretical (latent) variables that are imperfectly measured by any single observed (indicator) variable. As an example, the size of an individual, be it plant or animal, is often influential in its reproduction, survival, and behavior. However, the functional significance of the concept of "size" may involve many specific attributes such as height, dry mass, muscle mass, wing length, energy reserves, genetic potential, etc. that can be intercorrelated. Thus, any individual measure we make can act only as a partial indicator of the conceptual variable "size."

In SEM we explicitly recognize the distinction between theoretical and observed variables, which forces us to evaluate the validity of concepts as well as the adequacy of our measurements. A complete structural equation model, then, contains two components; (1) a measurement model that specifies the hypothesized relationships between specific indicator variables and general latent variables and (2) a structural model that specifies the hypothesized set of dependencies among latent variables.

A few ecological applications of SEM have been performed (e.g., Johnson et al. 1991, Wesser and Armbruster 1991, Mitchell 1992, Walker et al. 1994, Grace et al. 2000a). Two recent treatments of the subject have sought to facilitate the application of SEM to ecology and evolutionary biology (Shipley 2000, Pugesek et al. 2002). Shipley provides a good introduction to various philosophical issues relating to testing multivariate causal hypotheses and broadly addresses the question of why one might want to use SEM. Pugesek's book is a very comprehensive reference on the practical application of SEM to ecological and evolutionary biology and serves as a reference guide for ecologists interested in applying the methodology.

As a generalized procedure for covariance analysis, SEM includes many different statistical procedures as special cases including; multiple regression, factor analysis, path analysis, ANOVA, MANOVA, and MANCOVA. A number of classic references remain essential resources for the serious student of modern SEM (Hayduk 1987, Bollen 1989).

SEM's development has been strongly influenced by the availability of computer programs that implement the analysis; the three most widely used being LISREL (Jöreskog and Sörbom 1996), AMOS (Arbuckle 1997), and EQS (Bentler and Wu 1996). Additional programs have become available in recent years, each with their own special features: including CALIS, LISCOMP, TETRAD, and others (see Schumacker and Lomax 1996 for a listing of sources). Training courses are regularly offered around the world. Finally, the interdisciplinary journal *Structural Equation Modeling* was launched in 1994, providing a source for current information on methods and applications.

Box 30.1 Terminology Associated with Structural Equation Modeling

Competing models strategy - The comparison of two or more a priori models.

Confirmatory analysis - Evaluation of a predefined hypothesis using an empirical test.

Construct validity - The degree to which the indicators for a latent variable accurately measure the intended concept.

Endogenous variable - Variable that is the dependent or outcome variable in at least one causal relationship. In a path diagram, there are one or more arrows leading into an endogenous variable (see also exogenous variable).

Exogenous variable - Variable that acts only as a predictor for other variables in the model.

Exploratory analysis - Development of a hypothesis based on the empirical examination of data.

Identification - Degree to which there is a sufficient number of known parameters to solve for each of the unknown coefficients (see Underidentification and Overidentification).

Indicator - Observed variable used as a measure of a concept or latent variable.

Latent variable - A variable that cannot be measured directly without error but only estimated indirectly using indicator variables.

Loadings - Bivariate correlations between indicator variables and latent variables.

Manifest model - Type of path model involving only indicator variables and no latent variables.

Measurement model - Submodel in SEM that specifies the indicators for each latent variable and assesses the reliability of each latent variable.

Modification indices - Indices calculated for each unestimated relationship possible in a specified model. Modification indices are helpful in finding where the current model is inconsistent with the data.

Nested models - A special case of competing models in which all models being compared have the same variables but differ in terms of the number or types of dependency relationships specified.

Overfitting - The ill-advised act of changing a model based on results that are actually due to chance relationships in the data at hand rather than based on a revised mechanistic assumption.

Overidentification - When there is a sufficient surplus of known parameters to create degrees of freedom for a test of model fit.

Path diagram - Graphical description of the complete set of relationships among a model's variables, also sometimes known as the path model or structural equation model.

Structural model - The relationships among latent variables.

Total effects - The impact of a predictor variable on a predicted variable taking into account the sum of all pathways between variables.

Underidentification - Occurs when there are insufficient known parameters to permit a unique solution to be obtained.

Validation - The process whereby the accuracy of a model is evaluated, typically through comparison with other models or additional data.

Table 30.1. Example of correlations among environmental and axis variables (Pearson correlation coefficients). Numbers that are underlined represent significant correlations at the $p = 0.05$ level. Variables are elevation, soil composition (calcium, magnesium, manganese, zinc, potassium, phosphorus, acidity, total carbon, and total nitrogen), and ordination axis scores.

	elev	Ca	Mg	Mn	Zn	K	P	pH	C	N	Axis1	Axis2
elev	1.0	--	--	--	--	--	--	--	--	--	--	--
Ca	.42	1.0	--	--	--	--	--	--	--	--	--	--
Mg	.21	.79	1.0	--	--	--	--	--	--	--	--	--
Mn	.71	.66	.42	1.0	--	--	--	--	--	--	--	--
Zn	-.10	.47	.45	.39	1.0	--	--	--	--	--	--	--
K	-.20	.40	.35	.25	.69	1.0	--	--	--	--	--	--
P	-.32	.32	.26	.08	.72	.74	1.0	--	--	--	--	--
pH	.43	-.02	-.14	.28	-.26	-.46	-.48	1.0	--	--	--	--
C	-.44	.04	.15	-.19	.49	.55	.62	-.63	1.0	--	--	--
N	-.46	-.05	.11	-.27	.31	.41	.51	-.53	.81	1.0	--	--
Axis1	.77	.30	.03	.55	-.20	-.23	-.37	.47	-.55	-.47	1.0	--
Axis2	.005	-.26	-.35	-.14	-.28	-.16	-.25	.11	-.21	-.13	.17	1.0

A comparison of regression, principal components, and SEM

We have found that one of the best ways to explain structural equation modeling is through comparison to other, more familiar techniques commonly used by ecologists. Users of mainstream methods, such as regression and principal components analysis, are often only vaguely aware of the major limitations of these methods for interpretative studies. It is much easier for one to appreciate the significance of SEM in light of the limitations of other methods. For this reason, here we present comparative analyses of an example dataset from a fairly typical plant community characterization (Grace et al. 2000b). The data are from a study of 107 quarter-meter-squared plots in a native tallgrass prairie in southern Louisiana. Data collected include plant species cover, microelevation, and a number of soil properties: pH; total carbon; total nitrogen; and extractable phosphorus, calcium, magnesium, potassium, manganese, and zinc.

The question of interest in this example, "How does the plant community relate to environmental conditions," is a fairly standard sort of question for plant community studies. In this case, the vegetation data were summarized as a community similarity matrix and species abundance data were ordinated in two dimensions using nonmetric multidimensional scaling (Ch. 16) in PC-ORD (McCune and Mefford 1995) as described in Grace et al. 2000b. To simplify interpretation, the resulting ordination was rotated to maximize the relationship between Axis 1 and elevation, the strongest individual environmental correlate. The focus in this methodological comparison is on the analysis of relationships among environmental variables and with ordination axis scores. The methods used in this comparison are bivariate regression, standard multiple regression, stepwise multiple regression, principal components analysis, and SEM. Regressions and principal components analyses were conducted using SAS (SAS Institute 1989) while SEM was conducted using LISREL (Jöreskog and Sörbom 1996).

Regression analyses of environmental relationships

The raw information used in this comparison is summarized in Table 30.1 as a half-matrix of correlations. This table shows two features of the underlying data. First, the simplest expressions of the complex relationships in the data are the bivariate correlations. These correlations, of course, represent the relationships between pairs of variables irrespective of any other variables. All ten of the environmental variables examined had significant correlations with one or both ordination axes. The strongest relationship was between elevation and Axis 1 ($r = 0.77$), while several other variables were also related to Axis 1. Five variables were correlated with Axis 2, with Mg having the strongest correlation ($r = -0.35$). Another characteristic of the correlation matrix in Table 30.1 is the large number of significant correlations among environmental variables. Thirty-six of the 45 correlations among the 10 environmental variables were statistically significant, indicating a potentially complex set of interconnections among parameters.

To facilitate comparisons among methods, the relationships among environmental variables and with Axis 1 and Axis 2 scores are presented graphically in Figure 30.2. The columns of correlations presented in Figure 30.2 contrast results from bivariate correlation, multiple regression, and stepwise regression. This figure represents multiple regression as a path model. Double-headed arrows represent the correlations among variables, which influence both the results and their interpretation. Other kinds of correlations, such as correlated errors, can also be important and are discussed later in this chapter.

Differences in coefficients resulting from the various types of regression (Fig. 30.2) come from differences in how the correlations among independent variables are addressed. In bivariate correlation analysis, which is very commonly used in overlaying environmental correlations on ordination diagrams and in evaluating environmental relations, the correlations among predictors are ignored. The maximum variance in Axis 1 scores explained by a single indicator is 59%. Standard regression seeks to evaluate the relationships between predictors and the independent variable simultaneously using partial correlations. In partial correlations, each path coefficient quantifies the strength of a relationship while the effects of the other predictor variables are held constant (see Box 30.2 below).

As can be seen in Figure 30.2, the results of a standard multiple regression yield partial correlation coefficients suggesting that only 4 variables have significant relationships with Axis1, in contrast to the results from bivariate analysis. One additional result from multiple regression that would not have been expected from just looking at the bivariate correlations is that now there is a significant pathway between Magnesium and Axis1. Table 30.1 shows that the raw bivariate correlation between Magnesium and Axis1 is +0.03, quite a departure from the -0.31 estimate derived from the partial correlation from standard multiple regression! How can this happen?

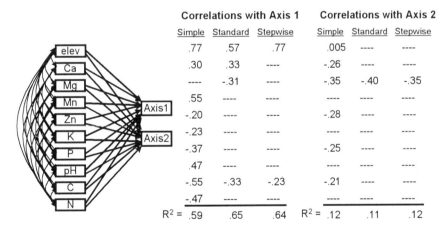

Figure 30.2. Illustration of the regression relationships between environmental parameters and Axes 1 and 2 of the ordination. Simple bivariate regression, multiple regression, and stepwise regression results are shown for comparison. "----" denotes nonsignificant coefficients. Double-headed arrows represent correlations between independent variables, which are dealt with differently in the three methods of correlation analysis. R^2 in the simple correlation column represents the highest R^2 obtained for any single variable; for other columns it is the variance explained for the whole model.

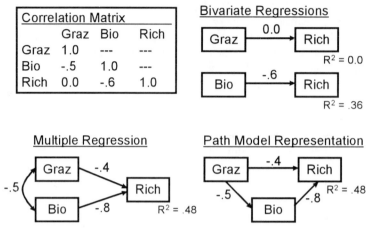

Figure 30.3. Offsetting pathways are represented differently by bivariate correlation and regression, multiple regression, and path models. Graz = grazing (yes or no), Bio = standing community biomass, and Rich = plant species richness. In this example, the path model shows how offsetting negative and positive effects of grazing on richness can result in a zero bivariate correlation.

Box 30.2 What is a partial correlation?

Partial correlations help to make sense of the mathematical interrelationships among a set of intercorrelated variables. As the name implies, partial correlations are calculated to remove the effect of a third variable and then recalculate the relationship between two variables. This is also referred to as "controlling for the third variable," "partialling out" the third variable, or "holding the third variable constant."

Consider, for example, three intercorrelated varibles, A, B, and C. We can represent the partial correlation between A and B, $r_{AB \cdot C}$, as

$$r_{AB \cdot C} = \frac{(r_{AB} - r_{AC} r_{BC})}{\sqrt{(1 - r_{AC}^2)(1 - r_{BC}^2)}}$$

where r_{AB}, r_{AC}, and r_{BC} represent the bivariate correlations between pairs of variables. If, for example, $r_{AB} = 0.5$, $r_{AC} = 0.8$, and $r_{BC} = 0.6$,

$$r_{AB \cdot C} = \frac{0.5 - (0.8)(0.6)}{\sqrt{(1 - 0.8^2)(1 - 0.6^2)}} = 0.04$$

To better understand this equation, consider what is accomplished by the numerator and denominator. First, the numerator subtracts out the indirect correlation between A and B that passes through their joint correlations with C. Second, the denominator adjusts the total standardized variance of A and B by subtracting the portions of each explained by C.

Digression on partitioning correlations

Figure 30.3 presents a simple example that shows how partial correlations, whether implemented through multiple regression or SEM, can give a substantially different impression than that obtained from bivariate correlations. Imagine a case where it is observed that grazing rates in a grassland are uncorrelated with plant species richness (for a related example, see Grace and Jutila 1999). Suppose that another measured predictor, community biomass, is negatively correlated with grazing. In this example, we also find that high levels of biomass are negatively related to plant species richness due to competitive exclusion. When richness is regressed on both grazing and biomass, we now observe a significant negative pathway between grazing and richness. Further, 48% of the observed variance in richness is now explained by two variables that independently explained 0% and 36%. This result illustrates how combinations of variables can explain variance that the individual variables cannot.

The interpretation of the example in Figure 30.3, both statistically and biologically, is enhanced by expressing the relationships as a path model. In *structuring* the relationships among predictor variables based on our knowledge of the situation (hence the name 'structural' equation modeling), we now specify that biomass is a variable that is dependent upon grazing. Note that our decision to recognize the path from grazing to biomass as directional (causal) is not based on the data, but is based on pre-existing knowledge. In the form of a path model, it becomes easier to recognize that grazing can have two kinds of influences on richness in this case, one mediated through effects of biomass and one that is more direct (the direct elimination of species). The strength of the path mediated through biomass is calculated based on the formula

$$\frac{\text{strength of a}}{\text{compound path}} = \frac{\text{product of path}}{\text{components}}$$

which in the case of Graz → Bio → Rich is

$$-0.5 \times -0.8 = +0.40$$

Further, the total effect of grazing on richness is the sum of the various paths that connect a predictor variable with a response variable. In this case,

$$\text{total effect} = \text{sum of individual paths}$$
$$= -0.40 + 0.40 = 0.0$$

Thus, we see that partial correlation analysis enhances our ability to partition correlations into components and to structure our understanding of the various ways in which variables interrelate using a combination of logical hypothesis formulation (developing the path model) and statistical analysis (deriving the various estimates).

A remaining question about our example in Figure 30.3 is how the values for R^2 (a measure of the variance in a response variable that was explained) are calculated (Box 30.3).

Return to our discussion of regression

Now that we have shown a simple example of how partial correlation analysis can give such a different picture from bivariate correlation analysis, you can appreciate how difficult it is to interpret a multiple regression when there are many intercorrelated predictors. Returning to our discussion of Figure 30.2, the wide range of differences between the column of bivariate correlations and standard multiple regression results is driven by the 45 correlations among predictors that factor into the partial correlation calculations. Since we have specified only 10 of the 55 interconnections among variables, the 10 that go from independent variables to our dependent variable, we are claiming no knowledge of the interrelations among predictors other than "they might be correlated, either positively or negatively."

One approach that is used to simplify the process of evaluating regression relationships is stepwise regression, which seeks the minimum set of predictors required to explain observed variation in a response variable. In our case, two variables, elevation and soil carbon, are able to explain almost as much variation in Axis 1 scores as any greater number of variables. Since stepwise procedures add variables into the

Box 30.3. Calculation of R^2 in a multiple regression or path model.

In multiple regression and path models, results compensate for the correlations among predictors. For the example in Figure 30.3, variation in richness is explained by both variation in grazing and biomass. If grazing and biomass were uncorrelated, the variance explained (R^2) would simply be the sum of the squared bivariate correlations. In such a case, the R^2 for richness would be

$$R^2 = (0.0)^2 + (-0.6)^2 = 0.36$$

However, when predictors are correlated, the variance explained different markedly from that estimated by a simple addition.

To simplify our discussion of R^2, we will use the formulae for zero-transformed variables (= z scores; all means are zero and variances are 1.0). This permits us to directly relate our calculations to the correlation matrix and partial regression coefficients in Figure 30.3. In this context, R^2 can be defined as the squared correlation between a z-transformed response variable Y_z, and its predicted values, $\hat{y}_z$. Estimates of the predicted values can be calculated as follows:

$$\hat{y}_z = \beta_1(x_{1z}) + \beta_2(x_{2z})$$

where β_1 and β_2 are standardized partial regression coefficients (see Box 30.2) and x_{1z} and x_{2z} are "z-transformed" predictor variables. As in bivariate regression, the values of the betas are those that satisfy the least squares criterion,

$$\min \sum (y - \hat{y})^2$$

Now, the R^2 for our example can be calculated using the formula

$$R^2_{y_{12}} = \beta_1(r_{y1}) + \beta_2(r_{y2}),$$

where r_{y1} and r_{y2} are the bivariate correlations between y and x_1 and y and x_2. For our example in Figure 30.3,

$$R^2_{y_{12}} = -0.4(0.0) - 0.8(-0.6) = 0.48$$

To come full circle in this illustration, if we had the case where grazing (x_1) and biomass (x_2) were uncorrelated as first mentioned in this Box, then the bivariate and partial correlations would be equal (i.e., $\beta_1 = r_{y1}$ and $\beta_2 = r_{y2}$) and the above equation would reduce to

$$R^2_{y_{12}} = (r_{y1})^2 + (r_{y2})^2 = (0.0)^2 + (-0.6)^2 = 0.36$$

multiple regression based on the ability to explain variance, elevation is added to the model first and its correlation coefficient is the same as its bivariate value. The coefficient presented for soil carbon, -0.23, represents the sequential improvement in variance explained by this variable after the effects of elevation have been taken into account. As a method for selecting the minimum set of predictors, stepwise regression is a good technique. If, however, the purpose is to develop a meaningful interpretation of the relationships among variables, stepwise regression is the least effective approach.

Principal components analysis

In Chapter 14 we discussed the problems of using principal components analysis (PCA) as an ordination method for community data. Data matrices of species abundance possess many characteristics, such as nonlinear relationships among species, that make PCA an inappropriate method for ordination of community data in most cases. However, when the task is to examine the relationships among a set of intercorrelated environmental variables, PCA is often appropriate. Relationships among environmental variables will often come reasonably close to meeting the statistical assumptions of PCA (with prior transformation as needed); the matrix of interest contains correlations among environmental factors, quite an appropriate subject for PCA.

The goal in using PCA to study environmental variables is to reduce a set of intercorrelated predictors to a small number of summary variables, and in so doing, to simplify the interpretation of environmental gradients. The mathematical principles behind PCA are described in detail in Chapter 14. Here we employ those principles to analyze and summarize the relationships among environmental factors in Table 30.1 presented earlier.

Principal components analysis was performed using the orthogonal procedure, which seeks to extract axes that are uncorrelated with one another (nonorthogonal rotations that may also be used). Following the extraction of principal components from the environmental matrix, components were regressed against the ordination axis scores obtained using NMS to determine the principal components that would be retained in the final model.

Principal components analysis extracted ten components from the environmental matrix. The first five components were judged to be important based on the criterion that additional components extracted less than five percent of the variance and because subsequent regression analysis showed that components

Table 30.2. Principal component loadings for the first five principal components (PC1 to PC5). Loadings greater than 0.3 are shown in bold to highlight patterns.

Variable	PC1	PC2	PC3	PC4	PC5
elev	-.18	**.44**	-.04	**.55**	-.26
Ca	.19	**.47**	-.29	-.14	.04
Mg	.22	**.36**	**-.60**	-.29	.25
Mn	.04	**.52**	.23	**.32**	-.11
Zn	**.38**	.20	**.38**	-.14	.22
K	**.41**	.10	.29	-.11	-.26
P	**.42**	.01	**.32**	-.13	-.01
pH	**-.33**	.20	**.35**	.03	**.77**
C	**.40**	-.19	-.09	**.41**	.12
N	**.35**	-.24	-.22	**.52**	**.37**

beyond component five explained little additional variance in the dependent variables. These first five components accounted for 91% of the total variance in the matrix of environmental correlations.

Loadings express the degree to which the extracted components correlate with environmental factors (Table 30.2). Component 1 was found to load most strongly on Zn, K, pH, C, N, and P, while component 2 loaded primarily on elev, Ca, Mg, and Mn. Component 3 loaded strongly with Mg, with lesser affinities for Zn, pH, and P. Component 4 loaded strongly with elevation and N, while also loading with C and Mn. Finally, component 5 was primarily related to pH.

Regression analysis using principal components as predictors revealed that four of the first five components were significantly related to axis scores ($p < 0.05$; Fig. 30.4). Component 3 did not contribute to the multiple regression equation for either axis. NMS Axis 1 was strongly related to components 1 and 2, and weakly related to components 4 and 5. NMS Axis 2 was related only to component 1. Sixty-two percent and 10 percent of the variance were explained for axes 1 and 2 respectively, approximately the same as by regression (Fig. 30.2)

When considering the merits and limitations of PCA as a method of multivariate data analysis, we should understand that what it provides is an empirical characterization of the correlation structure. Principal components analysis is based on the premise that a matrix of correlations among variables can be reduced to a subset of higher-order variables or components. As reflected by the direction of arrows from component variables to observed variables in Figure 30.4, principal components are sometimes interpreted in terms of shared underlying causes or macrovariables. Their substantive meaning is usually derived post facto, based on the patterns of loadings (note that these loadings are simple bivariate correlations between principal components and observed variables).

In PCA, the process of identifying components is directed by the data, though certainly you can choose rotation options, use discretion in choosing the number of principal components, and select the variables to include in the analysis. While orthogonal rotations (those that create uncorrelated principal components) are the norm in ecological studies, oblique rotations (which allow components to be correlated) can represent intercorrelated ecological factors.

In general, PCA can be useful in exploratory studies that seek to develop hypotheses about higher-order influences. The primary limitations of PCA by itself include: (1) limited flexibility in specifying the structure of the model; (2) limited ability to estimate underlying latent parameters; (3) limited capacity for hypothesis testing; and (4) an inability to account for measurement error (discussed later in this chapter).

Structural equation models

In this section we present some of the general features of SEM without getting into all the details of how the analyses are performed. Our objective is to give just enough of a representation of SEM results to familiarize you with the methodology. Later in the chapter we will present a more detailed description of the steps involved in performing a structural equation analysis from start to finish.

We have already seen a very simple application of SEM in Figure 30.3. This exercise represented the simplest form of a structural equation model as a structured regression, represented as a path model. When moving from a multiple regression to a structural equation model, the underlying mathematics changes from

$$y = a_1 + b_1 x_1 + b_2 x_2$$

where y is Rich and x_1 and x_2 are Graz and Bio, respectively, to a structured set of simultaneous regression equations (hence, "structural equation" modeling)

$$y_1 = a_1 + b_{11} x_1 + b_{12} y_2$$
$$y_2 = a_2 + b_{21} x_1$$

where y_1 is Rich, y_2 is Bio, and x_1 is Graz. In the example shown in Figure 30.3, the path model is of a type referred to as a <u>manifest model</u>. In a manifest

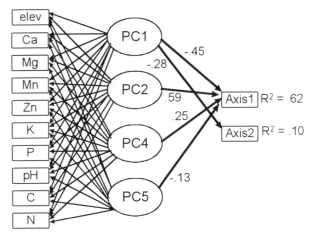

Figure 30.4. Use of principal components in combination with multiple regression. PC1 through PC5 represent principal components of the environmental variables at left. Axis1 and Axis2 represent scores on NMS ordination axes of community data. Numbers along arrows are partial correlation coefficients.

model, only observed or indicator variables are used. This is the case typically employed in methods that predate modern SEM, which are known as "Path Analysis" (Box 30.4).

Figure 30.5 depicts how a full structural equation model is developed relative to a multiple regression model. It is generally desirable to have the SEM model formulated before data are collected. However, in this example, we will assume that we are developing an exploratory SEM with preexisting data (the issue of exploratory versus confirmatory analyses will be considered later in the chapter). In this example, there are seven observed variables that were measured with one, y, being of prime interest as a response variable. In the regression model, we presume no knowledge about the relationships among predictors. In contrast, in developing the structural equation model, we have incorporated more of our complete scientific knowledge about our system into the path model formulation, which serves as a multivariate hypothesis. In developing the path model in Figure 30.5, we have reason to believe that variables x_1, x_2, and x_3 are really different facets of some common feature (A) of our system. For example, the observed variables may be tree height, forest canopy density, and gap fraction, all different reflections of overstory development. In formulating our model, we have organized different measures of forest overstory around a latent variable, which could be something like "canopy closure." The outward direction of the arrows from A to x_1, x_2, and x_3 reflects the premise that the observed variables are simply indicators or measures of the underlying causal property, canopy closure.

A similar logic is used in this example to organize variables x_4, x_5, and x_6 around a second latent variable *B*. Perhaps in this case, variables x_4, x_5, and x_6 are measures of sunlight reaching the forest herbaceous layer such as calculated solar penetration (based on canopy photography systems), average light penetration (using a radiometer), and sunfleck density (also determined from a radiometer). Finally, in a full SEM, even our single response variable *y* (e.g., herbaceous understory biomass) is considered an indicator of a latent variable, *C* (e.g., site favorability). As you can see, a full structural equation model expresses your ideas about the causal relationships in your data much more completely than does multiple regression.

Latent variables and measurement error

The concept of a latent variable requires some elaboration for two reasons: (1) it is a fundamental aspect of SEM, and (2) it is an important philosophical and statistical concept that is rarely discussed in the ecological and evolutionary literature.

The main premise behind the specification of latent variables is the recognition that we are rarely able to measure with absolute accuracy the underlying properties of conceptual interest. In the example just given, our structural model (the relationships among latent variables) specifies what we are primarily interested in: how the forest canopy affects light reaching the subcanopy and the growth of the herbaceous understory. Ecologists customarily think about this matter as one of degrees of abstraction rather than in terms of latent variables and measurement error. We could conceivably break down the processes by which forest canopy properties affect light into a large number of specific subprocesses, which themselves could be decomposed into finer and finer causal interactions, ultimately all the way to the level of quantum physics. The statistical field of "measurement theory" takes a different approach to this problem by using the twin concepts of latent variables and measurement error.

In our example, we explicitly recognize not only that we are really interested in the latent variable canopy closure, but also that we cannot make a single, completely accurate measurement of canopy closure that will encapsulate all the functional properties of the canopy that might influence the herbaceous understory. Instead, we recognize that we are faced with measurement error; the degree to which our measured variables

---

Box 30.4. SEM versus Path Analysis.

A point of considerable confusion in ecology and evolutionary biology has to do with the relationship between Structural Equation Modeling and what is referred to as Path Analysis (e.g., Sokal and Rohlf 1969, Mitchell 1993, Legendre and Legendre 1998). SEM evolved out of a desire to develop a statistical approach that would encompass factor analysis and Wright's path analysis. The fundamental structure of SEM was developed in the early 1970's and at that time, statisticians abandoned the pursuit of refining the old-style "Path Analysis" in favor of working on the more generalized methodology of SEM. However, because SEM has been largely applied to the social sciences, ecologists and evolutionary biologists have continued to identify with the familiar examples used by Wright and continue to rely on the old pre-SEM literature to analyze simultaneous regression equations (path models or path diagrams).

The accompanying change in names from Path Analysis to Structural Equation Modeling has continued to isolate ecological and evolutionary biologists from the modern methods that have been developed (Grace and Pugesek 1998, Pugesek and Grace 1998). A trip to the University library will reveal a couple of out-of-date books on Path Analysis in the biological sciences section of the library while over in the social sciences section there will be a wall of modern books on SEM describing the analysis of path relationships. The same holds true for biometric training. If students in ecology want to learn about SEM, they usually have to head across campus, for example to the Psychology or Marketing Department, to find the courses relating to SEM. Recent treatments of the fundamentals of SEM by biologists (Shipley 2000, Pugesek et al. 2002, as well as the present treatment) aim to break down the barriers that have isolated ecologists from these methods over the past several decades.

---

fail to accurately reflect the underlying causal property of interest. Obtaining a completely accurate estimate of measurement error is, in itself, extremely difficult. However, by having multiple indicators of canopy traits, we can derive an estimate of measurement error through a determination of how well the different indicators correlate with each other.

While we are discussing the concept of measurement error, we should make it clear that even simple ecological concepts are usually measured with error. In our example, we may be quite content to use dry biomass in August as the response variable of interest.

No matter how careful we are, some of the variation among replicate samples comes from the errors associated with measurement. Also, in most cases, we are interested in the general applicability of our response variable. Therefore, it is beneficial to make explicit the distinction between latent and indicator variables if for no other reason than to keep us conscious of the distinction.

Measurement error is not represented in multiple regression. This is a limitation of regression, because it assumes that the predictors are measured without error, even though this assumption is rarely if ever met. The result is that partial correlations are underestimated (biased) and model R^2 values are incorrect (Bollen 1989, ch. 5).

Commonly, the formulation of an SEM does not try to specify the magnitude of error with latent variables connected with single indicator variables. Instead, for single-indicator latent variables, the value of the path between the latent and indicator variable is set to a value of 1.0. It is possible, however, to use preexisting estimates of measurement error to set such a path to a different value (see Fig. 30.14 for such an example). One statistical benefit of identifying measurement error is that some of the variance that is otherwise ascribed to the pathways is partitioned out as error, resulting in more accurate estimates of path coefficients and higher (and more accurate) R^2 values for the dependent variables.

Results and interpretation of a structural equation model

Structural equation models were built for the data presented in Table 30.1 using LISREL (Jöreskog and Sörbom 1996). Step-by-step procedures for a complete SEM analysis will be presented later in this chapter in the section entitled, "How it Works." In this section, we present the initial measurement model (Fig. 30.6), modified measurement model (Fig. 30.7), and final full model (Fig. 30.8). Also presented are the measurement model factor loadings (standardized coefficients for paths from latent variables to indicators; Table 30.3) and the correlations among latent variables (Table 30.4).

The initial measurement model (Fig. 30.6) was hypothesized based on our knowledge of the relationships between major pedogenic forces such as elevation (ELEV), mineral content (MINRL), and hydric influences (HYDR) without reference to the correlational structure of our particular data set. As shown in this figure, we initially proposed that ELEV would be indicated by the measured surface microelevation (determined by laser level surveying), MINRL would be indicated by soil mineral components (Ca, Mg, Mn, Zn, K, and P), and hydric influences on soil formation would be reflected in values of pH, soil total carbon,

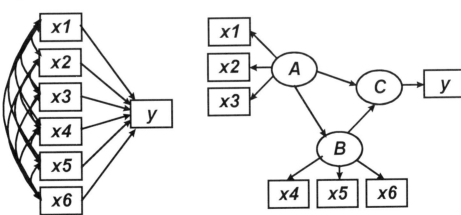

Figure 30.5. Development of a structural equation model in contrast to a regression model. Boxes represent measured or indicator variables while ellipses represent conceptual or latent variables. In the structural equation model, the indicator variables are organized around the hypothesis that x_1, x_2, and x_3 are different facets of a single underlying causal variable, A, while x_4, x_5, and x_6 are different facets of the causal variable, B. Further, y represents an available estimate of the latent variable C.

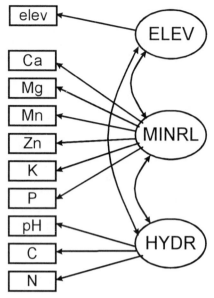

Figure 30.6. Initial (hypothesized) measurement model relating three latent variables and ten indicator variables.

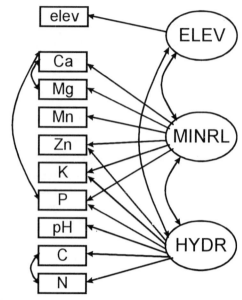

Figure 30.7. Modified measurement model.

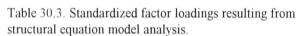

Table 30.3. Standardized factor loadings resulting from structural equation model analysis.

Measured	Latent factors				
Variable	ELEV	MINRL	HYDR	AXIS1	AXIS2
elev	1.00	---	---	---	---
Ca	---	0.66	---	---	---
Mg	---	0.43	---	---	---
Mn	---	0.99	---	---	---
Zn	---	0.66	0.81	---	---
K	---	0.53	0.84	---	---
P	---	0.39	0.93	---	---
pH	---	---	-0.66	---	---
C	---	---	0.78	---	---
N	---	---	0.67	---	---
axis1	---	---	---	1.00	---
axis2	---	---	---	---	1.00

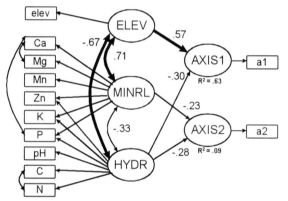

Figure 30.8. Final full model. Here a1 and a2 represent the measured ordination axis scores.

Table 30.4. Correlations among latent variables.

	ELEV	MINRL	HYDR	AXIS 1	AXIS 2
ELEV	1.00	---	---	---	---
MINRL	0.71	1.00	---	---	---
HYDR	-0.67	-0.33	1.00	---	---
AXIS1	0.77	0.50	-0.67	1.00	---
AXIS2	0.03	-0.14	-0.21	0.07	1.00

and soil total nitrogen. It was further assumed that the latent variables ELEV, MINRL, and HYDR might be correlated with one another; especially ELEV and HYDR.

Fitting our data to the measurement model in Figure 30.6 revealed that there was a significant difference between the data structure and the initial model. When a significant deviation between data and model are found, LISREL presents <u>modification indices</u>, which evaluate the potential consequences of all single parameter changes for model fit. In this case, the resulting modification indices led us to consider changes to our initial model, leading to the modified model shown in Figure 30.7. Once an acceptable measurement model was determined, the new full model was assessed; results of this analysis are shown in Figure 30.8.

As for interpreting the results summarized in Figure 30.8 and Tables 30.3 and 30.4, we can say that the covariance relations in the data are consistent with the interpretation that the complex of environmental effects can be summarized as relating to three major influences — microelevation, mineral soil, and hydric soil conditions. The first axis of the ordination is primarily related to elevational influences (the direct path from ELEV to AXIS1 = 0.57) and secondarily related to influences from hydric conditions (-0.30). Variation in the vegetation along axis 2 of the ordination was more or less equally related to mineral and hydric soil influences, though neither relationship was strong (-0.23 and -0.28 respectively).

The correlations among latent variables (Table 30.4) in combination with path coefficients (Fig. 30.8) make it possible for us to examine the relationships among latent variables more fully. For example, by comparing correlations among latent variables to path coefficients, we can show that the total correlation between ELEV and AXIS1 (0.77) may be partitioned into influences mediated through the relationship between elevation and hydric conditions (-0.67 × -0.30 = +0.20) as well as the remaining direct path (0.57), which represents elevational effects that have not been fully explained. The sum of these paths equals the net correlation (0.20 + 0.57 = 0.77).

Elevation is related to axis 2 through a pair of offsetting influences. One is mediated through mineral soil influences (0.71 × -0.23 = -0.16) and the other through hydric soil influences (-0.67 × -0.28 = +0.19). Summing these two offsetting paths results in a nonsignificant net correlation between elevation and axis 2 that belies the underlying effects of elevation in the model.

Overall, the results of this SEM analysis are consistent with the hypothesis that the vegetation at the study site is responding primarily to hydric soil influences, mineral soil variations, and additional influences of elevation (the direct pathway). Because our initial measurement model was rejected, we must view the measurement portion of the model in Figure 30.8 as tentative and in need of further testing. As always with structural equation modeling, it is important to keep in mind that consistency between the statistical model and data does not mean our scientific interpretations are correct, only that the data are consistent with our interpretations.

When to use it

As a statistical procedure, SEM is very versatile (see Variations section below). Here, we attempt to (1) establish the basic requirements for the most common types of analyses that community ecologists might perform, and (2) consider some of the broader issues relative to its application.

Conditions that favor the use of SEM for analyzing multivariate data

Desired conditions for SEM include:

- sufficient a priori information to support a mechanistic interpretation of a data set;

- an interest in, or need to understand relationships within a multivariate context where the relative strengths of different relationships can be compared;

- the desire to use all the correlations or covariances among variables in an analysis;

- an interest in studying the network of relationships among a set of correlated variables;

- a desire to work toward the eventual testing of a priori hypotheses about how a system works;

- a situation where one wishes to develop the most power possible for evaluating a priori hypotheses with nonexperimental data;

- field or other experiments that involve a complex set of covariates that make the results conditional;

- any case where predictor variables are measured with error, especially when there are either multiple types of measurements of a parameter of interest or some other estimate of measurement error; and

- situations where the investigator wishes to detect and study macroecological factors that are of broad conceptual interest.

Required sample sizes

Several kinds of recommendations have been proposed regarding desirable sample sizes in SEM studies. Perhaps the simplest rule of thumb might be: 200 sample units are desirable, 100 are adequate, 50 are tolerated, and less than 50 can lead only to tentative results. Recommendations on sample sizes are sometimes stated in terms of the optimum number of sample units per variable in the model, say 20 sample units per variable, which would lead to a recommendation of 240 sample units for a model that contains 12 variables.

Four points put the issue of sample size into context:

(1) SEM is typically no more sensitive to small sample size effects than are multiple regression, principal components analysis, and many other inferential methods. The desirability of having a large sample size when performing more conventional regression analyses is generally not appreciated, though it exists to the same degree.

(2) Bootstrapping methods have been found to support reasonably consistent conclusions with sample sizes as low as 50. However, for many ecological situations, small sample sizes really do not represent the broader population of samples being studied.

(3) Methods exist for conducting certain types of SEM that have very low sample size requirements. For the Partial Least Squares method of path model analysis (Falk and Miller 1992), the minimum sample size required is one sample more than the maximum number of variables associated with a latent variable. Theoretically, the model shown in Figure 30.8 could be analyzed with 7 samples, though the likelihood of the results having any generality would be extremely small.

(4) In the initial phase where a field such as ecology begins to use SEM methods, early applications may rely on inadequate sample sizes due to a failure to anticipate or appreciate the need for larger samples sizes. As investigators begin to become more familiar with SEM, they adjust data collection to a more reasonable standard. As always, conclusions must be reasonable for the data that are available. In some cases even 200 sample units may not be adequate for supporting a broad-based conclusion about an ecological phenomenon.

Data types

Initially the theory as well as the software used to analyze structural equation models was limited to certain types of data. Considerable effort has been made over the years to expand the range of conditions under which SEM can be used appropriately. Software is available to enable the user to work with both correlation matrices and covariance matrices without statistical bias. Both continuous and categorical variables can be analyzed using modern methods as well. Distributional assumptions are no longer a barrier to analysis due to the variety of model fitting procedures that are available. Methods for the analysis of hierarchically-structured data have recently become more diversified. Finally, experimental as well as non-experimental data are amenable to analysis using SEM.

Model formulation, testing, refinement, and validation

SEM is considered by practitioners to be a confirmatory method. What this means is that SEM is meant to be used to evaluate relationships that have been proposed a priori based on scientific judgement and experience. This is in contrast to exploratory analyses where the researcher selects a number of variables for examination and then decides the importance of variables based purely on statistical criteria (e.g., R^2 values).

One of the major strengths of SEM is its confirmatory nature, since it allows the researcher to contribute his or her full knowledge of the subject to model design and evaluation. Thus, when the results of an analysis suggest that a relationship exists in the data that we do not have reason to believe is generally true, the model does not have to be modified to include that relationship (fitting a model to the nongeneral features of a data set is referred to as overfitting).

Some people may be uncomfortable with the latitude afforded the researcher to decide which statistically significant relationships are scientifically meaningful and which are not. The ultimate guide will be additional studies that determine if this unexpected relationship is general and recurring, or just a chance feature of a particular data set. When Gough et al. (1994) examined the relationship between species richness and abiotic factors, they found indications of a weak but significant effect of soil organic matter on the size of the species pool. They did not expect this result and they did not fully trust its generality. However, subsequent studies have repeatedly shown that relationship to exist, leading to a greater confidence that it is a real phenomenon.

"Theory" versus scientific judgement

The field of ecology would seem to be a good candidate for confirmatory analyses because of the potential for experimental verification of relationships and because of fewer problems in establishing the direction of dependence than some other fields. However, the lack of agreement about terms and concepts within ecology is a considerable barrier to progress at times (e.g., Grace 1991). SEM practitioners typically use the phrase "theoretical justification" as the criterion for development and refinement of a model. In ecology, the meaning of "theory" is not well established (Pickett 1994). We believe that what is at issue here is what ecologists might better understand as "scientific judgement." When stated in this way, we believe that it is clear that ecologists will very often be able to use scientific judgement to establish a priori relationships among variables that can make up a model for confirmatory analysis. Some ecologists may not believe that we have an adequate "theory" linking plant growth to soil salinity, but most will agree that scientific judgement suggests that it is reasonable to presume a directional dependence.

A common misunderstanding about confirmatory analyses by ecologists is the relative roles of statistical analysis and scientific judgement. In the case of SEM, accepting a model based on statistical criteria does not validate the causality of the underlying relationships. The responsibility for establishing and defending the causality (dependence) of relationships in the model is solely based on scientific judgement. Statistical analysis only assesses the degree to which the data are consistent with our ideas. We see this as a healthy relationship between statistical analysis and scientific judgement in which we do not base our interpretations entirely on statistical criteria nor do we base them entirely on our preconceptions. In fact, this interdependency should be recognized when using any statistical technique, though it has been a chronic source of confusion (Yoccoz 1991).

Exploration, confirmation, and scientific progress

A central feature of SEM is its ability to permit the rejection of multivariate hypotheses with complex structure. On the surface, this would seem to be at odds with our discussion (above) of the role of statistical analysis in model acceptance. Keep in mind, however, that our goal is scientific progress: the improvement of our understanding and our ability to predict. Model evaluation and refinement test our prior experience and understanding against additional data. In cases where our under- standing is advanced, new data should lead to only small refinements. In cases where our understanding is weak, new data will suggest major changes in our models.

SEM offers two approaches to model evaluation and refinement. One approach is to test a priori models against a collected dataset. This can be done with either a strictly confirmatory approach, in which a single model is compared to data, or a competing models analysis that tests among predetermined alternatives. The most commonly used approach for competing models is a nested analysis in which a series of related models are evaluated. As in all analyses, the strategy used will depend on the nature of the problem and the state of knowledge for the problem.

The second approach to model evaluation offered by SEM lies in validation. By validation we mean the process of finding additional information with which to test our ideas. In many experimental fields a finding is held to be true only if it can be replicated in an independent laboratory. Ecology, like many other fields that deal with complex and long-term processes, is not always amenable to such overt validation (though much progress in this direction could be made). However, it is part of SEM philosophy to continue the evaluation of models using new data (experimental or nonexperimental) until consistency is obtained. For example, Gough and Grace (1999) experimentally evaluated predictions made from the SEM model of Grace and Pugesek (1997). The take home message is simple: we must always recognize that the evaluation of an idea with a single data set is of unknown validity. This is just as true for analysis of variance as it is for SEM.

Validation methods and ecological problems

Models can be validated at several levels. The first would be at the level of a particular data set. Larger data sets can be subdivided, with half used in an initial analysis and the other half reserved for model validation. In many circumstances, however, it is simply the process of competing model analysis that is used to arrive at validation for that data set. Validation at the level of the data set is easily achieved in many cases, though problems with equivalent models can arise (Spirtes et al. 1993).

A second level at which validation can be sought is at the level of the population, community, or ecosystem being sampled. Here, independent data sets and/or experimental studies can be used with great effect (e.g., Wootton 1994). Since ecological systems vary in time and space, we wish to evaluate the general validity of a model over time or space. In this sense, temporal validity would imply that the model held over time for a population. Spatial validity, in contrast, would refer to a model that could be shown to apply across space. For example, a model for a salt marsh might be shown to apply to other salt marshes in the region or even in other regions.

Finally, a model might be shown to apply to a variety of ecosystems, for example a salt marsh, a prairie, and an abandoned agricultural field. The level of validation being sought will be determined by the objectives of the research. If the objective is to obtain the maximum understanding of an ecosystem and its behavior, then a detailed model will be sought that will not apply to other ecosystems in its entirety. However, if the goal is to obtain a model that can apply to a wide range of conditions, a more general model will be required. As Grace and Pugesek (1997) have shown, it is possible to evaluate both specific and general models using a single data set.

An example of how validation may be approached can be taken from our studies of species richness. In Grace and Pugesek (1997) the final model was validated only at the level of the dataset itself (i.e., this was the best fitting model), not at any higher level. Therefore, these results must be viewed with caution as the authors noted in the following statement: "Our use of SEM was, in part, exploratory in that we accepted models other than the initial models evaluated. The structural models that we estimated and found to fit the data well should, therefore, be considered as being in need of further confirmation."

Several independent efforts to validate that initial study have been conducted (Gough and Grace 1999, Grace and Guntenspergen 1999, Grace and Pugesek unpublished). Finally, a different kind of validity for the model is now being assessed by using the general model to evaluate the factors controlling species richness in different types of grasslands (Grace et al. unpublished).

Conditions that interfere with the use of SEM

A number of factors restrict the use of SEM. First, there are practical limits to the model complexity. Hair et al. (1995) suggest that more than 20 latent variables can be difficult to manage and interpret, though much will have to do with the degree of uncertainty about relationships; the greater the certainty, the more complex the model can be. In our experience, models with more than about 10 latent variables become increasingly difficult to resolve and also difficult to explain to the reader. Second, for problems where it is difficult to test assumptions about the directionality of relationships, equivalent models will limit the confidence that can be achieved. Thus, SEM will not be fruitful in some situations.

When it can be applied, SEM can further our understanding and help us to quantify relationships in multivariate systems. For problems where there is insufficient information for establishing a priori hypotheses, exploratory regression analyses may still be the most profitable avenue for progress. For very general or deterministic topics, logic and analytical mathematical analysis will continue to be the most appropriate methods. However, for the application of general theories to real world problems, SEM can provide a framework for translating abstract theory into operational form. As an illustration, the example presented in the next section of this chapter was stimulated by a desire to operationalize the general theories of Connell (1978) and Grime (1979) for use in predicting the consequences of environmental change on species diversity.

Overall, SEM should prove to be a useful tool for ecologists and will complement existing methods. The application of this methodology to our field, however, will require some patience and persistence as we determine how best to adapt it to our needs and special problems.

How it works

Up to this point we have focused primarily on why one might use SEM and the kinds of information produced by this methodology. While there are many general texts on the subject, there are few sources available where the ecologist can see a step-by-step example of an SEM analysis for familiar material. For this reason, in this section we present another example of an ecological application of SEM, this time with a focus on the chronological sequence of steps in the analysis.

Overview of the structural equation modeling process

The process of developing and evaluating a structural equation model can be broken down into a seven-step process: (1) develop a theoretically based model, (2) collect data for the analysis, (3) screen the data for suitability, (4) evaluate the fit between the data and the initial measurement model, (5) create a modified measurement model if theoretically justified, (6) evaluate the fit between the data and the full model, and (7) draw the appropriate interpretations from the results.

Step 1: Develop a theoretically based model

A fundamental feature of SEM is that we begin with an a priori hypothesis based on underlying causal relationships. Clearly, many philosophical issues come to mind when we begin to talk about "theoretically based models" and "causal relationships" (Hayduk 1987, 1996). While SEM is often described as dealing with causal relationships, we can avoid a certain amount of controversy over what constitutes causation by focusing our attention on the "dependence" of a relationship.

Bollen (1989) offers three criteria for use in determining whether a relationship between two variables is a dependency or a simple correlation, (1) association, (2) isolation, and (3) directionality. In the simple case, if the manipulation of variable A results in variation in variable B while other variables are held constant, then we say that B depends on A and the relationship is one of directional dependence. Confirmation of directionality can be obtained by manipulating variable B with the expectation that A will not consequently vary.

In ecological situations we often have a notion of the direction of dependence of a relationship (e.g., the addition of saltwater may reduce a plant's growth). Feedback processes can certainly increase the complexity of our system, though they often operate on a much longer time scale (e.g., plant effects on soil salinity). What is important in Step 1 of the SEM process is that the relationships among variables are determined by the researcher based on the accumulated knowledge of the field of science, not based on the statistical analysis of the dataset at hand. Thus, when we say that our initial model must be theoretically based, we mean that it is not derived from the data used to test it and we must have justification for the paths hypothesized.

A fundamental component of developing an initial SEM model is the specification of the path diagram. An example is presented later in this chapter in Figure 30.10. The path diagram specifies the latent variables, in this case abiotic conditions (**Abiotic**), disturbance (**Disturb.**), community biomass (**Biomass**), and species richness (**Rich**), as well as the indicator variables abiotic indices (**abio#1**, **abio#2**), **dist**, **mass**, **lt-hi**, **lt-lo**, and **rich**. In general there is at least one indicator variable associated with each latent variable, though there can exist "second order" latent variables that are estimated only from two or more latent variables.

To analyze a model, the relationships in the path diagram must be converted to a set of equations that specify (1) the structural equations linking latent variables, (2) the measurement model specifying how latent variables will be estimated from indicator variables, and (3) any hypothesized correlations among variables.

Different software packages use different conventions for specifying models. Some, such as LISREL allow the user three different methods of specifying the model to be estimated. Whatever the software package, the above three pieces of information are required.

Additional issues must be considered to specify the complete set of equations at the mathematical level. In certain cases, we may expect correlated errors among variables. For example, if a carbon/nitrogen analyzer is used for soil samples, we could expect the errors for these two parameters to be correlated.

We may also at times wish to set the values for a relationship. For example, we may wish to let the loadings between **Abiotic** and both **abio#1** and **abio#2** be equal as a way of simplifying the model.

Another example of a specification would be to include an estimate of the reliability of the relationship between a latent variable and indicator variable, (e.g., **Disturb.** and **dist**) using a previously available estimate of the measurement error. Finally, the scale of the measurement must be determined, which is usually accomplished by setting the path between one indicator and the latent variable to a value of one (thus, scaling the latent variable to that indicator variable) or setting the variance of a latent variable to one (thus, zero transforming the latent variable). For more discussion of these issues refer to Hair et al. (1995).

SEM uses degrees of freedom to estimate parameters and test hypotheses. This means we must assess the identification level of the model. A model is said to be underidentified if there are more unknown than known parameters specified. In this case it is not possible to obtain a unique solution to the estimation process. Having degrees of freedom available for model estimation is a requirement but not necessarily a sufficient condition to ensure appropriate model identification. In addition, every model parameter must have a unique solution. The most common cause of underidentification is estimation of a nonparsimonious model (trying to examine all possible relationships among variables) while possessing a small number of indicators for each latent variable. Some data sets and models have a finite limit to the number of relationships that can be examined. Typically, however, various strategies can be used to reduce the number of unknown parameters that must be estimated, permitting the analysis to proceed (Hair et al. 1995).

Step 2: Collect data for the analysis

The initial model should ideally be specified prior to data collection (e.g., Grace and Pugesek 1997). Specifying the model prior to data collection lends credibility to the contention that SEM is a confirmatory procedure rather than an exploratory one. (For comparison, one does not generally decide prior to data collection the number of principal components and the loading structure they will accept in PCA; thus, PCA is usually used as an exploratory technique.) We must consider, however, that in the early stages of application of SEM a rigid adherence to this requirement will likely stifle its application. For example, in some cases, models are developed for preexisting data (e.g., Grace and Jutila 1999). In such cases, to the degree possible it is important to try to formulate the initial model from preexisting ideas about the subject rather than from results of regressions and other analyses of the data set at hand. This "naive" approach to model formulation may be facilitated by involving someone familiar with SEM but not involved in any initial analyses that were performed in the model specification process. More importantly, it is necessary in these initial experiences with SEM not to overstate the interpretation of the results but to view the experience as initial model development, and hence, the first step in an ongoing process of model refinement and testing.

Step 3: Screen data for suitability

Software packages such as LISREL offer modules (such as PRELIS) for prescreening of data prior to SEM analysis. At this stage in the process, certain key statistical issues can be addressed. The primary issues are (1) coding data so as to set a similar range of variation for variables, (2) distributional properties of variables, and (3) linearity of relationships, which may precipitate the need for transformations.

On the first of these issues, it is simply important to realize that model fitting can be hampered when variables are on widely different scales (say variable x_1 ranges from 0.001 to 0.01 while x_2 ranges from 10,000 to 100,000). The simple solution is to transform all variables to be of a similar range (Ch. 9), preferably numbers less than 100, to prevent problems.

The distributional properties of variables are important because of their impact on the validity of the significance tests. Standard parametric procedures such as maximum likelihood estimation (as well as least squares estimation) depend on the assumption of multivariate normality. What is often misunderstood is that these distributional assumptions do not apply to exogenous (independent variables), only endogenous (dependent) variables. In most modern software packages, estimation methods can be chosen to compensate for either categorical or nonnormal dependent variables in the model.

One of the trickier issues dealing with data has to do with curvilinear and nonlinear relationships, which can pose inherent problems for models that contain optima or threshold relationships. This complex topic is beyond the scope of our treatment (see Schumacker and Marcoulides 1998). However, the simplest approach is to transform the data to improve linearity (e.g., Table 2 in Grace and Pugesek 1997), which can provide an effective solution to the statistical problem (though it can create some difficulties in communicating the relationships).

Step 4: Evaluate the fit between data and the initial measurement model

The evaluation of model fit is a two-stage process that involves first, assessment of the measurement model and second, assessment of the full model. This two-stage process allows the investigator to assess misspecifications in the measurement model separately from those that might occur in the structural model. Here we separate these stages into Steps 4-7. All of these steps require a common process, the assessment of model fit.

The process of assessing model fit, whether it is a measurement model or structural model, initial model or final model, is one whereby the correspondence between observed covariances (or correlations) and the hypothesized model is evaluated. In essence, a hypothesized model implies a covariance structure that can be compared to the observed covariance structure. A number of fit indices have been derived to facilitate this evaluation process. The most basic fit index is the chi-square estimate for observed versus predicted. A nonsignificant chi-square would generally be interpreted as indicative of a good fit of model to the data. However, model parsimony and sample size are ignored in the use of the overall chi-square statistic and, therefore, a number of additional indices (such as the Goodness of Fit and Root Mean Square Error of Approximation) have been developed to deal with these influencing factors. A more complete discussion of this topic can be found in Jöreskog and Sörbom (1996).

Step 5: Create a modified measurement model if theoretically justified

If a model turns out to deviate significantly from the data, the problem arises as to how to determine the causes of such deviation and the remedy, if any, to this situation. Software packages such as LISREL automatically estimate the model fit for alternative models and generate a set of modification indices that show whether changes to the model could improve the fit. Often, researchers will specify a set of alternative models that they particularly wish to compare (e.g., Grace and Guntenspergen 1999), which helps to focus the evaluation process. In situations where it is particularly difficult to establish the logical superiority of one model over others (e.g., where experimental manipulation of relationships is not possible), evaluation of alternative and equivalent models can be made using programs such as TETRAD II (Scheines et al. 1994, Spirtes et al. 1994). The goal in most cases, however, is to compare the fit of the originally specified model(s) to the data.

An important point about SEM that is often misunderstood in practice is that the goal is to evaluate the general validity of our model. This has been mentioned earlier but warrants repeating. If our model is valid, it may still deviate from the data in our sample because of chance characteristics of our sample. In evaluating a model using SEM it will often be the case that slight improvements in model fit can be made by changing the model to represent all details of our data. However, if our data are not a perfect representation of the covariance structure of the population about which we wish to generalize, we risk overfitting the model and losing our generality. For this reason, the process of modifying an initial model based on the results of the analysis requires scientific justification (example below).

Step 6: Evaluate the fit between data and the full model

Once an acceptable measurement model is obtained, the full model is now tested against the data using the previously described procedures. This test then reports deviations between expected and observed covariances that reflect properties primarily in the structural model. Often we ascribe a different level of importance to the structural model than the measurement model. The two-stage fitting process allows us to explore the fit of the structural model with some degree of separation from the measurement model. Procedurally, the same approach is taken to the full model as was described for the measurement model.

Step 7: Draw the appropriate interpretations from the results

It is important to describe the precise steps that were performed in SEM analyses so that the audience can understand the true degree to which data fit the initial model and how changes were performed. The standard for this full disclosure is much higher than in normal multivariate statistical analyses, where an investigator may choose to drop a few variables out of the analysis because they did not seem to add much to the results.

Aside from proper description of the methods, an important issue is the appropriate statement of inference. A useful approach to this problem is defined by Jöreskog and Sörbom (1993):

We distinguish among three situations:

SC *Strictly Confirmatory.* In this case the researcher has formulated one single model and has obtained empirical data to test it. The model should be accepted or rejected.

AM *Alternative Models.* Here, the researcher has specified several models and on the basis of an analysis of a single set of empirical data, one of the models should be selected.

MG *Model Generating.* Here, the researcher has specified a tentative initial model. If the initial model does not fit the given data, the model should be modified and tested again using the same data. Several models may be tested in this process. The goal may be to find a model that not only fits the data well from a statistical point of view, but also has the property that every parameter of the model can be given a substantively meaningful interpretation. The re-specification of each model may be theory-driven or data-driven. Although a model may be tested in each round, the whole approach has more to do with generating than testing models.

What to report

The example presented in the next section illustrates what information should be reported. In general terms, the several pieces of information that should be reported in a scientific publication include the following (in addition to the methodological procedures):

☐ The logical basis for the initial model.

☐ The logical basis for any changes that were made to the model.

- Model fit parameters, particularly for the final model.
- Total variance explained in the endogenous variables.
- Relative strengths of relationships in the model as indicated by the standardized path coefficients.
- Partitioning of the total effects of predictor latent variables on dependent latent variables into different pathways.
- Correlations among latent variables.

Examples

Grace and Pugesek (1997) applied SEM to the problem of identifying the direct and indirect influences of abiotic factors on plant species richness. We use selected aspects of this analysis to illustrate the step-by-step application of SEM to an ecological problem.

In an earlier study, Gough et al. (1994) proposed a general conceptual model of the direct and indirect influences of abiotic conditions on species richness (Fig. 30.9). The primary emphasis of their model was the proposition that abiotic factors can have two distinct kinds of influences on species diversity. First, abiotic factors can regulate community biomass, which in turn can control richness through density-dependent processes such as competitive exclusion. Second, abiotic factors can also determine the diversity of a plot by regulating the pool of species that are physiologically capable of living there (potential richness).

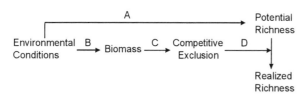

Figure 30.9. Initial conceptual model (from Gough et al. 1994).

The intent of Grace and Pugesek was to evaluate this conceptual model and to estimate the relative importance of indirect (density mediated) and direct (species pool) pathways between abiotic conditions and richness. To accomplish this, the authors collected data on abiotic variables, vegetation biomass, and species richness from 190 plots across a coastal marsh landscape.

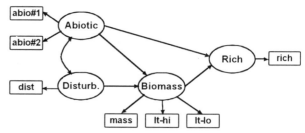

Figure 30.10. Initial structural equation model.

Figure 30.10 represents the general a priori model that was developed by Grace and Pugesek. The authors also evaluated a specific model that explicitly considered the individual abiotic factors as separate latent variables as well as the general model presented here. For the purposes of illustrating SEM, we present only the general model in which measured abiotic factors were aggregated into two indices, abio#1 and abio#2.

We first ascertained the adequacy of the measurement model with a confirmatory factor analysis using LISREL (Fig. 30.11). Assessment of the measurement model in this fashion allows the latent variables to freely correlate and focuses on determining the degree to which the hypothesized latent variables explain the covariances among indicator variables. The results of this analysis (Fig. 30.11) showed that factor loadings were high and measurement errors low. Despite this, however, this initial model fit the data rather poorly. The chi-square for this model was 50.6 with 12 degrees of freedom (p = 0.000015), indicating a significant lack of fit between data and model. The RMSEA (root mean square error of approximation), another index of model fit, was 0.1304, much higher than the recommended value of 0.06.

Examination of the modification indices produced by LISREL suggested that measured biomass and light data did not combine well into a single latent variable. Further, examination of the data for biomass and light revealed that the morphological variability found among vegetation types results in a variable correlation between biomass and light penetration. While not initially anticipated, the authors felt that there was conceptual justification for the recognition of **Biomass** and **Light** as separate latent variables. A subsequent confirmatory factor analysis was performed with these as separate latent variables (Fig. 30.12) and the results showed a good fit between model and data (chi-square = 10.6 with 8 degrees of freedom, p = 0.22; RMSEA = 0.042, p = 0.52).

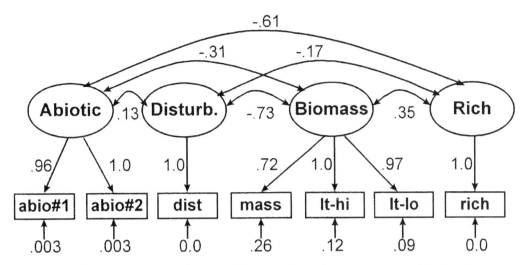

Figure 30.11. Initial results of confirmatory factor analysis of measurement model. Numbers are path coefficients, represent partial regression coefficients and correlation coefficients.

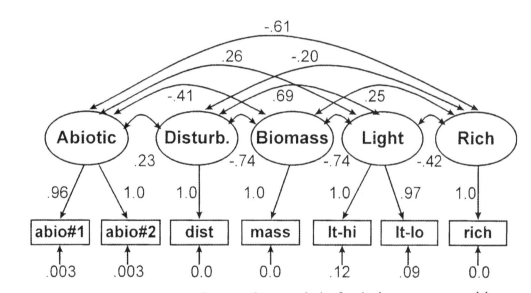

Figure 30.12. Final results for confirmatory factor analysis of revised measurement model.

Based on the results from the factor analyses, the original SEM (Fig. 30.10) was modified as shown in Figure 30.13. Evaluation of this model using LISREL suggested further refinements. One refinement was to recognize that there could be a significant path from **Disturb.** to **Light** due to an influence of disturbance on the relationship between community biomass and light capture per unit biomass. Inspection of the data indicated that the reason for this relationship was that plant canopies were vertically arranged in disturbed plots while more horizontally arranged in less disturbed sites. Thus, again there was conceptual justification for including a new relationship, this one being a path from **Disturb.** to **Light**. With this path included, LISREL produced revised estimates (Fig. 30.14). This final model was found to have a chi-square of 11.4 with 10 degrees of freedom (p = 0.33) and a RMSEA of 0.027 (p = 0.67), indicating a very good fit between model and data (Hair et al. 1995).

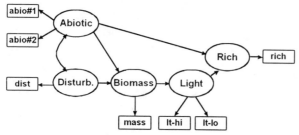

Figure 30.13. Revised structural equation model.

LISREL provides a number of results from the analysis, some of which are summarized in Figure 30.14. First, a high correspondence between indicators and latent variables was found for **Abiotic** and **Light** as shown by the high loading values. Note that for **Biomass**, a loading of 0.86 was specified by the investigators based on independent estimates of the precision of **mass** samples. Second, R^2 values for the endogenous variables represent the variance explained for those variables by the model and were found to be 0.42, 0.59, and 0.61 for **Rich**, **Light**, and **Biomass** respectively. Third, standardized partial regression coefficients illustrate the relative amounts of variance explained by different pathways. Interpreting the diagram from right to left, it can be seen in this example that the predominant pathways explaining variance in **Rich** were from **Abiotic** and **Light** with the direct pathway from **Biomass** being of somewhat less importance. A similar approach can be used to examine the relative importance of pathways explaining variance in **Light** and **Biomass**.

Now, in addition to reading the diagram from right to left, it is possible to examine the total effects of predictor variables on explained variables, which relates variables on the left to those on the right. Table 30.5 presents standardized total effects, which include all pathways connecting a pair of variables. For example, a total of three different pathways between **Abiotic** and **Rich** are included within the total effect, one direct, one through **Biomass**, and one through **Biomass** and then **Light**. As mentioned earlier in this chapter, the importance of compound pathways (those that pass through intermediate variables) can be determined by multiplying the coefficients in the compound path.

Total effects represent the sum of all the directed pathways. Standardized total effects values can also be used to perform "sensitivity analyses" (*sensu* Swartzman and Kaluzny 1987) that show the relative importance of various variables in the model.

For completeness it should be noted that the net correlation between latent variables includes both the directed pathways (those included in total effects) as well as the undirected pathways (those that include a correlation). An example of an undirected pathway in Figure 30.14 is from **Abiotic** to **Disturb.** to **Biomass** to

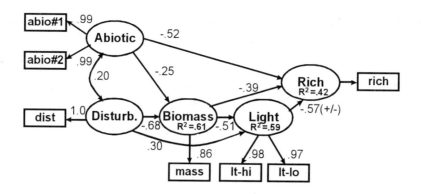

Figure 30.14. Final structural equation model. Path coefficients shown represent partial regression and correlation coefficients. R^2 values specify the amount of variance explained for the associated endogenous variable.

Table 30.5. Standardized total effects of predictors on predicted variables for the model in Figure 30.14. Total effects include both indirect and direct effects, and represent the sum of the strengths of all pathways between two variables. Numbers in parentheses are standard errors. Numbers in brackets are t values. Reprinted with permission from Grace and Pugesek (1997).

Predicted variables	Predictors			
	abiotic	disturbance	biomass	light
biomass	-0.2533 (0.0510) [-4.97]	-0.6846 (0.0743) [-9.21]	--	--
light	0.1303 (0.0361) [3.86]	0.6553 (0.0648) [10.11]	-0.5144 (0.0980) [-5.25]	--
richness	-0.4971 (0.0602) [-8.25]	-0.1046 (0.0568) [-1.84]	-0.0981 (0.0971) [-1.01]	-0.5665 (0.0761) [-7.44]

Light to **Rich**. Thus, while the total effect of **Abiotic** on **Rich** is -0.4971 (Table 30.5), the correlation between **Abiotic** and **Rich** is the total effect plus the total effect of **Disturb**. and **Rich** multiplied times the correlation between **Abiotic** and **Rich** (-0.4971 + 0.20 × -0.1046) = -0.5180

SEM analysis provides additional useful results: estimates of precision for various coefficients, standard errors and significance levels for all coefficients, and error terms. Another important kind of information is the unstandardized total effects of predictor variables. When the data analyzed are from a variance/covariance matrix, the unstandardized total effects can be used for prediction/projection purposes. Altogether, SEM generates a rather amazing amount of useful information about the information contained in a covariance matrix. The exact portions of this information to be presented in a paper often depend on what aspects are to be emphasized (Pugesek et al. 2002).

Variations

There are at least three different threads of ideas that have become woven together into the modern field of path model analysis, or SEM *sensu lato*. We refer to these as (1) mainstream, (2) exploratory and causal search methods, and (3) partial least squares. Other observers of the topic might easily classify the conglomeration of ideas in a different fashion.

The most fully developed of these three subdivisions is mainstream SEM, which has been emphasized in this chapter. As alluded to earlier, there are many variations on the mainstream methods. The interested reader is referred to Pugesek et al. (2002) for ecological and evolutionary examples using many of the more commonly used variations of mainstream SEM.

Along a different line, Spirtes et al. (1993) and Shipley (2000) present examples from the tradition of exploratory SEM analyses, a set of techniques that seek evidence for causal structure from the data itself. Related to this is the methodology of linear chain models, which searches for conditional independence among variables (Whittaker 1990).

The third subdivision of methods, known as partial least squares (PLS), is an alternative approach to path model analysis that is designed to rely on fewer assumptions and yields less robust analyses of the validity of path models. We have found PLS particularly useful in dealing with nonlinearities and interaction terms, as well as situations where sample sizes are too small for traditional methods. One example of the use of PLS as a preconditioning method for SEM is presented in Grace and Jutila (1999). A more general treatment of the method is in Falk and Miller (1992).

Methods that are concerned with the analysis of path models continue to evolve. Much effort is aimed at resolving some of the thornier major limitations, such as nonlinearities. Other efforts are concerned with extending SEM techniques to accommodate measurement error in regression, ANOVA, MANOVA, and MANCOVA analyses. All of these extensions will be of potential value to the ecologist.

One final variation we will mention has to do with the interplay between SEM and system simulation models. SEM and dynamic systems modeling represent different approaches to dealing with complex data, each with their own purpose, strengths and weaknesses. We have been involved in two kinds of approaches to integrating these two techniques into a comprehensive system. For example, the use of SEM analyses to generate coefficients that can be used for predicting the future behavior of ecological systems (Gough and Grace 1999) represents a linkage of potential value to ecologists. On the other side of the equation, Grace (2001) used simulation models to generate expected path models based on different mechanistic scenarios. We have little doubt that SEM methods will continue to be expanded and elaborated, providing a useful set of tools and approaches to analyzing complex systems.

Appendix 1. Elementary Matrix Algebra

Definitions

1. A **matrix** of dimension $r \times c$ is a rectangular array of numbers with r rows and c columns.

$$\mathbf{A} = \begin{bmatrix} a_{11} & a_{12} & \cdots & a_{1c} \\ a_{21} & a_{22} & \cdots & a_{2c} \\ \vdots & \vdots & & \vdots \\ a_{r1} & a_{r2} & \cdots & a_{rc} \end{bmatrix}$$

Each number in the matrix is called an **element**. The general term for an element is a_{ij}, which refers to the number in the ith row and the jth column. The row subscript is typically given before the column subscript. Matrices are symbolized by bold-faced (or wavy-underlined) capital letters. The array of elements is enclosed in square brackets. The matrix can also be written as

$$\mathbf{A} = \begin{bmatrix} a_{ij} \end{bmatrix}$$

2. A matrix with a one column (dimension $r \times 1$) or a one row (dimension $1 \times c$) is a **vector**. A vector is symbolized by a bold-faced (or wavy-underlined) *lower case* letter.

$$\mathbf{a} = \begin{bmatrix} a_1 \\ a_2 \\ \vdots \\ a_r \end{bmatrix} \text{ is a column vector}$$

$\mathbf{b} = \begin{bmatrix} b_1 & b_2 & \cdots & b_c \end{bmatrix}$ is a row vector

3. A **scalar** is a "matrix" with dimension 1×1 (having only one element). It is symbolized by a lower case letter in italic type.

4. A **square matrix** is a matrix with $r = c$.

5. The **diagonal** of a square matrix consists of elements from the upper left corner to the lower right corner. If the diagonal has n elements, then those elements are

$$a_{11}, a_{22}, \ldots a_{nn}.$$

6. The **trace** is the sum of diagonal elements.

7. A **symmetrical matrix** is a square matrix that has $a_{ij} = a_{ji}$ for all i and j.

8. A **diagonal matrix** is a square matrix in which all elements but the diagonal are zero.

9. An **identity matrix**, **I**, is a diagonal matrix in which all diagonal elements are one. For example,

$$\mathbf{I} = \begin{bmatrix} 1 & 0 & 0 \\ 0 & 1 & 0 \\ 0 & 0 & 1 \end{bmatrix}$$

Operations

1. **Transposing a matrix**: A matrix can be transposed by changing the rows to columns and the columns to rows. The transpose of **A** (indicated as **A'**) is

$$\mathbf{A} = \begin{bmatrix} 10 & 0 & 4 \\ 7 & 1 & 2 \end{bmatrix} \quad \mathbf{A'} = \begin{bmatrix} 10 & 7 \\ 0 & 1 \\ 4 & 2 \end{bmatrix}$$

The prime mark (') signifies that a matrix has been transposed. Note that transposing is like grabbing the upper left and lower right corners of the rectangle and flipping it over.

A vector in its normal position is a column vector. A row vector is the transform of a column vector and it is indicated by the prime mark.

The transpose of a symmetrical matrix is identical to the original: e.g., $\mathbf{I'} = \mathbf{I}$.

2. **Augmenting**. Matrices can be augmented by adding single rows or columns or multiple rows or columns. If rows are augmented then the number of columns must match. If columns are augmented then the number of rows must match. In the first example below, **A** is augmented on the right with **B**. In the second example below, **A** is augmented below with **B**.

$$[\mathbf{AB}] \quad \begin{bmatrix} \mathbf{A} \\ \mathbf{B} \end{bmatrix}$$

3. **Partitioning**. A matrix can be partitioned into submatrices. For example,

$$\mathbf{A} = \begin{bmatrix} a_{11} & a_{12} & a_{13} \\ a_{21} & a_{22} & a_{23} \\ a_{31} & a_{32} & a_{33} \end{bmatrix} = \begin{bmatrix} a_{11} & \mathbf{a}_{1j} \\ \mathbf{a}_{i1} & \mathbf{A}_{22} \end{bmatrix}$$

4. **Addition and subtraction.** The sum of two $r \times c$ matrices **A** and **B** is a third $r \times c$ matrix **S**:

$$S = A + B$$

To do this the corresponding individual elements are added:

$$s_{ij} = a_{ij} + b_{ij}$$

Subtraction is similarly performed on corresponding elements. To be added or subtracted, matrices must have the same dimensions. For addition, note that

$$A+B = B+A \quad \text{and} \quad A+(B+C) = (A+B)+C$$

5. **Multiplication or division by a scalar.** A matrix **A** is multiplied or divided by a scalar d, by multiplying or dividing each element of **A** by d:

$$d\mathbf{A} = [da_{ij}] \quad (1/d)\mathbf{A} = [a_{ij}/d]$$

6. **Multiplication of two matrices.** Two matrices **A** and **B** can be multiplied if the number of columns of **A** is equal to the number of rows of **B**. Let **A** be an $r \times m$ matrix and **B** an $m \times c$ matrix.

$$\mathbf{A}\mathbf{B} = \mathbf{P}$$

In this case **A** is called the premultiplicand matrix and **B** is the postmultiplicand matrix. Each element is obtained by:

$$p_{ij} = \sum_{k=1}^{m} a_{ik} b_{kj}$$

The product matrix **P** has dimensions $r \times c$. *Important: Multiplication requires that the number of columns of the premultiplicand matrix equal the number of rows of the postmultiplicand matrix.* A helpful way to see if two matrices can be multiplied, and what the dimensions of the product will be, is to write the number of rows and columns below the matrices:

$$\begin{array}{ccc} \mathbf{A} & \mathbf{B} & = \mathbf{P} \\ 2\times3 & 3\times5 & 2\times5 \end{array}$$

The adjacent numbers under the multiplicands must be equal (they are 3's in the example). The dimensions of the resulting matrix are the first and last numbers under the multiplicands.

Note that:

$$A(B+C) = AB + AC \quad \text{and} \quad A(BC) = (AB)C$$

and almost always:

$$AB \neq BA$$

Two vectors with the same number of elements can be multiplied if one is a row vector and the other a column vector. The product is a scalar, called the inner product of the two vectors.

$$\mathbf{x'y} = \begin{bmatrix} x_1 & x_2 & \cdots & x_n \end{bmatrix} \begin{bmatrix} y_1 \\ y_2 \\ \vdots \\ y_n \end{bmatrix} = \sum_{i=1}^{n} x_i y_i$$

The **length** of a vector **x** is the square root of the inner product with itself:

$$\text{length} = \sqrt{\mathbf{x'x}} \quad \text{where} \quad \mathbf{x'x} = \sum_{i=1}^{n} x_i^2$$

This is the *n*-dimensional version of the Pythagorean theorem, measuring the distance between the point specified by the vector and the origin.

7. **Inverting a matrix.** To quote Tabachnik and Fidell (2001), "If you liked matrix multiplication you'll love matrix inversion." There is no division of one matrix by another, but an analogous operation is finding the inverse of a matrix so that

$$\mathbf{AA}^{-1} = \mathbf{I}$$

The superscript "-1" indicates the inverse of a matrix. Only square matrices have an inverse and not all of them do. Note that

$$\mathbf{AA}^{-1} = \mathbf{A}^{-1}\mathbf{A} \quad \text{and} \quad (\mathbf{A}^{-1})^{-1} = \mathbf{A}$$

Finding the inverse of a matrix is typically difficult to impossible by hand. For very small matrices, such as 2×2 matrices, a number of techniques are available to find the inverse. One method involves making a series of operations that convert an augmented matrix into a specific form. This involves a variable number of steps, but the outcome is always the same, as long as the rules for the operations are strictly followed (below). The first step is to augment a matrix by attaching the same sized identity matrix to the right:

$$\begin{bmatrix} a_{11} & a_{12} & 1 & 0 \\ a_{21} & a_{22} & 0 & 1 \end{bmatrix}$$

One then selects operations from below, and operates on the augmented matrix one row at a time, until the identity matrix comes out on the left:

$$\begin{bmatrix} 1 & 0 & ? & ? \\ 0 & 1 & ? & ? \end{bmatrix}$$

At this point the values on the right hand side (the question marks) will be the inverse of **A**. In other words,

[**AI**]   is converted to   [**IA**$^{-1}$]

An operation can be performed on a row (treating one row at a time) in one of the following ways:

(1) Division or multiplication of each element in a given row by a constant.

(2) Adding to or subtracting from the row, another row or a constant times another row.

This is a fun puzzle until you start trying to do larger matrices.

Other special features of a square matrix

1. The **determinant**. Every square matrix has a determinant, symbolized by vertical parallel lines, such as $|\mathbf{A}|$. A determinant is a scalar derived from a polynomial of the elements of the matrix. It is as intuitively abstruse as it is mathematically powerful. A verbal definition of the determinant is "the sum of all the products that can be formed by taking as factors one element in succession from each row and column and giving to each product a positive or negative sign depending upon whether the number of permutations necessary to place the indices representing each factor's position in its row or column in the order of natural numbers is odd or even" (Webster's Dictionary 1969).

Tabachnik and Fidell give a more descriptive, but still non-intuitive, description: if the matrix is a variance-covariance matrix (variances in the diagonal, otherwise covariances), the determinant can be conceived as the variance of a matrix less the covariance. The result is sometimes called a "generalized variance."

For a 2 × 2 matrix, calculation is straightforward:

$$\begin{bmatrix} a_{11} & a_{12} \\ a_{21} & a_{22} \end{bmatrix} = a_{11}a_{22} - a_{12}a_{21}$$

For a 3 × 3 matrix:

$$\begin{bmatrix} a_{11} & a_{12} & a_{13} \\ a_{21} & a_{22} & a_{23} \\ a_{31} & a_{32} & a_{33} \end{bmatrix} = \begin{matrix} a_{11}a_{22}a_{33} + a_{12}a_{23}a_{31} + a_{13}a_{23}a_{32} \\ - a_{11}a_{23}a_{32} - a_{12}a_{21}a_{33} - a_{13}a_{22}a_{31} \end{matrix}$$

For higher dimension matrices, calculation becomes more complicated, involving cofactors of each element in a row. The determinant consists of $n!$ terms. Thus a 7 × 7 matrix has 5,040 terms in its determinant. But the determinant is critical to the simultaneous solution of many equations. Note that a matrix for which $|\mathbf{A}| = 0$ has no inverse.

2. **Eigenvalues** and **eigenvectors**. Eigenvalues (λ_i), also known as latent roots, of a square matrix ($n \times n$) are scalars such that

$$|\mathbf{A} - \lambda \mathbf{I}| = 0$$

There are n values of λ. Latent roots are comparable to the roots of a quadratic equation, but they can pertain to polynomials of a higher degree than two. For each eigenvalue λ_i, there is an eigenvector, also known as a latent vector, $\mathbf{y}_i$, such that

$$[\mathbf{A} - \lambda_I \mathbf{I}]\mathbf{y}_i = \mathbf{0}$$

See Chapter 14 on principal components analysis for more on eigenvalues and eigenvectors. They are important in many multivariate methods.

An application

Now let's do something useful, combining several operations of matrix algebra. Calculate the multiple regression of **y** on two independent variables (stored in **X**) with a sample size of n. The first column of **X**, an $n \times 3$ matrix, is filled with ones. It is used to calculate the intercept. The column vector **y** contains the n observations of the dependent variable. The equation for calculating the coefficients in the least-squares regression equation (call it **B** for betas) is given below. The resulting vector **B** contains the regression coefficients: b_1 = intercept, b_2 = coefficient for $\mathbf{x}_2$, and b_3 = coefficient for $\mathbf{x}_3$.

$$(\mathbf{X'X})^{-1}\mathbf{X'y} = \mathbf{B}$$

Some folks love this economy of expression. Once you have obtained **B**, you can solve the equation for a particular set of observations, represented by the vector **z**, with the equation:

$$\hat{y} = \mathbf{zB}$$

Bon voyage!

References

Abbott, E. A. 1950. Flatland; a romance of many dimensions. Blackwell, Oxford.

Affourtit, J., J. P. Zehr & H. W. Paerl. 2001. Distribution of nitrogen-fixing microorganisms along the Neuse River estuary, North Carolina. Microbial Ecology 41:114-123.

Allen, T. F. H., S. M. Bartell & J. F. Koonce. 1977. Multiple stable configurations in ordination of phytoplankton community change rates. Ecology 58:1076-1084.

Allen, T. F. H. & J. F. Koonce. 1973. Multivariate approaches to algal stratagems and tactics in systems analysis of phytoplankton. Ecology 54:1234-1246.

Arbuckle, J. L. 1997. Amos User's Guide. Smallwaters Press, Chicago.

Anderson, A. J. B. 1971. Ordination methods in ecology. Journal of Ecology 59:713-726.

Anderson, M. J. 2001. A new method for non-parametric multivariate analysis of variance. Austral Ecology 26:32-46.

Andersson, P.-A. 1988. Ordination and classification of operational geographic units in southwest Sweden. Vegetatio 74:95-106.

Andrew, N. L. & B. D. Mapstone. 1987. Sampling and the description of spatial pattern in marine ecology. Oceanography & Marine Biology, Annual Review. 25:39-90.

Angermeier, P. L. & R. A. Smogor. 1995. Estimating number of species and relative abundances in stream-fish communities: effects of sampling effort and discontinuous spatial distributions. Canadian Journal of Fisheries and Aquatic Sciences 52:936-949.

Arkin, A., P. Shen & J. Ross. 1997. A test case of correlation metric construction of a reaction pathway from measurements. Science 277:1275-1279.

Armitage, P. D., D. Moss, J. F. Wright & M. T. Furse. 1983. The performance of a new biological water quality score system based on macroinvertebrates over a wide range of unpolluted running-water sites. Water Research 17:333-347.

Arnott, S. E. & M. J. Vanni. 1993. Zooplankton assemblages in fishless bog lakes: influence of biotic and abiotic factors. Ecology 74:2361-2380.

Arrhenius, O. 1921. Species and area. Journal of Ecology 9:95-99.

Auclair, A. N. & F. G. Goff. 1971. Diversity relations of upland forests in the western Great Lakes area. American Naturalist 105:499-528.

Austin, M. P. 1968. An ordination study of a chalk grassland community. Journal of Ecology 56:739-57.

Austin, M. P. 1972. Models and analysis of descriptive vegetation data. Proceedings of the 12th Symposium of the British Ecological Society, pp. 61-86.

Austin, M. P. 1977. Use of ordination and other multivariate descriptive methods to study succession. Vegetatio 35:165-175.

Austin, M. P. 1979. Current approaches to the non-linearity problems in vegetation analysis. Pp. 197-210 in G. P. Patil & M. L. Rosenzweig (eds.), Contemporary quantitative ecology and related ecometrics. International Co-operative Publishing House, Fairland, Maryland.

Austin, M. P. 1981. Permanent quadrats: An interface for theory and practice. Vegetatio 46:1-10.

Austin, M. P. 1987. Models for the analysis of species response to environmental gradients. Vegetatio 69:35-45.

Austin, M. P. & L. Belbin. 1982. A new approach to the species classification problem in floristic analysis. Australian Journal of Ecology 7:75-89.

Austin, M. P. & R. B. Cunningham. 1981. Observational analysis of environmental gradients. Proceedings of the Ecological Society of Australia. 11:109-119.

Austin, M. P., R. B. Cunningham & P. M. Flemming. 1984. New approaches to direct gradient analysis using environmental scalars and statistical curve-fitting procedures. Vegetatio 55:11-27.

Austin, M. P. & P. Greig-Smith. 1968. The application of quantitative methods to vegetation survey. II. Some methodological problems of data from rain forest. Journal of Ecology 56:827-844.

Austin, M. P., A. O. Nicholls & C. R. Margules. 1990. Measurement of the realized qualitative niche: environmental niches of five *Eucalyptus* species. Ecological Monographs 60:161-177.

Austin, M. P., A. O. Nicholls, M. D. Doherty & J. A. Meyers. 1994. Determining species response functions to an environmental gradient by means of a b-function. Journal of Vegetation Science 5:215-228.

Austin, M. P. & I. Noy-Meir. 1971. The problem of non-linearity in ordination: experiments with two-gradient models. Journal of Ecology 59:763-774.

Austin, M. P. & L. Orloci. 1966. Geometrical models in ecology. II. An evaluation of some ordination techniques. Journal of Ecology 54:217-227.

Ayyad, M. A. & R. E. M. El-Ghareeb. 1982. Salt marsh vegetation of the western Mediterranean desert of Egypt. Vegetatio 49:3-19.

Batcheler, C. L. 1971. Estimation of density from a sample of joint point and nearest-neighbor distances. Ecology 52:703-709.

Beals, E. W. 1960. Forest bird communities in the Apostle Islands of Wisconsin. Wilson Bulletin 72:156-181.

Beals, E. W. 1965. Species pattern in a Lebanese Poterietum. Vegetatio 13: 69-87.

Beals, E. W. 1969. Vegetational change along altitudinal gradients. Science 165:981-985.

Beals, E. W. 1973. Ordination: mathematical elegance and ecological naivete. Journal of Ecology 61:23-35.

Beals, E. W. 1984. Bray-Curtis ordination: an effective strategy for analysis of multivariate ecological data. Advances in Ecological Research 14:1-55.

Beers, T. W., P. E. Dress & L. C. Wensel. 1966. Aspect transformation in site productivity research. Journal of Forestry 64:691.

Belbin, L. & C. McDonald. 1993. Comparing three classification strategies for use in ecology. Journal of Vegetation Science 4:341-348.

Bell, G. 2001. Neutral macroecology. Science 293:2413-2418.

Bellwood, D. R. & T. P. Hughes. 2001. Regional-scale assembly rules and biodiversity of coral reefs. Science 292:1532-1534.

Bentler, P. M. & E. J. C. Wu. 1996. EQS for Windows User's Guide. Multivariate Software, Encino, California.

Benzécri, J.-P. 1973. L'Analyse des Données. Vol. 1: La Taxinomie. Vol. 2: L'Analyse des Correspondance. Dunod, Paris.

Bernhard, A. E., Colbert, D., McManus, J. & Field, K. G. 2002. Microbial community dynamics based on 16S rDNA profiles in a Pacific Northwest estuary. Unpublished manuscript.

Berry, K. J., K. L. Kvamme & P. W. Mielke, Jr. 1983. Improvements in the permutation test for the spatial analysis of the distribution of artifacts into classes. American Antiquity 48: 547-553.

Biondini, M. E., C. D. Bonham & E. F. Redente. 1985. Secondary successional patterns in a sagebrush (*Artemisia tridentata*) community as they relate to soil disturbance and soil biological activity. Vegetatio 60: 25-36.

Biondini, M. E., P. W. Mielke, Jr. & K. J. Berry. 1988. Data-dependent permutation techniques for the analysis of ecological data. Vegetatio 75:161-168.

Bock, J. H., C. E. Bock & Randall J. Fritz. 1981. Biogeography of Illinois reptiles and amphibians: a numerical analysis. American Midland Naturalist 106:258-270.

Böhm, M., B. McCune & T. Vandetta. 1991. Diurnal curves of tropospheric ozone in the western United States. Atmospheric Environment 25A:1577-1590.

Böhm, M., B. McCune & T. Vandetta. 1995a. Ozone regimes in or near forests of the western United States. I. Regional patterns. Journal of the Air and Waste Management Association 45:477-489.

Böhm, M., B. McCune & T. Vandetta. 1995b. Ozone regimes in or near forests of the western United States. II. Factors influencing regional patterns. Journal of the Air and Waste Management Association 45:235-246.

Bollen, K. A. 1989. Structural Equations with Latent Variables. John Wiley and Sons, New York. 514 pp.

Boyce, R. L. & P. C. Ellison. 2001. Choosing the best similarity index when performing fuzzy set ordination on binary data. Journal of Vegetation Science 12:711-720.

Bradfield, G. E. & N. C. Kenkel. 1987. Nonlinear ordination using flexible shortest path adjustment of ecological distances. Ecology 68:750-753.

Brady, W. W., J. E. Mitchell, C. D. Bonham & J. W. Cook. 1995. Assessing the power of the point-line transect to monitor changes in plant basal cover. Journal of Range Management 48:187-190.

Bratton, S. P. 1975. A comparison of the beta diversity functions of the overstory and herbaceous understory of a deciduous forest. Bulletin of the Torrey Botanical Club 102:55-60.

Braun-Blanquet, J. 1965. Plant Sociology: The study of plant communities. Hafner, London. 439 pp. (Translation by G.D. Fuller & H.S. Conard).

Bray, J. R. 1955. The savanna vegetation of Wisconsin and an application of the concepts order and complexity to the field of ecology. Ph.D. Thesis. University of Wisconsin, Madison.

Bray, J. R. 1956. A study of mutual occurrence of plant species. Ecology 37: 21-28.

Bray, J. R. & J. T. Curtis. 1957. An ordination of the upland forest communities in southern Wisconsin. Ecological Monographs 27:325-349.

Brazner, J. C. & E. W. Beals. 1997. Patterns in fish assemblages from coastal wetland and beach habitats in Green Bay, Lake Michigan: a multivariate analysis of abiotic and biotic forcing factors. Canadian Journal of Fisheries and Aquatic Sciences 54:1743-1761.

Breiman, L., J. H. Friedman, R. A. Olshen & C. J. Stone. 1984. Classification and regression trees. Chapman & Hall, New York.

Briggs, J. M. & A. K. Knapp. 1991. Estimating biomass in tallgrass prairie with the harvest method: determining proper sample size using jackknifing and monte carlo simulations. Southwestern Naturalist 36:1-6.

Brown, T. C. & T. C. Daniel. 1990. Scaling of ratings: concepts and methods. USDA Forest Service Research Paper RM-293.

Brunner, I., S. Brudbeck, U. Büchler & C. Sperisen. 2001. Molecular identification of fine roots of trees from the Alps: reliable and fast DNA extraction and PCR-RFLP analysis of plastid DNA. Molecular Ecology 10:2079-2087.

Brzeziecki, B. 1987. Analysis of vegetation-environment relationships using a simultaneous equations model. Vegetatio 71:175-184.

Buckland, S. T., D. R. Anderson, K. P. Burnham & J. L. Laake. 1993. Distance Sampling. Chapman & Hall, New York. 446 pp.

Buffo, J., L. J. Fritschen & J. L. Murphy. 1972. Direct solar radiation on various slopes from 0 to 60 degrees north latitude. USDA Forest Service Research Paper PNW-142.

Burgman, M. A. 1987. An analysis of the distribution of plants on granite outcrops in southern western Australia using Mantel tests. Vegetatio 71:79-86.

Burnham, K. P. & W. S. Overton. 1979. Robust estimation of population size when capture probabilities vary among animals. Ecology 60:927-936.

Buys, M. H., J. S. Maritz, C. Boucher & J. J. A. Van Der Walt. 1994. A model for species-area relationships in plant communities. Journal of Vegetation Science 5:63-66.

Cao, Y., D. D. Williams & N. E. Williams. 1998. How important are rare species in aquatic community ecology and bioassessment? Limnology and Oceanography 43:1403-1409.

Carleton, T. J. 1980. Non-centered component analysis of vegetation data: a comparison of orthogonal and oblique rotation. Vegetatio 42:59-66.

Carleton, T. J. 1990. Variation in terricolous bryophyte and macrolichen vegetation along primary gradients in Canadian boreal forests. Journal of Vegetation Science 1:585-594.

Carleton, T. J., R. H. Stitt & J. Nieppola. 1995. Constrained indicator species analysis (COINSPAN): an extension of TWINSPAN. Journal of Vegetation Science 7:125-130.

Carnes, B. A. & N. A. Slade. 1982. Some comments on niche analysis in canonical space. Ecology 63:888-893.

Carpenter, S. R. 1990. Large-scale perturbations: opportunities for innovation. Ecology 71: 2038-2043.

Carpenter, S. R., T. M. Frost, D. Heisey & T. K. Kratz. 1989. Randomized intervention analysis and the interpretation of whole-ecosystem experiments. Ecology 70:1142-1152.

Causton, D. R. 1988. An Introduction to Vegetation Analysis: Principles, Practice and Interpretation. London: Unwin Hyman.

Cavigelli, M. A., G. P. Robertson & M. J. Klug. 1995. Fatty acid methyl ester (FAME) profiles as measures of soil microbial community structure. Plant Soil 170:99-113.

Chardy, P, M. Glemarec & A. Laurec. 1976. Application of inertia methods to benthic marine ecology: practical implications of the basic options. Estuarine and Coastal Marine Science 4:179-205.

Chardy, P, M. Glemarec & A. Laurec. 1976. Application of inertia methods to benthic marine ecology: practical implications of the basic options. Estuarine and Coastal Marine Science. 4:179-205.

Cheal, P, J. A. Davis, J. E. Growns, J. S. Bradley & F. H. Whittles. 1993. The influence of sampling method on the classification of wetland macroinvertebrate communities. Hydrobiologia 257:47-56.

Clarke, K. R. 1993. Non-parametric multivariate analyses of changes in community structure. Australian Journal of Ecology 18:117-143.

Clarke, K. R. & M. Ainsworth. 1993. A method of linking multivariate community structure to environmental variables. Marine Ecology Progress Series 92:205-219.

Clarke, K. R. & R. H. Green. 1988. Statistical design and analysis for a 'biological effects' study. Marine Ecology Progress Series 46:213-226.

Clay, K. & J. Holah. 1999. Fungal endophyte symbiosis and plant diversity in successional fields. Science 285:1742-1744.

Clements, F. E. 1905. Research methods in ecology. University Publishing Co., Lincoln, Nebraska.

Clymo, R. S. 1980. Preliminary survey of the peat-bog Hummel Knowe Moss using various numerical methods. Vegetatio 42:129-148.

Coddington, J. A., L. H. Young, & F. A. Coyle. 1996. Estimating spider species richness in a southern Appalachian cove hardwood forest. Journal of Arachnology 24:111-128.

Cohen, J. 1988. Statistical Power Analysis for the Behavioral Sciences. 2nd Edition. Lawrence Erlbaum Assoc., Hillsdale, New Jersey. 567 pp.

Colwell, R. K. & J. A. Coddington. 1994. Estimating terrestrial biodiversity through extrapolation. Philosophical Transactions of the Royal Society of London (Series B) 345:101-118.

Condit, R., N. Pitman, E. G. Leigh, Jr., J. Chave, J. Terborgh, R. B. Foster, P. Núnez V., S. Aguilar, R. Valencia, G. Villa, H. C. Muller-Landau, E. Losos & S. P. Hubbell. 2002. Beta-diversity in tropical forest trees. Science 295:666-669.

Connell, J. H. 1978. Diversity in tropical rain forests and coral reefs. Science 199:1302-1309.

Connor, E. F. & E. D. McCoy. 1979. The statistics and biology of the species-area relationship. American Naturalist 113:7910833.

Connor, E. F. & D. Simberloff. 1979. The assembly of species communities: chance or competition? Ecology 60:1132-1140.

Cornelius, J. M. & J. F. Reynolds. 1991. On determining the statistical significance of discontinuities within ordered ecological data. Ecology 72:2057-2070.

Cottam, G. & J. T. Curtis. 1956. The use of distance measures in phytosociological sampling. Ecology 37:451-460.

Crovello, T. J. 1981. Quantitative biogeography: an overview. Taxon 30:563-575.

Curtis, J. T. 1959. The Vegetation of Wisconsin. University of Wisconsin Press, Madison.

Curtis, J. T. & R. P. McIntosh. 1951. An upland forest continuum in the prairie-forest border region of Wisconsin. Ecology 32:476-496.

Czekanowski, J. 1913. Zarys Metod Statystiycnck. E. Wendego, Warsaw.

Dale, M. R. T. 1986. Overlap and spacing of species' ranges on an environmental gradient. Oikos 47:303-308.

Dale, M. R. T. 1988. The spacing and intermingling of species boundaries on an environmental gradient. Oikos 53:351-356.

Dargie, T. C. D. 1984. On the integrated interpretation of indirect site ordinations: a case study using semi-arid vegetation in southeastern Spain. Vegetatio 55:37-55.

Dargie, T. C. D. 1986. Species richness and distortion in reciprocal averaging and detrended correspondence analysis. Vegetatio 65:95-98.

Daubenmire, R. 1959. A canopy coverage method of vegetation analysis. Northwest Science 33:43-64.

De'ath, G. 1999a. Extended dissimilarity: a method of robust estimation of ecological distances from high beta diversity data. Plant Ecology 144:191-199.

De'ath, G. 1999b. Principal curves: a new technique for indirect and direct gradient analysis. Ecology 80:2237-2253.

De'ath, G. 2002. Multivariate regression trees: a new technique for modeling species-environment relationships. Ecology 83:1105-1117.

de Swart, E. O. A. M., A. G. van der Valk, K. J. Koehler & A. Barendregt. 1994. Experimental evaluation of realized niche models for predicting responses of plant species to a change in environmental condition. Journal of Vegetation Science 5:541-552.

del Moral, R. 1980. On selecting indirect ordination methods. Vegetatio 42:75-84.

del Moral, R. & M. F. Denton. 1977. Analysis and classification of vegetation based on family composition. Vegetatio 42:75-84.

Delincé, J. 1986. Robust density estimation through distance measurements. Ecology 67:1576.

Derr, C. C. 1994. Lichen biomonitoring in southeast Alaska and western Oregon. M. S. Thesis, Oregon State University, Corvallis. 98 pp.

Dethier, M. N., E. S. Graham, S. Cohen & L. M. Tear. 1993. Visual versus random-point percent cover estimations: 'objective' is not always better. Marine Ecology Progress Series 96:93-100.

Deuser, R. D. & H. H. Shugart, Jr. 1979. Niche pattern in a forest-floor small-mammal fauna. Ecology 60:108-118.

Deuser, R. D. & H. H. Shugart, Jr. 1982. Reply to comments by Van Horne and Ford and by Carnes and Slade. Ecology 63:1174-1175.

Diaz, S., A. Acosta & M. Cabido. 1992. Morphological analysis of herbaceous communities under different grazing regimes. Journal of Vegetation Science 3:689-696.

Diaz, S. & M. Cabido. 1997. Plant functional types and ecosystem function in relation to global change. Journal of Vegetation Science 8:463-474.

Diaz, S., M. Cabido, M. Zak, E. Martinez Carretero & J. Aranibar. 1999. Plant functional traits, ecosystem structure and land-use history along a climatic gradient in central-western Argentina. Journal of Vegetation Science 10:651-660.

Diaz Barradas, M. C., M. Zunzunegui, R. Tirado, F. Ain-Lhout & F. Garcia Novo. 1999. Plant functional types and ecosystem function in Mediterranean shrubland. Journal of Vegetation Science 10:709-716.

Dietz, E. J. 1983. Permutation tests for association between two distance matrices. Systematic Zoology 32:21-26.

Dietz, H. & T. Steinlein. 1996. Determination of plant species cover by means of image analysis. Journal of Vegetation Science 7:131-136.

Dolédec, S. & D. Chessel. 1994. Co-inertia analysis: an alternative method for studying species-environment relationships. Freshwater Biology 31:277-294.

Douglas, M. E. & J. A. Endler. 1982. Quantitative matrix comparisons in ecological and evolutionary investigations. Journal of Theoretical Biology 99:777-795.

Dufrêne, M. & P. Legendre. 1997. Species assemblages and indicator species: the need for a flexible asymmetrical approach. Ecological Monographs 67:345-366.

Eberhardt, L. L. & J. M. Thomas. 1991. Designing environmental field studies. Ecological Monographs 61:53-73.

Efron, B. 1982. The jackknife, the bootstrap and other resampling plans. SIAM Publications, Philadelphia.

Efron, B. & R. Tibshirani. 1991. Statistical data analysis in the computer age. Science 253:390-395.

Ellis, R. J., I. P. Thompson & M. J. Bailey. 1995. Metabolic profiling as a means of characterizing plant-associated microbial communities. Microbial Ecology 16:9-18.

Elzinga, C. L. & A. G. Evenden. 1997. Vegetation monitoring: an annotated bibliography. USDA Forest Service, General Technical Report INT-GTR-352. 184 pp.

Emlen, J. T. 1972. Size and structure of a wintering avian community in southern Texas. Ecology 53:317-329.

Emlen, J. T. 1977. Land bird communities of Grand Bahama Island: the structure and dynamics of an avifauna. Ornithological Monographs 24:1-129.

Enright, N. J. 1982. Recognition of successional pathways in forest communities using size-class ordination. Vegetatio 48:133-140.

Everson, T. M., G. P. Y. Clarke & C. S. Everson. 1990. Precision in monitoring plant species composition in montane grasslands. Vegetatio 88:135-141.

Faith, D. P., P. R. Minchin & L. Belbin. 1987. Compositional dissimilarity as a robust measure of ecological distance. Vegetatio 69:57-68.

Faith, D. P. & R. H. Norris. 1989. Correlation of environmental variables with patterns of distribution and abundance of common and rare freshwater macroinvertebrates. Biological Conservation 50:77-98.

Falk, R. F. & N. B. Miller. 1992. A primer for soft modeling. The University of Akron Press, Akron, Ohio.

Fasham, M. J. R. 1977. A comparison of nonmetric multidimensional scaling, principal components and reciprocal averaging for the ordination of simulated coenoclines and coenoplanes. Ecology 58:551-561.

FAUNMAP Working Group. 1996. Spatial response of mammals to late Quaternary environmental fluctuations. Science 272:1601-1606.

Federal Interagency Committee for Wetland Delineation. 1989. Federal manual for identifying and delineating jurisdictional wetlands. U.S. Army Corps of Engineers, U.S. Environmental Protection Agency, U.S. Fish and Wildlife Service, and U.S.D.A. Soil Conservation Service, Washington, D.C. Cooperative technical publication. 76 pp. plus appendices.

Feigen, P. D. & A. Cohen. 1978. On a model for concordance between judges. Journal of the Royal Statistical Society B 12:171-181.

Feoli, E. 1977. On the resolving power of principal component analysis in plant community ordination. Vegetatio 33:119-125.

Feoli, E. & L. Orloci. 1985. Species dispersion profiles of anthropogenic grasslands in the Italian eastern pre-Alps. Vegetatio 60:113-118.

Feoli, E. & M. Scimone. 1984. A quantitative view of textural analysis of vegetation and examples of application of some methods. Arch. Bot. Biogeogr. Ital. 60:72-94.

Fewster & L. Orloci. 1983. On choosing a resemblance measure for non-linear predictive ordination. Vegetatio 54:27-35.

Field, J. G., K. R. Clarke & R. M. Warwick. 1982. A practical strategy for analysing multispecies distribution patterns. Marine Ecology-Progress Series 8:37-52.

Field, J. G. & G. McFarlane. 1968. Numerical methods in marine ecology. I. A quantitative "similarity" analysis of rocky shore samples in False Bay, South Africa. Zoologica Afrricana 3:119-137.

Fischer, M. & J. Stöcklin. 1997. Local extinctions of plants in remnants of extensively used calcareous grasslands 1950-1985. Conservation Biology 11:727-737.

Fisher, R. A. 1940. The precision of discriminant functions. Annals of Eugenics. London 10:422-429.

Floyd, D. A. & J. E. Anderson. 1987. A comparison of three methods for estimating plant cover. Journal of Ecology 75:221-228.

Forman, R. T. T. 1969. Comparison of coverage, biomass and energy as measures of standing crop of bryophytes in various ecosystems. Bull. Torrey Bot. Club 96:582-591.

Fortier, J. J. 1966. Simultaneous linear prediction. Psychometrika 31:369-381.

Foster, M. S., C. Harrold & D. D. Hardin. 1991. Point vs. photo quadrat estimates of the cover of sessile marine organisms. Journal of

Experimental Marine Biology and Ecology. 146:193-203.

Franklin, J. 1998. Predicting the distribution of shrub species in southern California from climate and terrain-derived variables. Journal of Vegetation Science 9:733-648.

Franklin, S. B., D. J. Gibson, P. A. Robertson, J. T. Pohlman & J. S. Fralish. 1995. Parallel analysis: a method for determining significant principal components. Journal of Vegetation Science 6:99-106.

Franquet, E., S. Dolédec & D. Chessel. 1995. Using multivariate analyses for separating spatial and temporal effects within species-environment relationships. Hydrobiologia 300/301:425-531.

Fresco, L. F. M. 1972. A direct quantitative analysis of vegetational boundaries and gradients. Pp. 99-111 in E. van der Maarel & R.Tuxen, eds: Grundfragen und Methoden in der Pflanzensoziologie, Berichte des Symposiums der Internationalen Vereinigung für Vegetationskde, Rinteln 1970. The Hague.

Friedel, M. H. & K. Shaw. 1987a. Evaluation of methods for monitoring sparse patterned vegetation in arid rangelands. I. Herbage. Journal of Environmental Management 25:297-308.

Friedel, M. H. & K. Shaw. 1987b. Evaluation of methods for monitoring sparse patterned vegetation in arid rangelands. II. Trees and shrubs. Journal of Environmental Management 25:309-318.

Frontier, S. 1976. Étude de la decrosissance des valeurs propres dans une analyze en composantes principales: comparison avec le modèle de baton brisé. Journal of Experimental Marine Biology & Ecology 25:67-75.

Fulton, M. R. 1996. The digital stopwatch as a source of random numbers. Bulletin of the Ecological Society of America 77:217-218.

Gabriel, K. R. 1971. The biplot graphical display of matrices with application to principal component analysis. Biometrika 58:453-467.

Garland, J. L. & A. L. Mills. 1991. Classification and characterization of heterotrophic microbial communities on the basis of patterns of community-level sole-carbon source utilization. Applied and Environmental Microbiology 57:2351-2359.

Gauch, H. G., Jr. 1973. The relationship between sample similarity and ecological distance. Ecology 54:618-622.

Gauch, H. G., Jr. 1977. ORDIFLEX: a flexible computer program for four ordination techniques. Cornell University, Department of Ecology and Systematics.

Gauch, H. G., Jr. 1982a. Multivariate Analysis in Community Ecology. New York: Cambridge Univ. Press.

Gauch, H. G., Jr. 1982b. Noise reduction by eigenvector ordination. Ecology 63:1643-1649.

Gauch, H. G., Jr. & W. M. Scruggs. 1979. Variants of polar ordination. Vegetatio 40:147-153.

Gauch, H. G. & R. H. Whittaker. 1972. Coenocline simulation. Ecology 53:446-451.

Gauch, H. G. & R. H. Whittaker. 1981. Hierarchical classification of community data. Journal of Ecology 69:537-557.

Gauch, H. G., Jr. & T. R. Wentworth. 1976. Canonical correlation analysis as an ordination technique. Vegetatio 33:17-22.

Gauch, H. G., Jr., R. H. Whittaker & T. R. Wentworth. 1977. A comparative study of reciprocal averaging and other ordination techniques. Journal of Ecology 65:157-174.

Gehlbach, F. R. 1988. Avian biotic provinces of the Texas-Mexican borderlands: new techniques for synthetic resource assessment and mapping. Southwestern Naturalist 33:129-136.

Gerdol, R., C. Ferrari & F. Piccoli. 1985. Correlation between soil characters and forest types: a study in multiple discriminant analysis. Vegetatio 60:49-56.

Gilbert, G. S., J. L. Parke, M. K. Clayton & J. Handselman. 1993. Effects of an introduced bacterium on bacterial communities on roots. Ecology 74:840-854.

Gittins, R. 1965. Multivariate approaches to a limestone grassland community. III. A comparative study of ordination and association analysis. Journal of Ecology 53:411-425.

Gleason, H. A. 1922. On the relation between species and area. Ecology 3:158-162.

Gnanadesikan, R. & M. B. Wilk. 1969. Data analytic methods in multivariate statistical analysis. Pp. 593-638 in Krishnaiah, P. R., ed., Multivariate Analysis II. Proceedings of the 2nd International Symposium on Multivariate Analysis, Wright State University, Dayton, Ohio, June 17-22, 1968. Academic Press, New York.

Goldsmith, F. B. 1973. The vegetation of exposed sea cliffs at South Stock, Anglesey. I. The multivariate approach. Journal of Ecology 61:787-818.

Goldstein, R. A. & D. F. Grigal. 1972. Definition of vegetation structure by canonical analysis. Journal of Ecology 60:277-284.

Goodall, D. W. 1952. Some considerations in the use of point quadrats for the analysis of vegetation. Australian Journal of Science Research B5:1-41.

Goodall, D. W. 1953a. Objective methods for the classification of vegetation. I. The use of positive interspecific correlation. Australian Journal of Botany 1:39-63.

Goodall, D. W. 1953b. Point-quadrat methods for the analysis of vegetation. Australian Journal of Botany 1:457-461.

Goodall, D. W. 1954. Objective methods for the classification of vegetation. III. An essay in the use of factor analysis. Australian Journal of Botany 2: 304-324.

Goodall, D. W. 1970. Statistical plant ecology. Annual Review of Ecology and Systematics. 1:99-124.

Goodall, D. W. 1973a. Sample similarity and species correlation. Handbook of Vegetation Science 5:107-156.

Goodall, D. W. 1973b. Numerical classification. Handbook of Vegetation Science 5:575-615.

Gough, L. & J. B. Grace. 1999. Predicting effects of environmental change on plant species density: experimental evaluations in a coastal wetland. Ecology 80:882-890.

Gough, L., J. B. Grace, & K. L. Taylor. 1994. The relationship between species richness and community biomass: the importance of environmental variables. Oikos 70:271-279.

Gower, J. C. 1966. Some distance properties of latent root and vector methods used in multivariate analysis. Biometrika 53:325-338.

Gower, J. C. 1967. A comparison of some methods of cluster analysis. Biometrics 23:623-627.

Gower, J. C. 1975. Generalized Procrustes analysis. Psychometrika 40:33-51.

Gower, J. C. & D. J. Hand. 1996. Biplots. Chapman & Hall, London. 277 pp.

Grace, J. B. 1991. A clarification of the debate between Grime and Tilman. Functional Ecology 5:583-587.

Grace, J. B. 2001. The roles of community biomass and species pools in the regulation of plant diversity. Oikos 92:191-207.

Grace, J.B. & G. R. Guntenspergen. 1999. The effects of landscape position on plant species density: evidence of past environmental effects in a coastal wetland. Ecoscience 6:381-391.

Grace, J.B. & H. Jutila. 1999. The relationship between species density and community biomass in grazed and ungrazed coastal meadows. Oikos 85: 398-408.

Grace, J. B. & B. H. Pugesek. 1997. A structural equation model of plant species richness and its application to a coastal wetland. American Naturalist 149:436-460.

Grace, J. B. & B. H. Pugesek. 1998. On the use of path analysis and related procedures for the investigation of ecological problems. American Naturalist 152: 151-159.

Grace, J. B., L. Allain & C. Allen. 2000a. Plant species richness in a coastal tallgrass prairie: the importance of environmental effects. Journal of Vegetation Science 11:443-452.

Grace, J.B., L. Allain & C. Allen. 2000b. Vegetation associations in a rare community type - coastal tallgrass prairie. Plant Ecology 147:105-115.

Green, R.H. 1971. A multivariate statistical approach to the Hutchinsonian niche: bivalve mollusks of central Canada. Ecology 52:543-556.

Green, R. H. 1980. Multivariate approaches in ecology: the assessment of ecologic similarity. Annual Review Ecology & Systematics 11:1-14.

Greenacre, M. J. 1984. Theory and Applications of Correspondence Analysis. London: Academic Press.

Greenacre, M. J. 1988. Correspondence analysis of multivariate categorical data by weighted least-squares. Biometrics 75:457-467.

References

Greig-Smith, P. 1964. Quantitative Plant Ecology, 2nd ed. Butterworth, London. 256 pp.

Greig-Smith, P. 1983. Quantitative Plant Ecology, 3rd ed. Blackwell Scientific, Oxford. 359 pp.

Grigal, D. F. & R. A. Goldstein. 1971. An integrated ordination-classification analysis of an intensively sampled oak-hickory forest. Journal of Ecology 59:481-492.

Grime, J. P. 1979. Plant strategies and vegetation processes. John Wiley & Sons, London.

Grossman, G. D., D. M. Nickerson & M. C. Freeman. 1991. Principal component analyses of assemblage structure data: utility of tests based on eigenvalues. Ecology 72:341-347.

Guissan, A. & F. E. Harrell. 2000. Ordinal response regression models in ecology. Journal of Vegetation Science 11:617-626.

Gurevitch, J. & S. T. Chester, Jr. 1986. Analysis of repeated measures experiments. Ecology 67:251-255.

Hagmeier, E. M. & C. D. Stults. 1964. A numerical analysis of the distributional patterns of North American mammals. Systematic Zoology 13:125-155.

Hair, J. F., Jr., R. E. Anderson, R. L. Tatham & W. C. Black. 1995. Multivariate Data Analysis. 4th ed. New York: Prentice-Hall.

Hamann, A., Y. A. El-Kassaby, M. P. Koshy, & G. Namkoong. 1998. Multivariate analysis of allozymic and quantitative trait variation in *Alnus rubra*: geographic patterns and evolutionary implications. Canadian Journal of Forest Research 28:1557-1565.

Harries, J. M. & H. J. H. MacFie. 1976. The use of a rotational fitting technique in the interpretation of sensory scores for different characteristics. Journal of Food Technology 11:449-456.

Harte, J. & R. Shaw. 1995. Shifting dominance within a montane vegetation community: results of a climate-warming experiment. Science 267:876-880.

Hastie, T. J. & W. Stuetzle. 1989. Principal curves. Journal of the American Statistical Association 84:502-516.

Hatheway, W. H. 1971. Contingency-table analysis of rain-forest vegetation. Pp. 271-313 in G. P. Patil, E. C. Pielou & W. E. Waters (eds.), Statistical Ecology. Vol. 3, Pennsylvania State University Press, University Park.

Hayduk, L. A. 1987. Structural Equation Modeling with LISREL: Essentials and Advances. Johns Hopkins University Press, Baltimore.

Hayduk, L. A. 1996. LISREL: Issues, debates, and strategies. Baltimore, Maryland: Johns Hopkins University Press,

Hayek, L. & M. A. Buzas. 1997. Surveying natural populations. Columbia University Press, New York. 563 pp.

Heiser, W. J. & J. Meulman. 1983. Analyzing rectangular tables by joint and constrained multidimensional scaling. Journal of Econometrics 22:139-167.

Hellmann, J. J. & G. W. Fowler. 1999. Bias, precision, and accuracy of four measures of species richness. Ecological Applications 9:824-834.

Heltshe, J. F. & J. DiCanzio. 1985. Power study of jack-knifed diversity indices to detect change. Journal of Environmental Management 21:331-341.

Heltshe, J. F. & N. E. Forrester. 1983. Estimating species richness using the jackknife procedure. Biometrics 39:1-12.

Hermy, M. 1988. Accuracy of visual cover assessments in predicting standing crop and environmental correlation in deciduous forests. Vegetatio 75:57-64.

Hill, J. L., P. J. Curran & G. M. Foody. 1994. The effect of sampling on the species-area curve. Global Ecology and Biogeography Letters 4:97-106.

Hill, M. O. 1973a. Diversity and evenness: a unifying notation and its consequences. Ecology 54:427-432.

Hill, M. O. 1973b. Reciprocal averaging: an eigenvector method of ordination. Journal of Ecology 61: 237-249.

Hill, M. O. 1979a. DECORANA -- a FORTRAN program for detrended correspondence analysis and reciprocal averaging. Ecology and Systematics, Cornell University, Ithaca, New York. 52 pp.

Hill, M. O. 1979b. TWINSPAN--A FORTRAN program for arranging multivariate data in an ordered two-way table by classification of the individuals and attributes. Ithaca, NY: Ecology and Systematics, Cornell University.

Hill, M. O. & H. G. Gauch. 1980. Detrended correspondence analysis: an improved ordination technique. Vegetatio 42: 47-58.

Hirschfield, H. O. 1935. A connection between correlation and contingency. Proceedings of the Cambridge Philosophical Society 31:520-524.

Hobbs, E. R. 1986. Characterizing the boundary between California annual grassland and coastal sage scrub with differential profiles. Vegetatio 65:115-126.

Hopf, F. A. & J. H. Brown. 1986. The bull's eye method for testing randomness in ecological communities. Ecology 67:1139-1155.

Hotelling, H. 1933. Analysis of a complex of statistical variables into principal components. Journal of Experimental Psychology 24:417-441, 493-520.

Huisman, J., H. Olff & L. F. M. Fresco. 1993. A hierarchical set of models for species response analysis. Journal of Vegetation Science 4:37-46.

Hurlbert, S. H. 1971. The nonconcept of species diversity: a critique and alternative parameters. Ecology 52:577-586.

Hurlbert, S. H. 1984. Pseudoreplication and the design of ecological field experiments. Ecological Monographs 54:187-211.

Husch, B., C. I. Miller & T. W. Beers. 1972. Forest Mensuration. Second edition. John Wiley & Sons, New York. 410 pp.

Hutchinson, J. 2001. Rare riparian lichens of northern Idaho. M.S. Thesis. Oregon State University. 174 pp.

Ivimey-Cook, R. B. & M. C. F. Proctor. 1967. Factor analysis of data from an East Devon heath: a comparison of principal component and rotated solution. Journal of Ecology 55:405-413.

Jaccard, P. 1901. Étude comparative de la distribution florale dans une portion des Alpes et des Jura. Bull. Société Vaudoise des Sciences Naturelles 37:547-579.

Jackson, D. A. 1993. Stopping rules in principal components analysis: a comparison of heuristical and statistical approaches. Ecology 74:2204-2214.

Jackson, D. A. & K. M. Somers. 1991. Putting things in order: the ups and downs of detrended correspondence analysis. American Naturalist 137:704-712.

Jackson, D. A., K. M. Somers & H. H. Harvey. 1989. Similarity coefficients: measures of co-occurrence and association or simply measures of occurrence? American Naturalist 133:436-453.

Jeffers, J. N. R. 1978. An Introduction to Systems Analysis: with Ecological Applications. University Park Press, Baltimore.

Jensen, S. 1978. Influences of transformation of cover values on classification and ordination of lake vegetation. Vegetatio 48:47-59.

Johansson, P. & L. Gustafsson. 2001. Red-listed and indicator lichens in woodland key habitats and production forests in Sweden. Canadian Journal of Forest Research 31:1617-1628.

Johnson, M. L, D. G. Huggins & F. deNoyelles, Jr. 1991. Ecosystem modeling with LISREL: a new approach for measuring direct and indirect effects. Ecological Applications 4:383-398.

Johnson, R. K. 1998. Spatiotemporal variability of temperate lake macroinvertebrate communities: detection of impact. Ecological Applications 8:61-70.

Johnson, W. C., T. L. Sharik, R. A. Mayes & E. P. Smith. 1987. Nature and cause of zonation discreteness around glacial prairie marshes. Canadian Journal of Botany 65:1622-1632.

Jonasson, S. 1988. Evaluation of the point intercept method for the estimation of plant biomass. Oikos 52:101-106.

Jongman, R. H. G., C. J. F. ter Braak & O. F. R. van Tongeren. 1987. Data analysis in community and landscape ecology. Wageningen: Pudoc.

Jongman, R. H. G., C. J. F. ter Braak & O. F. R. van Tongeren. 1995. Data analysis in community and landscape ecology. Cambridge University Press, Cambridge.

Jöreskog, K. G. 1973. A general method for estimating a linear structural equation system. In: A. S. Goldberger, & O. D. Duncan (eds.), Structural Equation Models in the Social Sciences. pp. 85-112. New York : Academic Press.

Jöreskog, K. G. & D. Sörbom. 1996. *LISREL 8: User's reference guide*. Chicago Illinois: Scientific Software International.

Kantvilas, G. & P. R. Minchin. 1989. An analysis of epiphytic lichen communities in Tasmanian cool temperate rainforest. Vegetatio 84:99-112.

Karr, J. R. 1991. Biological integrity: a long-neglected aspect of water resource management. Ecological Applications 1:66-84.

Keesling, J. W. 1972. Maximum Likelihood Approaches to Causal Analysis. Ph.D. dissertation. Department of Education, Chicago, Illinois: University of Chicago.

Kendall, M. G. 1975. Rank correlation methods. Charles Griffin, London.

Kenkel, N. C. & C. E. Burchill. 1990. Rigid rotation of nonmetric multidimensional scaling axes to environmental congruence. Abstracta Botanica 14:109-119.

Kenkel, N. C. & G. E. Bradfield. 1981. Ordination of epiphytic bryophyte communities in a wet-temperate coniferous forest, south-coastal British Columbia. Vegetatio 45:147-154.

Kenkel, N. C. & L. Orloci. 1986. Applying metric and nonmetric multidimensional scaling to ecological studies: some new results. Ecology 67:919-928.

Kenkel, N. C. & J. Podani. 1991. Plot size and estimation efficiency in plant community studies. Journal of Vegetation Science 2:539-544.

Kennedy, K. A. & P. A. Addison. 1987. Some considerations for the use of visual estimates of plant cover in biomonitoring. Journal of Ecology 75:151-157.

Kent, M. & P. Coker. 1992. Vegetation Description and Analysis: a Practical Approach. Belhaven Press, London.

Kerans, B. L. & J. R. Karr. 1994. A benthic index of biotic integrity (B-IBI) for rivers of the Tennessee Valley. Ecological Applications 4:768-785.

Kercher, J. R. & R. A. Goldstein. 1977. Analysis of an east Tennessee oak-hickory forest by canonical correlation of species and environmental parameters. Vegetatio 35:153-163.

Kessell, S. R. 1979. Gradient Modeling: Resource and Fire Management. Springer-Verlag, New York. 432 pp.

Ketchledge, E. H. & R. E. Leonard. 1984. A 24-year comparison of the vegetation of an Adirondack mountain summit. Rhodora 86:439-444.

Kinkel, L. L., M. Wilson & S. E. Lindow. 1995. Effect of sampling scale on the assessment of epiphytic bacterial populations. Microbial Ecology 29:283-297.

Kirby, K. J., T. Bines, A. Burn, J. Mackintosh, P. Pitkin & I. Smith. 1966. Seasonal and observer differences in vascular plant records from British woodlands. Journal of Ecology 74:123-131.

Kleyer, M. 1999. Distribution of plant functional types along gradients of disturbance intensity and resource supply in an agricultural landscape. Journal of Vegetation Science 10:697-708.

Knox, R. G. 1989. Effects of detrending and rescaling on correspondence analysis: solution stability and accuracy. Vegetatio 83:129-236.

Knox, R. G. & R. K. Peet. 1989. Bootstrapped ordination: a method for estimating sampling effects in indirect gradient analysis. Vegetatio 80:153-165.

Krajina, V. J. 1933. Die Pflanzengesellschaften des Mlynica-Tales in den Vysoke Tatry (Hohe Tatra). Mit besonderer Brucksichtigung der okologischen Verhaltnisse. Botan. Centralb., Beigh., Abt. II, 50:774-957; 51:1-224.

Krebs, C. J. 1989. Ecological Methodology. Harper & Row, New York.

Kruskal, J. B. 1964a. Multidimensional scaling by optimizing goodness of fit to a nonmetric hypothesis. Psychometrika 29:1-27.

Kruskal, J. B. 1964b. Nonmetric multidimensional scaling: a numerical method. Psychometrika 29:115-129.

Kruskal, J. B. & J. D. Carroll. 1969. Geometrical models and badness of fit functions. Pp. 639-671 in Krishnaiah, P. K., ed., Multivariate Analysis II. Proceedings of the 2nd International Symposium on Multivariate Analysis, Wright State University, Dayton, Ohio, June 17-22, 1968. Academic Press, New York.

Kruskal, J. B. & M. Wish. 1978. Multidimensional Scaling. Sage Publications, Beverly Hills, California. 93 pp.

Lambert, J. M. & M. B. Dale. 1964. The use of statistics in phytosociology. Advances in Ecological Research 2:59-99.

Lance, G. N. & W. T. Williams. 1967. A general theory of classification sorting strategies. I. Hierarchical systems. Computer Journal 9:373-380.

Lance, G. N. & W. T. Williams. 1968. A general theory of classification sorting strategies. II. Clustering systems. Computer Journal 10:271-277.

Lange, R. T. & A. D. Sparrow. 1985. Moving analysis of interspecific associations. Australian Journal of Botany 33:639-644.

Landsberg, J., S. Lavorel & J. Stol. 1999. Grazing response groups among understorey plants in arid rangelands. Journal of Vegetation Science 10:683-696.

Lavorel, S., S. McIntyre & K. Grigulis. 1999. Plant response to disturbance in a Mediterranean grassland: how many functional groups? Journal of Vegetation Science 10:661-672.

Leathwick, J. R. & Austin M. P. 2000. Competitive interactions between tree species in New Zealand's old-growth indigenous forests. Ecology 82: 2560-2573.

LeBlanc, F. & J. de Sloover. 1970. Relation between industrialization and the distribution and growth of epiphytic lichens and mosses in Montreal. Canadian Journal of Botany 48:1485-1496.

Lechowicz, M. J. & G. R. Shaver. 1982. A multivariate approach to the analysis of factorial fertilization experiments in Alaskan Arctic tundra. Ecology 63:1029-1038.

Lee, Y.-W. & D. B. Sampson. 2000. Spatial and temporal stability of commercial groundfish assemblages off Oregon and Washington as inferred from Oregon trawl logbooks. Canadian Journal of Fisheries and Aquatic Science 57:2443-2454.

Lees, B. G. & K. Ritman. 1991. Decision-tree and rule induction approach applied to integration of remotely sensed and GIS data in mapping vegetation in disturbed or hilly environments. Environmental Management 15:823-831.

Lefkovitch, L. P. 1984. A nonparametric method for comparing dissimilarity matrices, a general measure of biogeographical distance, and their application. American Naturalist 123:484-499.

Legendre, P. & M. J. Anderson. 1999. Distance-based redundancy analysis: testing multispecies responses in multifactorial ecological experiments. Ecological Monographs 69:1-24.

Legendre, P., R. Galzin & M. L. Harmelin-Vivien. 1997. Relating behavior to habitat: solutions to the fourth-corner problem. Ecology 78:547-562.

Legendre, P. & L. Legendre. 1983. Numerical Ecology. Elsevier. Amsterdam, The Netherlands.

Legendre, P. & L. Legendre. 1998. Numerical Ecology, 2nd English edition. Elsevier Science BV. Amsterdam, The Netherlands. 853 pp.

Legendre, P. & A. Vaudor. 1991. The R Package: Multidimensional Analysis, Spatial Analysis. Department of Biological Sciences, University of Montreal, Montreal, Canada.

Lesica, P. & B. M. Steele. 1997. Use of permanent plots in monitoring plant populations. Natural Areas Journal 17:331-340.

Lesica, P., B. McCune, S. Cooper & W. S. Hong. 1991. Differences in lichen and bryophyte communities between old-growth and managed second-growth forests. Canadian Journal of Botany 69: 1745-1755.

Limpert, E., W. Stahel & M. Abbt. 2001. Log-normal distributions across the sciences: keys and clues. BioScience 51:341-352.

Lindsey, A. A., J. D. Barton, Jr. & S. R. Miles. 1958. Field efficiencies of forest sampling methods. Ecology 39:428-444.

Lively, C. M., P. T. Raimondi & L. F. Delph. 1993. Intertidal community structure: space-time interactions in the northern Gulf of California. Ecology 74:162-173.

Loucks, O. L. 1962. Ordinating forest communities by means of environmental scalars and phytosociological indices. Ecological Monographs 32:137-166.

Lougheed, V. L. & P. Chow-Fraser. 2002. Development and use of a zooplankton index of wetland quality in the Laurentian Great Lakes Basin. Ecological Applications 12:474-486.

References

Ludwig, J. A. & J. F. Reynolds. 1988. Statistical Ecology. John Wiley & Sons, New York. 337 pp.

Ludwig, J. A. & J. M. Cornelius. 1987. Locating discontinuities along ecological gradients. Ecology 68:448-450.

Maarel, E. van der. 1979. Transformation of cover-abundance values in phytosociology and its effects on community similarity. Vegetatio 39:97-114.

MacArthur, R. H. 1960. On the relative abundance of species. American Naturalist 25-36.

MacArthur, R. H. 1965. Patterns of species diversity. Biological Review 40:410-533.

MacArthur, R. H. & E. O. Wilson. 1967. The Theory of Island Biogeography. Princeton University Press, Princeton. 203 pp.

MacArthur, R. H. & J. W. MacArthur. 1961. On bird species diversity. Ecology 42:594-598.

MacNally, R. C. 1990. Modelling distributional patterns of woodland birds along a continental gradient. Ecology 71:360-374.

MacNaughton-Smith, P., W. T. Williams, M. B. Dale & L. G. Mockett. 1964. Dissimilarity analysis: a new technique of hierarchical subdivision. Nature 202:1034-1035.

Magurran, A. E. 1988. Ecological diversity and its measurement. Princeton University Press, Princeton, NJ.

Manly, B. F. J. 1995. A note on the analysis of species co-occurrences. Ecology 76:1109-1115.

Mankin, J.B., R.V. O'Neill, H.H. Shugart, & B.W. Rust. 1975. The importance of validation in ecosystem analysis. Pp. 63-71 in G. S. Innis, editor. New Directions in the Analysis of Ecological Systems, Part 1. Simulation Councils Proceedings Series, Simulation Councils, Inc., LaJolla, California.

Mann, L. K. & H. H. Shugart. 1983. Discriminant analysis of some east Tennessee forest herb niches. Vegetatio 52:77-89.

Mantel, N. 1967. The detection of disease clustering and generalized regression approach. Cancer Research 27:209-220.

Marcoulides, G. A. & R. E. Schumacker. 1996. Advanced Structural Equation Modeling. Lawrence Erlbaum Associates, Mahwah, New Jersey.

Marks, P. L. & P. A. Harcombe. 1981. Forest vegetation of the Big Thicket, southeast Texas. Ecological Monographs 51:287-305.

Mather, P. M. 1976. Computational methods of multivariate analysis in physical geography. J. Wiley & Sons, London. 532 pp.

Matthews, J. A. 1979. A study of the variability of some successional and climax plant assemblage-types using multiple discriminant analysis. Journal of Ecology 67:255-271.

May, R. M. 1975. Island biogeography and the design of wildlife preserves. Nature 254:177-178.

McArdle, B. H. & M. J. Anderson. 2000. Fitting multivariate methods to community data: a comment on distance-based redundancy analysis. Ecology 82:290-297.

McAuliffe, J. R. 1990. A rapid survey method for the estimation of density and cover in desert plant communities. Journal of Vegetation Science 1:653-656.

McCoy, E. D., S. S. Bell & K. Walters. 1986. Identifying biotic boundaries along environmental gradients. Ecology 67:749-759.

McCune, B. 1988. Ecological diversity in North American pines. American Journal of Botany 75:353-368.

McCune, B. 1990. Rapid estimation of abundance of branch epiphytes. Bryologist 93:39-43.

McCune, B. 1992. Components of error in predictions of species compositional change. Journal of Vegetation Science 3:27-34.

McCune, B. 1994. Improving community analysis with the Beals smoothing function. Ecoscience 1:82-86.

McCune, B. 1997. Influence of noisy environmental data on canonical correspondence analysis. Ecology 78:2617-2623.

McCune, B. & T. F. H. Allen. 1985. Will similar forests develop on similar sites? Canadian Journal of Botany 63:367-376.

McCune, B. & J. A. Antos. 1981. Diversity relationships of forest layers in the Swan Valley, Montana. Bulletin of the Torrey Botanical Club 108:354-361.

McCune, B. & E.W. Beals. 1993. History of the development of Bray-Curtis ordination. Pp. 67-79 in J. S. Fralish, R. P. McIntosh & O. L.

Loucks (eds). John T. Curtis: Fifty Years of Wisconsin Plant Ecology. Wisconsin Academy of Science, Arts & Letters, Madison, Wisconsin. 339 pp.

McCune, B. & W. J. Daly. 1993. Consumption and decomposition of lichen litter in a temperate coniferous rainforest. Lichenologist 26:67-71.

McCune, B., J. Dey, J. Peck, D. Cassell, K. Heiman, S. Will-Wolf & P. Neitlich. 1997a. Repeatability of community data: species richness versus gradient scores in large-scale lichen studies. Bryologist 100:40-46.

McCune, B, J. Dey, J. Peck, K. Heiman, S. Will-Wolf. 1997b. Regional gradients in lichen communities of the southeast United States. Bryologist 100:145-158.

McCune, B. & P. Lesica. 1992. The trade-off between species capture and quantitative accuracy in ecological inventory of lichens and bryophytes in forests in Montana. Bryologist 95:296-304.

McCune, B. & M. J. Mefford. 1995. PC-ORD. Multivariate Analysis of Ecological Data, Version 2.0. MjM Software Design, Gleneden Beach, Oregon.

McCune, B. & M. J. Mefford. 1997. PC-ORD. Multivariate Analysis of Ecological Data. Version 3.0. MjM Software, Gleneden Beach, Oregon, USA.

McCune, B. & M. J. Mefford. 1999. PC-ORD. Multivariate Analysis of Ecological Data. Version 4.0. MjM Software, Gleneden Beach, Oregon, USA.

McCune, B. & E. S. Menges. 1986. Quality of historical data on midwestern old-growth forests. American Midland Naturalist 116:163-172.

McCune, B., R. Rosentreter, J. M. Ponzetti & D. C. Shaw. 2000. Epiphyte habitats in an old conifer forest in western Washington, USA. Bryologist 103:417-427.

McLaughlin, S. P. 1989. Natural floristic areas of the western United States. Journal of Biogeography 16:239-248.

McQuitty, L. L. 1966. Similarity analysis by reciprocal pairs for discrete and continuous data. Educational and Psychological Measurement 26:825.

Meese, R. J. & P. A. Tomich. 1992. Dots on the rocks: a comparison of percent cover estimation methods. Journal of Experimental Marine Biology and Ecology 165:59-73.

Menges, E. S., W. G. Abrahamson, G. T. Givens, N. P. Gallo & J. N. Layne. 1993. Twenty years of vegetation change in five long-unburned Florida plant communities. Journal of Vegetation Science 4:575-586.

Mielke, P. W., Jr. 1979. On the asymptotic non-normality of null distributions of MRPP statistics. Communications in Statistics A5:1409-1424.

Mielke, P. W., Jr. 1984. Meteorological applications of permutation techniques based on distance functions. Pp. 813-830. In P. R. Krishnaiah & P. K. Sen, eds., Handbook of Statistics, Vol. 4. Elsevier Science Publishers.

Mielke, P. W., Jr. 1991. The application of multivariate permutation methods based on distance functions in the earth sciences. Earth-Science Reviews 31:55-71.

Mielke, P. W., Jr. & K. J. Berry. 1982. An extended class of permutation techniques for matched pairs. Communications in Statistics. Part A - Theory and Methods 11:1197-1207.

Mielke, P. W., Jr. & K. J. Berry. 2001. Permutation Methods: A Distance Function Approach. Springer Series in Statistics. 344 pp.

Mielke, P. W., Jr., K. J. Berry & E. S. Johnson. 1976. Multiresponse permutation procedures for a priori classifications. Communications in Statistics A5:1409-1424.

Mielke, P. W., Jr., K. J. Berry & G. W. Brier. 1981. Application of multi-response permutation procedures for examining seasonal changes in monthly sea-level pressure patterns. Monthly Weather Review 109:120-126.

Mielke, P. W., K. J. Berry & J. G. Medina. 1982. Climax I and II: distortion resistant residual analyses. Journal of Applied Meteorology. 21:788-792.

Mielke, P. W. & H. K. Iyer. 1982. Permutation techniques for analyzing multiresponse data from randomized block experiments. Communications in Statistics. Part A - Theory and Methods 11:1427-1437.

Miles, J. 1979. Vegetation Dynamics. Chapman & Hall, London. 80 pp.

Minieka, E. 1978. Optimization algorithms for networks and graphs. Marcel Dekker, New York.

Minchin, P. R. 1987a. An evaluation of the relative robustness of techniques for ecological ordination. Vegetatio 69:89-107.

Minchin, P. R. 1987b. Simulation of multidimensional community patterns: towards a comprehensive model. Vegetatio 71:145-156.

Minchin, P. R. 1989. Montane vegetation of the Mt. Field Massif, Tasmania: a test of some hypotheses about properties of community patterns. Vegetatio 83:97-110.

Mitchell, J. E., Bartling, P. N. S. & O'Brien, R. 1988. Comparing cover-class macroplot data with direct estimates from small plots. American Midland Naturalist 120:70-78.

Mitchell, R. J. 1992. Testing evolutionary and ecological hypotheses using path analysis and structural equation modeling. Functional Ecology 6:123-129.

Mitchell, R. J. 1993. Path analysis: pollination. Pp. 211-231, *in* S. M. Scheiner and J. Gurevitch (eds.). Design and Analysis of Ecological Experiments. Chapman and Hall, New York.

Moore, D. M., B. G. Lee, & S.M. Davey. 1991. A new method for predicting vegetation distributions using decision tree analysis in a geographic information system. Environmental Management 15:59-71.

Moral, R. del. 1975. Vegetation clustering by means of isodata: revision by multiple discriminant analysis. Vegetatio 29:179-190.

Moral, R. del. & A. F. Watson. 1978. Gradient structure of forest vegetation in the central Washington Cascades. Vegetatio 38:29-48.

Morrison, R. G. & G. A. Yarranton. 1970. An instrument for rapid and precise point sampling of vegetation. Canadian Journal of Botany 48:293-297.

Moser, E. B., A. M. Saxton & S. R. Pezeshki. 1990. Repeated measures analysis of variance: application to tree research. Canadian Journal of Forest Research 20:524-535.

Mueller-Dombois, D. & H. Ellenberg. 1974. Aims and Methods of Vegetation Ecology. John Wiley & Sons, New York.

Muir, P. S. & B. McCune. 1987. Index construction for foliar symptoms of air pollution injury. Plant Disease 71:558-565.

Muir, P. S. & B. McCune. 1988. Lichens, tree growth, and foliar symptoms of air pollution: are the stories consistent? Journal of Environmental Quality 17:361-370.

Muir, P. S. & B. McCune. 1992. A dial quadrat for mapping herbaceous plants. Natural Areas Journal 12:136-138.

Myers, R. T., D. R. Zak, D. C. White & A. Peacock. Landscape-level patterns of microbial community composition and substrate use in upland forest ecosystems. Soil Science Society of America Journal 65:359-367.

Neilson, R. P. 1987. Biotic regionalization and climatic controls in western North America. Vegetatio 70:135-147.

Neitlich, P. & B. McCune. 1997. Hotspots of Epiphytic Lichen Diversity in Two Young Managed Forests. Conservation Biology 11:172-182.

Norris, J. M. & J. P. Barkham. 1970. A comparison of some Cotswold beechwoods using multiple-discriminant analysis. Journal of Ecology 58:603-619.

Noy-Meir, I. 1971. Multivariate analysis of the semi-arid vegetation in south-eastern Australia: nodal ordination by component analysis. Proceedings of the Ecological Society of Australia 6:159-193.

Noy-Meir, I. 1973a. Data transformations in ecological ordination. I. some advantages of non-centering. Journal of Ecology 61:329-341.

Noy-Meir, I. 1973b. Divisive polythetic classification of vegetation data by optimized division on ordination components. Journal of Ecology 61:753-760.

Noy-Meir & W. T. Williams. 1975. Data transformations in ecological ordination. II. On the meaning of data standardization. Journal of Ecology 63:779-800.

O'Brien, R. G. & M. K. Kaiser. 1985. MANOVA method for analyzing repeated measures designs: an extensive primer. Psychological Bulletin 97:316-333.

Økland, R. H. 1986. Rescaling of ecological gradients. I. Calculation of ecological distance between vegetation stands by means of their floristic composition. Nordic Journal of Botany 6:651-660.

Økland, R. H. 1992. Studies in SE Fennoscandian mires: relevance to ecological theory. Journal of Vegetation Science 3:279-284.

Økland, R. H. 1994. Patterns of bryophyte associations at different scales in a Norwegian boreal spruce forest. Journal of Vegetation Science 5:127-138.

Økland, R. H. 1996. Are ordination and constrained ordination alternative or complementary strategies in general ecological studies? Journal of Vegetation Science 7:289-292.

Økland, R. H. 1999. On the variation explained by ordination and constrained ordination axes. Journal of Vegetation Science 10:131-136.

Økland, R. H. & O. Eilertsen. 1994. Canonical correspondence analysis with variation partitioning: some comments and an application. Journal of Vegetation Science 5:117-126.

Økland, R. H., O. Eilertsen & T. Økland. 1990. On the relationship between sample plot size and beta diversity in boreal coniferous forests. Vegetatio 87:187-192.

Oksanen, J. 1983. Ordination of boreal heath-like vegetation with principal components analysis, correspondence analysis, and multidimensional scaling. Vegetatio 52:181-189.

Oksanen, J. 1985. Cluster seeking with non-centred component analysis and rotation in forested sand dune vegetation in Finland. Annales Botanici Fennici 22:263-273.

Oksanen, J. 1988. A note on the occasional instability of detrending in correspondence analysis. Vegetatio 74:29-32.

Oksanen, J. 1997. Why the beta function cannot be used to estimate skewness of species responses. Journal of Vegetation Science 8:147-152.

Oksanen, J., E. Läärä, P. Huttunen & J. Meriläinen. 1988. Estimation of pH optima and tolerances of diatoms in lake sediments by the methods of weighted averaging, least squares and maximum likelihood, and their use for the prediction of lake acidity. Journal of Paleolimnology 1:39-49.

Oksanen, J. & P. R. Minchin. 1997. Instability of ordination results under changes in input data order: explanations and remedies. Journal of Vegetation Science 8:447-454.

Oksanen, J. & T. Tonteri. 1995. Rate of compositional turnover along gradients and total gradient length. Journal of Vegetation Science 6:815-824.

Oksanen, L. 1976. On the use of the Scandinavian type class system in coverage estimation. Annales Botanici Fennici 13:149-153.

Olhorst, S. L., W. D. Liddell, R. J. Taylor & J. M. Taylor. 1988. Evaluation of reef census techniques. Vol. 2:319-324 *in* J. H. Choat et al. (eds.) Proceedings of the Sixth International Coral Reef Symposium, Townsville, Australia, August 1988.

Olhorst, S. L., W. D. Liddell, R. J. Taylor & J. M. Taylor. 1992. An evaluation of coral reef census strategies: implications for monitoring programs. Pacific Science 46:380 (abstract only).

Orloci, L. 1966. Geometric models in ecology. I. The theory and application of some ordination methods. Journal of Ecology 54:193-215.

Orloci, L. 1967a. An agglomerative method for classification of plant communities. Journal of Ecology 55:193-206.

Orloci, L. 1967b. Data centering: a review and evaluation with reference to component analysis. Systematic Zoology 16:208-212.

Orloci, L. 1974. Revisions for the Bray and Curtis ordination. Canadian Journal of Botany 52:1773-1776.

Orloci, L. 1975. Multivariate Analysis in Vegetation Research. The Hague: Junk.

Orloci, L. 1978. Multivariate Analysis in Vegetation Research. The Hague: Junk.

Osenberg, C. W., R. J. Schmitt, S. J. Holbrook, K. E. Abu-Saba & A. R. Flegal. 1994. Detection of environmental impacts: natural variability, effect size, and power analysis. Ecological Applications 4:16-30.

Palmer, M. W. 1990. The estimation of species richness by extrapolation. Ecology 71:1195-1198.

Palmer, M. W. 1991. Estimating species richness: the second-order jackknife reconsidered. Ecology 72:1512-1513.

Palmer, M. W. 1993. Putting things in even better order: the advantages of canonical correspondence analysis. Ecology 74:2215-2230.

Palmer, M. W. 1995. How should one count species? Natural Areas Journal 15:124-135.

Palmer, M. W. & P. S. White. 1994. Scale dependence and the species-area relationship. American Naturalist 144:717-740.

Pan, V. Y. 1998. Solving polynomials with computers. American Scientist 86:62-69.

Payandeh, B. & A. R. Ek. 1986. Distance methods and density estimators. Canadian Journal of Forest Research 16:918-924.

Pearson, K. 1901. On lines and planes of closest fit to systems of points in space. Philosophical Magazine, Sixth Series 2:559-572.

Peck, J. E. & B. McCune. 1998. Commercial moss harvest in northwestern Oregon: biomass and accumulation of epiphytes. Biological Conservation 86:299-305.

Peet, R. K. 1974. The measurement of species diversity. Annual Review of Ecology & Systematics 5:285-307.

Peet, R. K. 1988. Putting things in order: the advantages of detrended correspondence analysis. American Naturalist 131:924-934.

Peterman, R. M. 1991. The importance of reporting statistical power: the forest decline and acidic deposition example. Ecology 71:2024-2027.

Peters, R. H. 1991. A Critique for Ecology. Cambridge University Press, Cambridge. 366 pp.

Peterson, B. J. & 16 other authors. 1993. Biological responses of a tundra river to fertilization. Ecology 74:653-672.

Peterson, E. B. & B. McCune. 2001. Diversity and succession of epiphytic macrolichen communities in low-elevation managed conifer forests in western Oregon. Journal of Vegetation Science 12:511-524.

Philippi, T. E., P. M. Dixon & B. E. Taylor. 1998. Detecting trends in species composition. Ecological Applications 8:300-308.

Pickett, S. T. A. 1994. Ecological Understanding. San Diego, CA, Academic Press.

Pielou, E. C. 1966. Shannon's formula as a measure of specific diversity: its use and misuse. American Naturalist 100:463-465.

Pielou, E. C. 1969. An Introduction to Mathematical Ecology. New York: John Wiley & Sons.

Pielou, E. C. 1975. Ecological Diversity. J. Wiley & Sons, New York.

Pielou, E. C. 1977. The latitudinal spans of seaweed species and their patterns of overlap. Journal of Biogeography 4:299-311.

Pielou, E. C. 1984a. Probing multivariate data with random skewers: a preliminary to direct gradient analysis. Oikos 42:161-165.

Pielou, E. C. 1984b. The Interpretation of Ecological Data: A Primer on Classificatio and Ordination. John Wiley & Sons, New York.

Pillar, V. D. & L. Orloci. 1996. On randomization testing in vegetation science: multifactor comparisons of relevé groups. Journal of Vegetation Science 7:585-592.

Pillar, V. D. 1999a. On the identification of optimal plant functional types. Journal of Vegetation Science 10:631-640.

Pillar, V. D. 1999b. The bootstrapped ordination re-examined. Journal of Vegetation Science 10:895-902.

Podani, J. & T. A. Dickinson. 1984. Comparison of dendrograms: a multivariate approach. Canadian Journal of Botany 62:2765-2778.

Pollard, J. E. 1981. Investigator differences associated with a kicking method for sampling macroinvertebrates. Journal of Freshwater Ecology 1:215-224.

Ponzetti, J. M. & B. McCune. 2001. Biotic soil crusts of Oregon's shrub steppe: community composition in relation to soil chemistry, climate, and livestock activity. Bryologist 104:212-225.

Post, W. M. & J. D. Sheperd. 1974. Hierarchical Agglomeration. Department of Botany, University of Wisconsin, Madison (unpublished typescript).

Prentice, I. C. 1977. Non-metric ordination methods in ecology. Journal of Ecology 65:85-94.

Prentice, I. C. 1980. Vegetation analysis and order invariant gradient models. Vegetatio 42:27-34.

Press, W. H., B. P. Flannery, S. A. Teukolsky & W. T. Vetterling. 1986. Numerical Recipes.

The Art of Scientific Computing. Cambridge University Press, Cambridge. 818 pp.

Preston, F. W. 1948. The commonness and rarity of species. Ecology 29:254-283.

Proctor, J. R. 1966. Systematic Zoology 15:131.

Proctor, M. C. F. 1974. Ordination, classification and vegetational boundaries. Pp. 1-16 in W. H. Sommer & R. Tuxen, eds. Tatsachen und Probleme der Grenzen in der Vegetation. Berichte des Symposiums der Internationalen Vereinigung für Vegetationskde, Rinteln 1968. J. Cramer, Lehre.

Pugesek, B., A. Tomer & A. von Eye (eds.). 2002. Structural Equations Modeling: Applications in Ecological and Evolutionary Biology Research. Cambridge, UK: Cambridge University Press.

Puroleit, R. & S. P. Singh. 1985. Submerged macrophytic vegetation in relation to eutrophication level in Kuruaun Himilaya. Environmental Pollution (Series A) 39:161-173.

Purvis, A. & A. Hector. 2000. Getting the measure of biodiversity. Nature 405:212-219.

Qian, H., K. Klinka & X. Song. 1999. Cryptogams on decaying wood in old-growth forests of southern coastal British Columbia. Journal of Vegetation Science 10:883-894.

Ramsey, F. L. & C. P. Marsh. 1984. Diet dissimilarity. Biometrics 40:707-715.

Rao, C. R. 1964. The use and interpretation of principal components analysis in applied research. Sankya A 26: 329-358.

Rényi, A. 1961. On measures of entropy and information. Pp. 547-561 in J. Neyman (ed.), Proceedings of the fourth Berkeley symposium on mathematical statistics and probability. University of California Press, Berkeley.

Rexstad, E. A., D. D. Miller, C. H. Flather, E. M. Anderson, J. W. Hupp & D. R. Anderson. 1988. Questionable multivariate statistical inference in wildlife habitat and community studies. Journal of Wildlife Management 52:794-798.

Ritchie, N. J., M. E. Schutter, R. P. Dick & D. Myrold. 2000. Use of length heterogeneity PCR and fatty acid methyl ester profiles to characterize microbial communities in soil. Applied and Environmental Microbiology 66:1668-1675.

Roberts, D. W. 1986. Ordination on the basis of fuzzy set theory. Vegetatio 66:123-131.

Robertson, P. A. 1978. Comparison of techniques for ordinating and classifying old-growth floodplain forests in southern Illinois. Vegetatio 37:43-51.

Rohlf, F. J. & R. R. Sokal. 1995. Statistical Tables, 3rd ed. W. H. Freeman, New York.

Root, T. 1988a. Energy constraints on avian distributions and abundances. Ecology 69:330-339.

Root, T. 1988b. Environmental factors associated with avian distributional boundaries. Journal of Biogeography 15:489-505.

Rosenweig, M. L. 1995. Species Diversity in Space and Time. Cambridge University Press, Cambridge.

Roux, G. & M. Roux. 1967. A propos de quelques methodes de classification en phytosociologie. Revue de Statistique Applique 15:59-72.

Rowell, J. G. & D. E. Walters. 1976. Analysing data with repeated observations on each experimental unit. J. Agric. Sci. 87:423-432.

Roweis, S. T. & L. K. Saul. 2000. Nonlinear dimensionality reduction by locally linear embedding. Science 290:2323-2269.

Sanderson, J. G. 2000. Testing ecological patterns. American Scientist 88:332-339.

Sanderson, J. G., M. P. Moulton & R. G. Selfridge. 1998. Null matrices and the analysis of species co-occurrences. Oecologia 116:275-283.

SAS Institute, 1989. *SAS/STAT User's Guide*, Version 6, Fourth Edition, Volume 2, Cary, NC: SAS Institute Inc.

Savary, S., N. Fabellar, E. R. Tiongco & P. S. Teng. 1993. A characterization of rice tungro epidemics in the Phillipines from historical survey data. Plant Disease 77:376-382.

Scheines, R., P. Spirtes, C. Glymour, & C. Meek. 1994. TETRAD II User's Manual. Mahwah, New Jersey: Lawrence Erlbaum Associates, Publishers.

Schmid, W. D. 1965. Distribution of aquatic vegetation as measured by line intercept with scuba. Ecology 46:816-823.

Schmida, A. & R. H. Whittaker. 1981. Pattern and biological microsite effects in two shrub communities, southern California. Ecology 62:234-251.

Schoenly, K. G. & W. Reid. 1987. Dynamics of heterotrophic succession in carrion arthropod assemblages: discrete series or a continuum of change? Oecologia 73:192-202.

Schoenly, K. G. & W. Reid. 1989. Dynamics of heterotrophic succession in carrion revisited. A reply to Boulton and Lake (1988). Oecologia 79:140-142.

Schönemann, P. H. 1966. A generalized solution of the orthogonal procrustes problem. Psychometrika 31:1-10.

Schopp-Guth, A., D. Maas & J. Pfadenhauer. 1994. Influence of management on the seed production and seed bank of calcareous fen species. Journal of Vegetation Science 5:569-578.

Schumacker, R. E. and R. G. Lomax. 1996. A Beginner's Guide to Structural Equation Modeling. Mahwah, New Jersey: Lawrence Erlbaum Associates, Publishers.

Schumacker, R.E. & Marcoulides, G.A. (eds.) 1998. Interaction and Nonlinear Effects in Structural Equation Modeling. Mahwah, New Jersey: Lawrence Erlbaum Associates.

Schutter, M. E. & R. P. Dick. 2000. Comparison of fatty acid methyl ester (FAME) methods for characterizing microbial communities. Soil Science Society of America Journal 64:1659-1668.

Schutter, M. E. & R. P. Dick. 2001. Comparison of fatty acid methyl ester (FAME) methods for characterizing microbial communities. Soil Biology & Chemistry 33:1481-1491.

Seagle, S. W., R. A. Lancia & D. A. Adams. 1987. A multivariate analysis of rangewide red-cockaded woodpecker habitat. Journal of Environmental Management 25:45-56.

Sheldon, A. L. 1969. Equitability indices: dependence on the species count. Ecology 50:466-467.

Shepard, R. N. 1962a. The analysis of proximities: Multidimensional scaling with an unknown distance function. I. Psychometrika 27:125-139.

Shepard, R. N. 1962b. The analysis of proximities: Multidimensional scaling with an unknown distance function. II. Psychometrika 27: 219-246.

Shipley, B. 2000. Cause and Correlation in Biology. Cambridge University Press, Cambridge, UK.

Sibson, R. 1972. Order invariant methods for data analysis. Journal of the Royal Statistical Society B. 34:311-349.

Simpson, E. H. 1949. Measurement of diversity. Nature 163:688.

Smilauer, P. 1990. Program CANODRAW, version 2.10. Trebon, Czechoslovakia, 33 pp.

Smith, E. P. 1998. Randomization methods and the analysis of multivariate ecological data. Environmetrics 9:37-51.

Smith, R. W., M. Bergen, S. B. Weisberg, D. Cadien, A. Dalkey, D. Montagne, J. K. Stull & R. G. Velarde. 2001. Benthic response index for assessing infaunal communities on the southern California mainland shelf. Ecological Applications 11:1073-1087.

Smith, S. D., S. C. Bunting & M. Hironaka. 1986. Sensitivity of frequency plots for detecting vegetation change. Northwest Science 60:279-286.

Smouse, P. E., J. C. Long & R. R. Sokal. 1986. Multiple regression and correlation extensions of the Mantel test of matrix correspondence. Systematic Zoology 35:627-632.

Sneath, P. H. A. & R. R. Sokal. 1973. Numerical Taxonomy: The Principles and Practice of Numerical Classification. W. H. Freeman & Co., San Francisco. 573 pp.

Sokal, R. R. 1979. Testing statistical significance of geographic variation patterns. Systematic Zoology 28:627-632.

Sokal, R. R. & C. D. Michener. 1958. A statistical method for evaluating systematic relationships. University of Kansas Science Bulletin 38:1409-1438.

Sokal, R. R. & F. J. Rohlf. 1969. Biometry. W.H. Freeman & Co., San Francisco

Sokal, R. R. & F. J. Rohlf. 1981. Biometry. W. H. Freeman & Co., New York.

Sokal, R. R. & F. J. Rohlf. 1995. Biometry. Third edition. W. H. Freeman & Co., New York.

Sokal, R. R. & P. H. A. Sneath. 1963. Principles of Numerical Taxonomy. Freeman.

Spirtes, P., C. Glymour & R. Scheines. 1993. Causation, Prediction, and Search. Springer-

Verlag Lecture Notes in Statistics 81, New York: Springer-Verlag.

Sprites, P., R. Scheines, C. Meek, & C. Glymour. 1994. TETRAD II Software. Mahwah, New Jersey: Lawrence Erlbaum Associates, Publishers.

Sprules, W. G. 1980. Nonmetric multidimensional scaling analyses of temporal variation in the structure of limnetic zooplankton communities. Hydrobiologia 69:139-146.

Sørensen, T. A. 1948. A method of establishing groups of equal amplitude in plant sociology based on similarity of species content, and its application to analyses of the vegetation on Danish commons. Biologiske Skrifter Kongelige Danske Videnskabernes Selskab 5:1-34.

Stahl, D. A. 1995. Application of phylogenetically based hybridization probes to microbial ecology. Molecular Ecology 4:535-542.

Stampfli, A. 1991. Accurate determination of vegetational change in meadows by successive point quadrat analysis. Vegetatio 96:185-194.

Stauffer, D. F., E. O. Garton & R. K. Steinhorst. 1985. A comparison of principal components from real and random data. Ecology 66:1693-1698.

Stein, R. A. & J. A. Ludwig. 1979. Vegetation and soil patterns on a Chihuahuan Desert bajada. American Midland Naturalist 101:28-37.

Stephenson, N.L. 1990. Climatic control of vegetation distribution: the role of the water balance. American Naturalist 135:649-670.

Stephenson, N.L. 1998. Actual evapotranspiration and deficit: biologically meaningful correlates of vegetation distribution across spatial scales. Journal of Biogeography 25:855-870.

Stephenson, W., W. T. Williams & G. N. Lance. 1970. The macrobenthos of Moreton Bay. Ecological Monographs 40:459-494.

Stewart-Oaten, A. & W. W. Murdoch. 1986. Environmental impact assessment: "pseudoreplication" in time? Ecology 67:929-940.

Stewart-Oaten, A., J. R. Bence & C. W. Osenberg. 1992. Assessing effects of unreplicated perturbations: no simple solutions. Ecology 73:1396-1404.

Stohlgren, T. J., M. B. Falkner & L. D. Schell. 1995. A modified-Whittaker nested vegetation sampling method. Vegetatio 117:113-121.

Strahler, A. H. 1978. Response of woody species to site factors of slope angle, rock type, and topographic position in Maryland as evaluated by binary discriminant analysis. Journal of Biogeography 5:403-423.

Stroup, W. W. & J. Stubbendieck. 1983. Multivariate statistical methods to determine changes in botanical composition. Journal of Range Management 36:208-212.

Sugihara, G. 1980. Minimal community structure: an explanation of species abundance patterns. American Naturalist 116:770-786.

Swan, J. M. A. 1970. An examination of some ordination problems by use of simulated vegetation data. Ecology. 51:89-102.

Swartzman, G. L. & S. P. Kaluzny. 1987. Ecological Simulation Primer. MacMillan Publishing Co., New York, USA

Sykes, J.M., Horrill, A.D. & M.D. Mountford. 1983. Use of visual cover assessments as quantitative estimators of some British woodland taxa. Journal of Ecology 71:437-450.

Tabachnik, B.G. & L. S. Fidell. 1983. Using Multivariate Statistics. New York: Harper & Row.

Tabachnik, B. G. & L. S. Fidell. 1989. Using multivariate Statistics. 2nd ed. Harper & Row, New York. 746 pp.

Tabachnik, B. G. & L. S. Fidell. 1996. Using Multivariate Statistics. 3rd ed. Harper Collins, New York. 880 pp.

Tabachnik, B. G. & L. S. Fidell. 2001. Using Multivariate Statistics. 4th ed. Allyn & Bacon, Boston. 966 pp.

Tausch, R. J., D. A. Charlet, D. A Weixelman & D. C. Zamudio. 1995. Patterns of ordination and classification instability resulting from changes in input data order. Journal of Vegetation Science 6:897-902.

Taylor, D. W. 1977. Floristic relationships along the Cascade-Sierran axis. American Midland Naturalist 97:333-349.

Ten Berge, J. M. F. 1977. Orthogonal Procrustes rotation for two or more matrices. Psychometrika 42:267-276.

Tenenbaum, J. B., V. de Silva & J. C. Langford. 2000. A global geometric framework for nonlinear dimensionality reduction. Science 290:2319-2322.

ter Braak, C. J. F. 1985. Correspondence analysis of incidence and abundance data: properties in terms of a unimodal response model. Biometrics 41:859-873.

ter Braak, C. J. F. 1986. Canonical correspondence analysis: a new eigenvector technique for multivariate direct gradient analysis. Ecology 67:1167-1179.

ter Braak, C. J. F. 1988. CANOCO. Agricultural Mathematics Group. Technical Report LWA-88-02. Wageningen, Netherlands. 95 pp.

ter Braak, C. J. F. 1990. Update Notes: CANOCO Version 3.10. Agricultural Mathematics Group. Wageningen, Netherlands. 35 pp.

ter Braak, C. J. F. 1992. Multidimensional scaling and regression. Statistica Applicata 4:577-586.

ter Braak, C. J. F. 1994. Canonical community ordination. Part I: Basic theory and linear methods. Ecoscience 1:127-140.

ter Braak, C. J. F. 1995. Canonical correspondence analysis and related multivariate methods in aquatic ecology. Aquatic Sciences 57:255-289.

ter Braak, C. J. F. & C. W. N. Looman. 1994. Biplots in reduced-rank regression. Biom. J. 36:983-1003.

ter Braak, C. J. F. & C. W. N. Looman. 1996. Biplots in reduced-rank regression. Weighted averaging, logistic regression and the Gaussian response model. Vegetatio 65:3-11.

Thaler, G. R. & R. C. Plowright. 1973. An examination of the floristic zone concept with special reference to the northern limit of the Carolinian zone in southern Ontario. Canadian Journal of Botany 51:1765-1789.

Therneau, T. M., and E. J. Atkinson. 1997. An introduction to recursive partitioning using the RPART routines. Technical report, Mayo Foundation (available at http://www.mayo.edu/hsr/techrpt.html).

Thrush, S. F., R. D. Pridmore & J. E. Hewitt. 1994. Impacts on soft-sediment macrofauna: the effects of spatial variation on temporal trends. Ecological Applications 4:31-41.

Tilman, D. 1993. Species richness of experimental productivity gradients: how important is colonization limitation? Ecology 74:2179-2191.

Titus, K., J. A. Mosher & B. K. Williams. 1984. Chance-corrected classification for use in discriminant analysis: ecological applications. American Midland Naturalist 111:1-7.

Tuomisto, H., K. Ruokolainen, R. Kalliola, A. Linna, W. Danjoy & Z. Rodriguez. 1995. Dissecting Amazonian biodiversity. Science 269:63-66.

Turkington, R. & E. Klein. 1993. Interactive effects of nutrients and disturbance: an experimental test of plant strategy theory. Ecology 74:863-878.

Underwood, A. J. 1978. The detection of non-random patterns of distribution of species along a gradient. Oecologia 36:317-326.

Urban, D. L., C. Miller, N.L. Stephenson, & P.N. Halpin. 2000. Forest pattern in Sierran landscapes: the physical template. Landscape Ecology 15:603-620.

Urban, D., S. Goslee, K. Pierce, & T. Lookingbill. 2002. Extending community ecology to landscapes. Ecoscience (in press).

van den Wollenberg, A. L. 1977. Redundancy analysis — an alternative to canonical correlation analysis. Psychometrika 42:207-219.

van der Maarel, E. 1974. Small-scale vegetational boundaries: on their analysis and typology. Pp. 75-80 in W. H. Sommer & R. Tuxen, eds., Tatsachen und Probleme der Grenzen in der Vegetation. Berichte des Symposiums der Internationalen Vereinigung für Vegetationskde., Rinteln 1968. J. Cramer, Lehre.

van der Maarel, E. 1976. On the establishment of plant community boundaries. Ber. Deutsch. Bot. Ges. 89:415-443.

van Groenewoud, H. 1992. The robustness of correspondence, detrended correspondence and TWINSPAN analysis. Journal of Vegetation Science 3:239-246.

Van Horne, B. 1982. Niches of adult and juvenile deer mice (*Peromyscus maniculatus*) in seral stages of coniferous forest. Ecology 63:992-1003.

Van Horne, B. & R. G. Ford. 1982. Niche breadth calculations based on discriminant analysis. Ecology 63:1172-1174.

Van Sickle, J. 1997. Using mean similarity dendrograms to evaluate classifications. Journal of Agricultural, Biological & Environmental Statistics 2: 370–388.

van Wijngaarden, R. P. A., P. J. van den Brink, J. H. O Voshaar & P. Leeuwangh. 1995. Ordination techniques for analysing response of biological communities to toxic stress in experimental ecosystems. Ecotoxicology 4:61-77.

Vayssieres, M. P., R. E. Plant & B. H. Allen-Diaz. 2000. Classification trees: an alternative non-parametric approach for predicting species distributions. Journal of Vegetation Science 11:679-694.

Vellend, M. 2001. Do commonly used indices of β-diversity measure species turnover? Journal of Vegetation Science 12:545-552.

Venables, W. N. & B. D. Ripley. 1999. Modern Applied Statistics with S-plus (3rd edition). Springer, New York.

Verbyla, D. 1986. Potential prediction bias in regression and discriminant analysis. Canadian Journal of Forest Research 16:1255-1257.

Waichler, W. S., R. F. Miller, & P. S. Doescher. 2001. Community characteristics of old-growth western juniper woodlands. Journal of Range Management 54:518-527.

Walker, M. D., P. J. Webber, E. H. Arnold & D. Ebert-May. 1994. Effects of interannual climate variation on aboveground phytomass in alpine vegetation. Ecology 75:393-408.

Warncke, W. M., Jr. 1998. The species composition, density, and distribution of the littoral zooplankton assemblage in Crater Lake, Oregon. M.S. Thesis, Oregon State University, Corvallis.

Ward, J. H. 1963. Hierarchical grouping to optimise an objective function. Journal of the American Statistical Association 58:236-244.

Wartenberg, D., S. Ferson & F. J. Rohlf. 1987. Putting things in order: a critique of detrended correspondence analysis. American Naturalist 129:434-448.

Webb, L. J., T. G. Tracey, W. T. Williams & G. N. Lance. 1967. Studies in the numerical analysis of complex rainforest communities. I. A comparison of methods applicable to site/species data. Journal of Ecology 55:171-191.

Webster, R. 1973. Automatic soil-boundary location from transect data. Mathematical Geology 5:27-37.

Webster's Dictionary. 1969. Webster's Seventh New Collegiate Dictionary. G. & C. Merriam, Springfield, Mass.

Werker, A. G. & E. R. Hall. 2001. Quantifying population dynamics based on community structure fingerprints extracted from biosolids samples. Microbial Ecology 41:195-209.

Wesser, S. D. & W. S. Armbruster. 1991. Species distribution controls across a forest-steppe transition: a causal model and experimental test. Ecological Monographs 61:323-342.

West, N. E. & G. A. Reese. 1991. Comparison of some methods for collecting and analyzing data on aboveground net production and diversity of herbaceous vegetation in a northern Utah subalpine context. Vegetatio 96:145-163.

Westman, W. E. 1978. Measuring the inertia and resilience of ecosystems. BioScience 28:705-710.

Westman, W. E. 1981. Factors influencing the distribution of species of Californian coastal sage scrub. Ecology 62:439-455.

Whitman, W. C. & E. I Siggeirsson. 1954. Comparison of line interception and point contact methods in the analysis of mixed-grass vegetation. Ecology 35:431-436.

Whittaker, J. 1990. Graphical Models in Applied Multivariate Statistics. Wiley, New York.

Whittaker, R. H. 1954. Plant populations and the basis of plant identification. In: Augewandte Pflanzensoziologie, Veroffentlichungen des Karntner Landesinstituts für augewandte Pflanzensoziologie in Klagenfurt, Fesschrift Aichinger, Vol. 1.

Whittaker, R. H. 1956. Vegetation of the Great Smoky Mountains. Ecological Monographs 26:1-80.

Whittaker, R. H. 1960. Vegetation of the Siskiyou Mountains, Oregon and California. Ecological Monographs 30:279-338.

Whittaker, R. H. 1965. Dominance and diversity in land plant communities. Science 147:250-260.

Whittaker, R. H. 1967. Gradient analysis of vegetation. Biological Reviews 42:207-264.

Whittaker, R. H. 1972. Evolution and measurement of species diversity. Taxon 21:213-251.

Whittaker, R. H. 1973. Direct gradient analysis: techniques. Handbook of Vegetation Science 5:9-31.

Whittaker, R. H. & H. G. Gauch, Jr. 1973. Evaluation of ordination techniques. Handbook of Vegetation Science 5:287-321.

Whittaker, R. H., L. D. Gilbert & J. H. Connell. 1979. Analysis of a two-phase pattern in a mesquite grassland, Texas. Journal of Ecology 67:935-952.

Whittaker, R. H., W. A. Niering & M. D. Crisp. 1979. Structure, pattern, and diversity of a mallee community in New South Wales. Vegetatio 39:65-76.

Whittington, H. B. & C. P. Hughes. 1972. Ordovician geography and faunal provinces deduced from trilobite distribution. Philosophical Transactions of the Royal Society of London, Series B. 263:235-278.

Whysong, G. I. & W. H. Miller. 1987. An evaluation of random and systematic plot placement for estimating frequency. Journal of Range Management 40:475-479.

Wiederholm, T. 1980. Use of benthos in lake monitoring. Journal of the Water Pollution Control Federation 52:537-547.

Wiegleb, G. 1980. Some applications of principal components analysis in vegetation: ecological research of aquatic communities. Vegetatio 42:67-73.

Wiegleb, G. 1981. Application of multiple discriminant analysis on the analysis of the correlation between macrophyte vegetation and water quality in running waters of central Europe. Hydrobiologia 79:91-100.

Wiemken, V., E. Laczko, K. Ineichen & T. Boller. 2001. Effects of elevated carbon dioxide and nitrogen fertilization on mycorrhizal fine roots and the soil microbial community in beech-spruce ecosystems on siliceous and calcareous soil. Microbial Ecology 42:126-135.

Wierenga, P. J., J. M. H. Hendrickx, M. H. Nash, J. A. Ludwig & L. A. Daugherty. 1987. Variation of soil and vegetation with distance along a transect in the Chihuahuan Desert. Journal of Arid Environments Arid Zone Environments 13:53-64.

Wijngaarden, R. P. A., P. J. Van den Brink, J. H. O Voshaar & P. Leeuwangh. 1995. Ordination techniques for analysing response of biological communities to toxic stress in experimental ecosystems. Ecotoxicology 4:61-77.

Wiley, D. E. 1973. The identification problem for structural equation models with unmeasured variables. Pp. 69-83 in: A. S. Goldberger & O. D. Duncan (eds.), Structural Equation Models in the Social Sciences. Academic Press, New York.

Williams, B. K. 1983. Some observations on the use of discriminant analysis in ecology. Ecology 64:1283-1291.

Williams, B. K. & K. Titus. 1988. Assessment of sampling stability in ecological applications of discriminant analysis. Ecology 69:1275-1285.

Williams, W. T. & J. M. Lambert. 1959. Multivariate methods in plant ecology. I. Association analysis. Journal of Ecology 47:83-101.

Williams, W. T., J. M. Lambert & G. N. Lance. 1966. Multivariate methods in plant ecology. V. Similarity analysis and information-analysis. Journal of Ecology 54:427-445.

Williamson, M. H. 1978. The ordination of incidence data. Journal of Ecology 66:911-920.

Williamson, M. H. 1983. The land-bird community of Skokholm: ordination and turnover. Oikos 41:378-384.

Will-Wolf, S. 1975. Multivariate analysis of foraging site selection by flower-feeding insects in a western South Dakota prairie. Ph.D. thesis, University of Wisconsin, Madison.

Wilson, J. B. 1987. Methods for detecting non-randomness in species co-occurrence: a contribution. Oecologia 73:579-582.

Wilson, M. V. 1981. A statistical test of the accuracy and consistency of ordinations. Ecology 62:8-12.

Wilson, M. V. & C. L. Mohler. 1983. Measuring compositional change along gradients. Vegetatio 54:129-141.

Wilson, M. V. & A. Shmida. 1984. Measuring beta diversity with presence-absence data. Journal of Ecology 72:1055-1064.

Winkworth, R. E. & D. W. Goodall. 1962. A crosswire sighting tube for point quadrat analysis. Ecology 43:342-343.

Wishart, D. 1969. An algorithm for hierarchical classifications. Biometrics 25:165-170.

Wishart, D. 1970. CLUSTAN 1A USER MANUAL, Computing Laboratory, University of St. Andrews, St. Andrews, Scotland.

Wishart, D. 1978. CLUSTAN Users Manual, 3rd edition. Edinburgh: Edinburgh University.

Wolda, H. 1981. Similarity indices, sample size and diversity. Oecologia 50:296-302.

Wootton, J. T. 1994. Predicting direct and indirect effects: an integrated approach using experiments and path analysis. Ecology 75:151-165.

Wright, J. F., M. T. Furse & P. D. Armitage. 1993. Use of macroinvertebrate communities to detect environmental stress in running waters. Pp. 15-34 in D. W. Sutcliffe, ed., Water Quality and Stress Indicators in Marine and Freshwater Ecosystems: Linking Levels of Organisation (Individuals, Populations, Communities). Freshwater Biological Association 1994.

Wright, S. 1918. On the nature of size factors. Genetics 3:367-374.

Wunderle, J. M., Jr., A. Diaz, I Valazquez & R. Scharron. 1987. Forest openings and the distribution of understory birds in a Puerto Rican rainforest. Wilson Bulletin 99:22-37.

Yarranton, G. A. 1967. Organismal and individualistic concepts and the choice of methods of vegetation analysis. Vegetatio 15:113-116.

Yoccoz, N.G. 1991. Use, overuse, and misuse of significance tests in evolutionary biology and ecology. Bulletin of the Ecological Society of America 106-111.

Zak, J. C., M. R. Willig, D. L. Moorhead & H. G. Wildman. 1994. Functional diversity of microbial communities: a quantitative approach. Soil Biology and Biochemistry 26:1101-1108.

Zimmerman, G. M., H. Goetz & P. W. Mielke, Jr. 1985. Use of an improved statistical method for group comparisons to study effects of prairie fire. Ecology 66: 606-611.

Index

A

abundance
   dust bunny, 38-43
   concentration in groups, 198
   measures, 13-16
   species, 1, 35-44
abundant species
   effect on stress, NMS, 133
accuracy, 13, 18-21, 23
   classification, 226
   defined, 19
   diversity, 32
   example, 21
   ordination, 178-181
      Wilson's method, 180
      pattern, 180
      SEM, 235
      species ranks, 180
      Wilson's method, 181
addition
   matrices, 258
agglomeration, 86
agglomerative cluster analysis, 81, **86-96**
   compared to TWINSPAN, 100
aggregate sample units, 77
aggregate univariate responses, 183
aggregation, 16
agreement within-group, 191
alignment of blocks, 194
AMOS, 234
analysis
   association, 83
   behavioral, 4
   community, 4
   canonical correspondence, 164-177
   correspondence, 152-158
   detrended correspondence, 159-163
   discriminant, 205-210
   documenting, 62
   functional groups, 4
   habitat-centered, 12
   indicator species, 97-101, 198-204
   logs, 62, 65-66
      blank form, 66
      example, 65
   niche-space, 4
   normal, 5-7
   principal components, 114-121
   taxonomic, 4
   transpose, 5-7
analysis of similarity, 195
analysis of variance, 83, 85
   multivariate, 184 (see also MANOVA)
ANOSIM, 188, 193, 195
   compared with MRPP, 195
   homogeneity of variance, 197
ANOVA, nested, 44, 218-221
arc distance, 46, 49
arch problem, 176
   CA, 152
   CCA, 176
arcsine-squareroot transformation, 13-14, 69, 79
artifacts
   CA, 178
   DCA, 178
   methodological, 178
   ordination, 178
   RA, 178
aspect, 22, 24
   transformation, 22
ASPT, 149
assemblage, 2
association
   analysis, 83
   measures, 37
   negative, 37
   positive, 37
   species, 39
assumptions
   MRPP, 188
   transformations to help, 67
asymptotic vs. randomization method
   Mantel test, 214
attributes, 3, 5
   ecological, 4
augmenting a matrix
   defined, 257
autopilot
   NMS, 130, 135
average distance function
   commensuration, 194
Average Score Per Taxon, 149
averaging sample units, 77

B

BACI designs, 185
bacteria, 3-4, 183-184
Bartlett's test of sphericity, 119, 179
basal area data, 15-16, 18, 37, 73
BATC designs, 185
Beals smoothing, 65, 67-68, **70-71**
   example, 71
bedrock, 3
before-after-control-impact, 185
before-after data, 77
before-after-treatment-control, 185
behavior, 3
behaviorial analysis, 4
Benthic Quality Index, 149
Benthic Response Index, 149
best fit line, 114-115
beta diversity, **27-31**, 38, 50-51, 78
   axis length, 30
   DCA, 28
   difficulty of ordination, 31
   half changes, 31
   MRPP, 193
   use of PCA, 119
   usefulness, 28
beta functions, 37
beta turnover, 28, 30
bias, 13, 17, 19-21, 23
   defined, 19
   diversity, 32
   example, 21
   partial correlation in regression, 243
   prediction, DA, 207
   weighted averaging, 150
bimodal species distributions, 36
binary by mean, 68
binary data, (*see also* presence) 3, 51, 70
   correlation coefficients, 108
   distances, 45-46
biodiversity, 3, 20
BIOLOG microplates, 4, 120
Biological Monitoring Working Party, 149
biomass, 13-14, 18, 25, 27, 43, 78, 185
   regression, 14
   transformation, 68
biplot, 109-110
   CCA, 109, 167
   example, 175
biplot scores
   CCA, 171-172
   example, 174, 175
bivariate distribution, 37, 42
   abundance, 37
   species abundance, 39
   scatterplot, 191
black spruce, 36
blocked design, 184, 194
blocked MRPP, 44, 185-186, 193-194
   example, 76, 195
blocks, 44, 184, 194-195
   alignment, 194
   example, 195
   distance between, 194
   MRPP, 193-194
BMWP, 149

Index

bootstrap
  CART, 226
  defined, 180
  parallel, 180
  PCA, 119
  sample, 180
  species richness, 32
boundaries
  community, 28
BQI, 149
Braun-Blanquet school, 14, 17
Bray-Curtis distance, 46, 48, 54
Bray-Curtis ordination, 143-148
  development, 143
  distance measures, 143
  endpoint selection, 145
  examples, 144
  geometry, 145-146
  how it works, 145-146
  outliers, 145
  reference points, 145
  refinements, 143
  residual distance, 145, 147-148
  scores, 147-148
  specific hypotheses, 145
  variance represented, 145
  variance-regression endpoints, 146
  what to report, 147
  when to use it, 143
broken-stick model
  PCA, 119

C

CA, *see* correspondence analysis, 152-157
CALIS, 234
canonical analysis, 165, 205
canonical coefficients, 169
  CCA, 169
  example, 173
canonical correlation, 54, 121-122, 164-165, 176-177
  interpretation, 176
  on axis scores, 122
canonical correspondence analysis, 49, 54, 77, 103, **164-177**, 210
  appropriate questions, 164
  arch problem, 176
  axis scaling, 166-167
  axis summary statistics, 168, 173
    example, 175
  basic method, 166
  basic questions, 164
  biplot, 167
    scaling, 167
    scores, 167, 171
    example, 175
  canonical coefficients, 169
  choice of scores, 169
  compared to RDA, 177

  compared with NMS, 175-176
  convergence, 167
  decision tree, 165
  detrended, 176
  distance measure, 166, 168
  eigenvalue, 168
  environmental constraints, 164
  environmental space, 164, 175
  example, 172
  geographic coordinates, 213
  Hill's scaling, 166-167, 172
  how it works, 166, 172
  hypothesis test, 172
  inertia, 168
  interset correlation, 171
    example, 174
  intraset correlation, 171
    example, 174
  iteration report, 167
  iterative method, 166
  joint plot, 167
  LC scores, 166, 169-174
    example, 175
  Monte Carlo test, 172
    example, 175
  multiple regression, 164, 168
    example, 175
  multiple regression, 166
  noise, effect of, 169
  null hypotheses, 172
  number of axes, 168
  percent variance, spp-envt, 168
  questions, 165
  randomization test, 172, 176
  rare species, 166
  regression, 166
    R-squared, 169
  relationship to DA, 210
  relativization, 176
  scaling constants, 167
  scaling for species or sites, 167
  scores, 167, 169-171
    centered with unit variance, 166
    choice of, 169
    example, 173, 174
  simple correlations, 167
  site scores, 166-167, 169-171
  software, 165
  species scores, 166-167
    example, 174
  species-environment correlation, 168
  tolerance, 167
  total variance, 168
  two methods, 166
  unimodal model, 164
  variance explained, 164, 168
    example, 175-176
  WA scores, 166, 169-174
  weighted averaging, 166
  weights, 171
  what to report, 172
  when to use it, 164-165
canonical variates, 205-206

CART, **222-232**
  accuracy, 226
  algorithm, 225-226
  bootstrap, 226
  choosing variables, 225
  classify new samples, 231
  confusion matrix, 226-227, 230
  cost-complexity, 226, 230-231
  cross-validation, 226, 230-231
  deviance, 225
  dichotomous key, 224
  examples, 223, 227-231
  GIS, 224
  habitat overlap, 224
  how it works, 224-227
  identifying the groups, 224-225
  impurity, 225
  misclassification, 224
  missing values, 224
  nonadditive effects, 224
  one-step look-ahead, 226
  overfitting, 226
  pruning, 226
  software, 231
  stopping rules, 225
  tree pruning, 226
  unimodal species response, 226
  validation, 226
  variations, 231
  what to report, 227
categorical variables, 3, 8, 225
  environmental, use with MRPP, 188
  ordination on, 204
categorization, 1
causal relationships
  in SEM, 242
CCA, *see* canonical correspondence analysis, 164-177
centroid, 21, 33, 46-47, 49, 77, 82, 114-115
  biplot, 109-110
  distances to, 196
  endpoint selection, 146
  group, 81
  joint plot, 108
  linkage, 87
  noncentered PCA, 121
  PCA, 117
centroid linkage clustering, 92
chaining, 84, 92-94
chance-corrected within-group agreement, 190
changes through time, 13, 77
characteristic equation
  CA, 154
  DA, 206
  PCA, 116
chi-square distance, 46, **49-51**, 155
  CA, 152, 155
  CCA, 168
  DCA, 159
  rare species, 50

chord distance, 46, 49, 195
city-block distance, 44, 47-49, 54-
        55, 57
    example, 57
    incompatibility in clustering, 88
    relativizing for, 73
    with MRPP, 189
city-block space, 46-47, 54
classification, 1, 7, 12, 44, 51, 54,
        80, 83, 104, 223
    accuracy, 226
    based on CA, 154
    DA, 206
        example, 209
    hierarchical, 177
    indicator species, 204
    multilevel, 82
    of unknown sample, 222
    one-way, 196
    refining with DA, 208
    summarizing with DA, 208
classification algorithm
    CART, 225
classification and regression trees,
        222-232
classification equation, DA, 206
classification power
    jack-knife, 231
classification table
    DA, 207
classification tree
    cost-complexity, 230
    example, 229
    multi-group, 224-225
    pruning, 226
classification trees, see also
        CART, 222-232
    multi-group, 224-225
climatic affinities, 151
climax adaptation values, 149
cluster analysis, 6, 10, 60, 80-85,
        86-96, 97, 100
    agglomerative, 81, 83, 86-96
    algorithm, 86
    before DA, 208
    chaining, 84, 92-94
    compared to TWINSPAN, 100
    compromise, 82
    constrained, 177
    dendrogram pruning, 203
    distance measures, 87-88
    divisive, 80-81
    ecological traits, 208
    example, 89-90
    flexible, 93
    hierarchical, 80-83, 86-96
    hierarchical divisive, 83
    indicator species analysis, 198, 201
    indicator species for stopping point,
        201
    k-means, 84
    large data sets, 84

linkage methods, 86
memory requirements, 84
monothetic, 80
nonhierarchical, 80, 84
optimum number of groups, 203
partitioning data set, 85
polythetic, 80
recommendations, 93
species groups, 6, 90-91
strategies, 86
supplement with DA, 207
coefficient of determination
    analogy in clustering, 82
    ordination, 111
        example, 111
    ordination distances, 111
coefficient of variation
    marginal totals, 70
COINSPAN, 100
column totals, CV, 70
columns, matrix, 5
combinatorial
    equation, 87-88
        example, 89
    hierarchical strategies, 86
    strategies, defined, 86
commensuration
    distance, 194
communalities, 120
community, 2
    abstract, 2
    analysis, 4
        questions, 43
    boundaries, 28
    change, 77
    concrete, 2
    data (see data)
        basic properties, 42-43
        frequency distribution, 206
    defined, 2
    ecology
        experimental, 182
        prehistoric, 2
    matrices, 3-12
    operational definition, 2
    organismal, 2
    questions, 2
    sampling, 13-24
    types, 9, 80
        CART, example, 227
        classification trees, 222
        from hierarchical clustering,
            example, 228
    indicator species, 198, 204
comparing groups, 182
compatible hierarchical strategies,
        86
    defined, 88
competing models strategy, 235
competing models, SEM, 247
complete linkage, 87, 92
complex response curves, 36

components of error
    direction, 186
    rate, 186
compositional change, 77
compositional vectors, 186
concentration of species
        abundance, 198
confirmatory analysis, SEM, **235**,
        247
confusion matrix, CART, 226-227,
        230
consistency
    defined, 179
    ordination, 179-181
    Wilson's method, 180-181
consistency of ordination, 179
constrained clustering, 177
constrained ordination, 102
construct validity, 235
contingency deviate relativization,
        75
contingency tables
    data, 152
    association, 83
continuous variables, 3
continuum index, 149
convergence, CA, 155
coordinate system, 8
correlation
    between matrices, 211
    Mantel test, 211
    matrix, 116, 236
    outliers, 116
    partial, 238
    structure, 1, 9, 115
correlation coefficient, 10, 36, 38,
        49, 116, 236
    as distance, 47
    binary data, 108
    convert to distance, 47
    effect of zeros, 42
    habitat, 12
    partitioning, 238
    noncentered, 47
    PCA loadings, 117
    permutation test, 195
    rank, Mantel test, 216
    with NMS, 136
    with ordination axes, 107
correlation distance, 46, 51
correspondence
    between ordinations, 180
    maximizing, 154
correspondence analysis, 28, 30,
        49, 103, 144, **152-158**,
        159, 165, 210
    arch, 152-153
    artifacts, 152-153, 178
    assumptions, 156
    basis in weighted averaging, 150
    CCA, 164

characteristic equation, 154
chi-square distance, 155
compared to PCA and NMS, 152-153, 156-157
covariance matrix, 154
cross-products matrix, 154
data transformation, 152
distance measure, 152, 155
eigenvalues, 154
eigenvectors, 155
examples
    compared to PCA and NMS, 152-153, 156-158
faults, 152
how it works, 154-156
linear model, 156
major steps, 155
quadratic distortions, 152
scaling factors, 155
TWINSPAN, 97
weights, 154
when to use it, 154
cost function, LLE, 139
cost-complexity
    CART, 226
    curve, CART, 230
    pruning table, 231
count data, 78
covariance analysis
    SEM, 234
covariance matrix, 50, 116, 259
    CA, 154
    canonical form, 165
    DA, 206
cover
    defined, 13
    percent, 3, 13
cover classes, 3, 13-14, 99
    midpoints, 13
cross-products matrix, 116
    CA, 154
cross-validation
    CART, 226
    importance in CART, 231
cryptic species, 23, 27
cube root transformation, 67
cut levels, 97, 99
Czekanowski distance, 46, 48

D

DA, *see* discriminant analysis, 205-210
data
    adjustments, 58-79
        general procedure, 78-79
        basic properties, 42-43
    before-and-after, 77
    matrices, examples, 5
    missing, 58-59
    partitioning, 85
    presence, (*see* presence)

reduction, **1**, 3, 6, 10, 23, 104, 111
    quality, 19
    screening, 58-61
    size class, 13
    transformation, 67-79
data reduction, **1**, 3, 6, 10, 23, 104, 111
    defined, 1
    goals, 3, 104
    ordination, 104
    quality of, 111
data transformation, 2, 9, 42, 67-79
dates, differences between, 77
db-MRT, 83
db-RDA, 177
DCA, *see* detrended correspondence analysis, 159-163
DECORANA, 159-160
deleting rare species, 75-77
delta, MRPP, 189
dendrogram, 80-82, 86
    as a mobile, 82-83
    comparing, 82
    examples, 81, 90-91, 94-96
    from TWINSPAN, 99
    interpretability, 85
    lack of dimensionality, 82
    mean similarity, 188
    pruning, 82, 88-99
        based on indicator species, 203
    scaling, 81-82
    where to cut, 82, 99, 203
density, 13-15, 18, 43
    defined, 14
    distance methods, 18
    pellet groups, 18
    relative, 14-15
descriptive statistics, 59, 78-79
design matrix
    as distance matrix, 216
    example, 217
    Mantel test, 216
designs
    blocked, 44, 184-186, 193-194
    factorial, 185, 196-197
    hierarchical, 218
    matched-pair, 184-185
    nested, 218-221
    paired, 184-185
    repeated measures, 22, 44, 185, 189
determinant of a matrix, 259
detrended CCA, 176
detrended correspondence analysis, 28, 30, 49-50, 77, 103, 144, **159-163**
    artifacts, 178
    axis length, 28, 30
    compared to NMS, 159, 161-162
    correcting faults of CA, 154
    detrending, 159-160

discontinuities, 162
distance measure, 159
eigenvalue, 161
eigenvector, 161
examples 161-163
    compared to CA, 161
    compared to NMS, 161-163
    two strong gradients, 162
Hill's scaling, 161
how it works, 160-161
instability, 159-160
output, 161
rescaling, 159-161
residual, 161
scores 160-161
segment length, 161
tolerance, 160-61
what to report, 161
when to use it, 160
detrending, 159
    by polynomials, 162
    by segments, 160
deviance
    CART, 225
    regression tree, 231
diagonal of matrix defined, 257
diagonal matrix defined, 257
dichotomous key, CART, 224
difference between dates, 77
    incompatibility with CA, 77
digital stopwatch, 17
dimensionality, 6, 65, 133, 136
    detection with PCA, 119
    lack of in dendrogram, 82
    NMS, 130-131, 134-135
        significance, 136
        vs. stress, NMS, 131
    of group differences, 205
    reduction, 104
direct DA, 210
direct and indirect effects
    response variables, 183
direct gradient analysis, 102-103
    CCA, 164-177
    multivariate, 103
discontinuous variables, 3
discriminant analysis, 12, 50, 54, 60, 82, 85, 121, 142, 164-165, 177, 183-184, 188, **205-210**, 214, 223
    assumptions, 205, 207
    characteristic equation 206
    classification, 206-207
        equation, 206
        score, 207
        table, 207
    coefficients, 207
    community types, 207
    compared to MANOVA, 205
    compared to PCA, 208
    complex designs, 184
    covariance matrix, 206

descriptive, 205
difference from MANOVA, 184
direct, 206
dominance types, 207
example of priors, 209
examples, 207, 209
F-ratio, 206
group membership, 205-207
habitat studies, 205
    example, 209
how it works, 206
interpretation, 207
linearity among variables, 206
misclassified items, 208
multivariate normality, 205
normality, 205
predicted group membership, 207
predictive, 205
prior probabilities, 206-207
relationship to CCA, 210
scores, 206
software, 205
stepwise, 206-207, 210
summary statistics, 207
traits, 208
uses in ecology, 205
what to report, 207
discriminant functions, 205-207
    coefficients, 207
dispersions, within-group, 191
dissimilarity (see also distance measures and distance matrices), 45-57
    correlation, 211
    clustering, 88
    matrix, 86, 88
distance function commensuration, 194
distance matrices, **45-57**
    additive partitioning, NPMANOVA, 196
    clustering, 81, 86
    comparison with Mantel test, 184
    congruence, 211
    correlation, 211
    example for MRPP, 191
    geographic coordinates, 213
    independence of elements, 211
    MRPP, 189
    NMS, 135
    NPMANOVA, 196
    partitioning variation in, 195
    rank transformation, 193, 216
    ranking, 125, 193
    total sum of squares, 196
    two-stage, 90
distance measures, 21, **45-57**
    affecting outliers, 59
    arc, 49
    average for detecting outliers, 60-61
    binary data, 51
    Bray-Curtis, 46, 48
    CA, 152, 155

chi-square, 46, **49-51**, 152, 155, 159, 168
chord, 46, 49
city-block, 47, 54
    example, 56-57
clustering, 87-89
compatibility, 54, 73-74, 87-88
correlation, 46-47, 49, 51, 211
Czekanowski, 48
DCA, 159
domains, 45
domains and ranges, 46
environmental distance, 50-51
Euclidean, 21, 46-47
    example, 56-57, 89
geodesic, 55, 139
geometry, 47
geographic, 213
Jaccard, 46
Kulczynski, 46, 48
Mahalanobis, 46, **50**, 60-61
Manhattan, 47
metric, defined, 46
MRPP, 189, 194
NMS, 125
    autopilot, 130
nonmetric, defined, 46
NPMANOVA, 196
PCoA, 120
performance, 50-55
proportion coefficients, 47-49, 56
Pythagorean, 46
quantitative symmetric, 46, 48
ranked, 51
relative Euclidean, 46, **48-49**, 61, 73
relative Sørensen, 46, 48
relativized Manhattan, 48
relativization, 73
semimetric, defined, 46
sensitivity, 42, 51
    to outliers, 54
slope correction, 24
sociological, 50
Sørensen, 46-**48**, 51-54, 56
    examples, 56-57
squared Euclidean, 46, 89
standardized Euclidean, 46
upper bound, 51
vs. similarities, 45
within-group, 191
distance-based MRT, 83
distance-based RDA, 177
distance-based sampling, 18
distribution
    bivariate, species, 37-39
    dust-bunny, **38-43**, 177, 191
    lognormal, 37, 43, 68
    multivariate normal, 41, 184
    normal, 43
    Poisson, 13
diversity, 2-3, 17, 19, 20, **25-34**, 38

accuracy, 32
alpha, 25
beta, 25, **27-31**, 51, 78
bias, 32
effect of rare species, 32
effective number of species, 26
evenness, 25-27, **32**, 34
gamma, 25, 31-34
Hill's index, 25
information theory, 26
landscape-level, 25, 31-32
measures, 25, 28, 31, 34
precision, 32
proportionate, 25
richness, 26
Shannon-Wiener, 26, 32
Simpson's, 26
software, 34
subsets of data, 34
divide data set, 85
divisive clustering, 80-81, 83, 177
documenting analyses, 62-66
    analysis log 62, 65-66
    flow charts, 62
dominance, 15, 18, 32
    relative, 15
    Simpson's index, 26
dominant species, 13
double relativization, 74
dummy variables, 3
dust bunny, 37, **38-43**, 54, 115, 157, 176-177, 184, 191

E

edge decisions, 18
effects
    nonadditive, 224
    size, 184
EI values, 149
eigenanalysis
    CA, 154
    CCA, 166
    DCA, 160
    DA, 205
    factor analysis, 120
    LLE, 139
    PCA, 114
    PCoA, 120
    rotations, 122
eigenvalue, 116-119, 172, 176, 259
    broken-stick, 119
    CA, 154
    CCA, 167-168, 172, 176
    DCA, 161
    defined, **116**, 259
    expected, 119
    negative, 120
    PCA, 116-119
eigenvector, 116-118
    CA, 155

defined, **117**, 259
  DA, 206
  DCA, 161
  PCA, 116-118
element, defined, 257
empty sample units, 38, 59
endogenous variable, SEM, **235**, 250
endpoint selection
  Bray-Curtis ordination, 145
  centroid, 146
  original method, 146
  subjective, 146
  variance-regression, 146
envelope, 36, 107
  based on standard deviation, 36
environmental categories
  indicator species, 199
environmental data, 9, 79, 122
  adjusting, 79
  noise, effect on CCA, 171
environmental gradients, 35-44
environmental matrix, 8
  example, 5
environmental ordination, 9
  drawbacks, 102
environmental space, 6-7, 10, 79, 102, 122
  analysis with DA, 206
  CCA, 164, 169, 171
  distances in, 50
environmental variables
  correlations, 236
  DA, 208
  relating to ordination, SEM, 233
  relativizing, 9
EQS, 234
equitability, 25-27, 32, 34
equivalent models, SEM, 247-248, 251
error
  bias, 19
  checking, 58
  classification, DA, depends on priors, 209
  distance measures, 51
  estimation, 178
  experimentwise, 193
  independent variable, 114
  influence of relativization, 73
  instrument, 178
  measurement, 27, 178, 241-243, 249
    effect on CCA, 171
    SEM, 242-243
  measures of, 178
  observer, 22
  out of range, 58
  predicted change, 186
  predictions from DA, 209
    example, 210
  predictor variables, 245

root mean square, 250, 252
sampling, 178
sources of, 178
species indication, 199
species-area, 33
standard, 19
error rate vs. complexity, CART, 226
error sum of squares
  clustering example, 89
  DA, 207
  defined, 92
  groups, 81
Euclidean distance, 21, 44, 46, **47**, 49, 51, 54-57, 87, 92
  example, 56
  in ordination space, 111
  relativized, 48-49, 61, 73
  relativization, 73
evenness, 25-27, **32**, 34
exclusive species, 198
exogenous variables, SEM, **235**, 250
experimental community studies, 182
experimental control, 182
experimental design
  reference point, 147
experiments
  aggregate response variables, 183
  defined, 182
  individual response variables, 183
  multivariate, 182-188
  natural, 182
  nonreplicated, 182
  randomized block, 193
  SEM, 245-246
  what to report, 184
exploratory analysis, SEM, 235
exposure, 110

F

factor analysis, 114, 120
factorial design, 196
  compared with nested, 218-220
  NPMANOVA, 185, 197
faithfulness of occurrence, 198
FAME, 4
  Mantel test, 213
families, 3
farthest neighbor linkage, 87, 92
Federal Wetlands Manual, 149
file names, 62
file types, 62
filtering noise, 76
fine roots, 3, 4
fixed-area sample units, 18
flexible clustering, 93
flexible-beta linkage, 87, 92-93

flowcharts, 62-64
  example, 63-64
focal species, 11-12
  DA example, 209
food, 3
forecasting, 2
fourth root transformation, 67
fourth-corner problem, 10
F-ratio, 50, 189
  DA, 206
  nested design, 218
  NPMANOVA, 196-197
frequency, 3, 13-15, 43
  defined, 13
  indicator species, example, 200
  relative, 15
frequency distribution
  abundance, 37
  average distance, 61
  community data, 206
  Mantel's Z, 216
  MRPP vs MANOVA, 188
  of delta, MRPP, 190
  outliers, 59
function
  beta, 37
  negative exponential, 37
functional groups, 3-4, 9-10, 151
fungi, 4
fuzzy sets, 48

G

GAM, 223
gamma diversity, 25, 31-34
Gaussian curves, 39
Gaussian response model
  least squares, 37
  maximum likelihood, 37
Gaussian responses, 35
genera, 3
general linear model, 37, 184, 223
general relativization, 73
generalized additive models, 223
geodesic distance, 55
  isomap, 139
geographic coordinates
  basis for distance matrix, 213
  in CCA, 213
  in RDA, 213
  Mantel test, 213
geographic distance, 213
geographic location, 213
GIS, 224
  CART models, 224
  environmental data, 224
gleasons, 28-30
GLM, 37, 184, 223
global minimum
  finding, 128
  NMS, 128, 134

gradient, 35-44
    analysis
        direct, 102-103
        indirect, 102-103
    beta diversity, 27-31
    compositional, 36
    curvature, 55
    environmental, 35-44
    length, 28, 30
    simulated 1-D, 153
    vector
        analogy, 128
        magnitude, 128
grain, 28
    influence on species-area, 34
graphical rotation, 122
grid overlay, 110
group average linkage, 87, 92-93
groups
    analyze variation within, 85
    average distance within, 188
    classification, 223
    comparison, 182-221
        MRPP, 184
    defining, 80-101
    describing, 85
    differences
        describing, 85
        Mantel test, 216
        testing, 85
    experimental, 182-187
    evaluating quality, 84
    finding, 80-101
    hierarchical, 218
    indicator species, 198
    maximizing separation, 205
    membership, 8, 10, 188, 223
        CART, 226
        DA, 207
        predicting, 205
    natural, 82
    separation, maximizing, 184
    test difference between, 188
    weights in MRPP, 189
guilds, 3, 4
    Mantel test, 213

H

H', 26-27, 32
habitat analysis, 11-12
habitat classification, 11-12, 224
habitat models, 11-12, 222-224
    CART, 223
habitat or not for species X, 223
habitat-centered analysis, 12
half changes, 28-31, 51-53
heat load, 22
heterogeneity, compositional, 31
heterogeneous data
    partitioning, 82

hierarchical, 80, 82-83, 177, 198, 201
hierarchical classification
    indicator species, 202
hierarchical cluster analysis 80-83, **86-96**
    before CART, 227
    example, 91
    of species, 91
    pros and cons, 82
hierarchical designs, 218, 221
hierarchical divisive clustering, 83
hierarchical grouping, 80, 92
    indicator species analysis, 201
hierarchical models
    habitat, 223
    CART, 226
hierarchical strategies, 86
hierarchy of indicator species, 202
Hill's scaling, 30, 161
    CCA, 166, 167, 172
    DCA, 161
homogeneity within sample units, 17
homogeneity of variance, 13, 67
    DA, 205
    in nonparametric methods, 197
    MRPP, 188
horseshoe pattern in PCA, 115, 152
hump-shaped species responses, 35, 43
hypercube, 38, 41
hyperellipsoid, 41
hypersphere, 49
    quarter, 49
hypotheses, evaluation of complex, 234

I

IBI, 149
identification, SEM, **235**, 249
identity matrix defined, 257
importance values, 15-16
impurity of groups, 83
    CART, 225
incompatibility
    city-block distance, 88
    strategy, 88
independent sample units, 189
independent-sample t-test, 186
index of atmospheric purity, 149
Index of Biotic Integrity, 149
indicator variable, SEM, 242
indicator species, 37, 85, 97, 99, 188, 198
    association analysis, 83
    classification, 204
    community types, 204

TWINSPAN, 99
indicator species analysis, 85, 188, **198-204**
    community types, 204
    companion to MRPP, 198
    examples, 199
    for pruning dendrogram, 203
    hierarchical grouping, 201
    how it works, 198
    indicator value, 199
    ordination, 204
        example, 204
    stopping point for clustering, 201
    two-way, 84, 97
    what to report, 199
    when to use it, 198
indicator values, 188, 198-199, 204
    cluster analysis, use with, 82
        example, 228
    dendrogram, 203
    example, 201
    maximum, 199
        example, 201
    properties, 199
    range, 199
    rare species, 199
    species, 151-152
    statistical significance, 199
    weighted averaging, 149
indicator variables, 3, 234-**235**, 243
    SEM, 234, 242
indirect gradient analysis, 102-103
individual tree data, 16
inertia, CCA, 168, 173, 176
information
    content, 26-27
    diversity, 26-27
    function of ubiquity, 68
    remaining from clustering, 81
    theory, 26, 74
        impurity, 225
    uncertainty, 27
inner product, 258
instability
    NMS, 128
        example, 134
interaction
    sum of squares, NPMANOVA, 197
    term
        Q_b method, 196
        NPMANOVA, 197
interpretability
    diversity, 26
    dendrogram, 85
    ordination, 112
    rotation for, 122
    transformation, 112
interset correlation, 171-172
    example, 174

intraset correlation, 171-172
  example, 174
inventory, 23
inversion of a matrix, 258
isomap, 103, **137**, 139
isometric feature mapping, 103, 139
isozymes, 4

J

Jaccard distance, 46-48, 51
jackknife,
  diversity, 32
  misclassification rate, CART, 230
joint absence, 6, 38
joint occurrences, 70-71
joint plot, 105, 108-110
  CCA, 167
  vector, 108
  vector fitting, 109

K

Kaiser-Guttman criterion, 119
Kendall's coefficient of
    concordance, 181
Kendall's rank correlation (tau)
  for Wilson's method, 181
  with ordination, 107
k-means, 84
knight's tour, 179
Kruskal-Wallis, 85
Kulczynski distance, 46, 48
kurtosis, 115

L

landscape, 2, 23, 34
  analogy, NMS, 128
  diversity, 25, 31-32
  habitat models, 224
  habitat predictions, 210
landscape-level diversity, 25, 31-32
large data sets, 128, 180
  clustering, 84
latent roots, 116, 259
  CA, 154
latent variables, SEM 234-**235**, 241-242, 252
latent vector, 259
LC scores, 166, 169, 171
  correlation with WA scores, 168
  example, 174-175
least squares, 114-115, 239
LH-PCR, 4
  Mantel test, 213
life history, 3, 9, 15, 151
  DA, 207

line intercept sampling, 18
linear chain models, 255
linear model, 67
  checking, 116
  DA, 206
  PCA, 116
  PCA vs. CA, 156
  RDA, 177
  regression, 177
  species responses, 35
linkage methods, 86-88, **92-93**
  centroid, 92
  complete linkage, 92
  farthest neighbor, 92
  flexible beta, 93
  group average, 92
  hierarchical grouping, 92
  McQuitty's, 93
  median, 92
  minimum variance, 92
  nearest neighbor, 92
  Orloci's method, 92
  properties, 87
  single linkage, 92
  unweighted pair-group, 92
  UPGMA, 92
  WPGMC, 92
LISCOMP, 234
LISREL, 234, 236, 243, 245, 249-254
LLE (*see* locally linear
    embedding), 103, 137, **139**
loading matrix, PCA, 117
loadings
  PCA, 240
  SEM, 235
local minima, NMS, 128
local NMS, 137
locally linear embedding, 103, 137, **139**
  neighborhood definition, 139
  weights, 139
log transformation, 37, 67, **68-69**, 73, 112, 135
  adding constant before, 69
  interpretability, 112
log(0), 68
logarithm of diversity, 26
logistic regression, 10, 12, 37, 150, 186, 205, 223
  compared to DA, 205
logit transformation, 69
log-likelihood estimator, CART, 225
lognormal distribution, 37, 43, 68

M

Mahalanobis distance, 46, **50**, 60-61
  for detecting outliers, 60-61

Manhattan distance, 47
manifest model, **235**, 241
MANOVA, 44, 85, 184-185, 188, 218-221, 223
  compared to DA, 205
  compared to MRPP, 188
  nested, 218-221
  nonparametric, 185
  nested, 221
Mantel test, 44, 184-185, **211-217**
  asymptotic approximation, 213
  asymptotic vs. randomization method, 214
  compare two species groups, 215
  compared with MRPP, 185, 211, 216
  design matrix, 216
    example, 217
  distance measures, 211
  examples, 213-215
  FAME, 213
  frequency distribution, 216
  geographic coordinates, 213
  group-contrast, 223
  how it works, 211
  large data sets, 213
  LH-PCR, 213
  Monte Carlo test, 213
  partial, 216
  permutation method, 212
  randomization, 211, 213
  rank transformed, 216
  sample size, 213
  spatial proximity, 213
  standardized Mantel statistic, 211
  what to report, 213
  when to use it, 211
  with Sørensen distance, 213
Mantel's asymptotic
    approximation, 213
Mantel's Z statistic, 211
matched case-control, 185
matched-pair designs, 185
  MRPP, 184
matrices
  addition, 258
  calculated, 7-8
  columns, 5
  community, 3-12
  correlation, 116, 236
  covariance, 116
  cross-products, 116
    CA, 154
  defined, 257
  examples, 5
  environmental, 8
  functional groups, 10
  identity, 116, **257**
  inversion, 258
  joint occurrences, 71
  multiplication, 7-11, 258
  rows, 5
  sparse, 8, 38-43

species × environment, 9
species composition, 8
subtraction, 258
symbols, 257
traits × environment, 10
transposed, 7-8, 257
variance-covariance, 116, 259
maximum likelihood species
 ranking, 181
maximum redundancy analysis,
 177
McQuitty's linkage, 87, 93, 96
MDS, 125
mean similarity dendrogram, 188
measurement 13-24
 abundance, 13-16
 error, 178
 model, SEM, 234-**235**, 242-243
 scales, making comparable, 67
 variables, 3
median
 alignment, 194
 blocked MRPP, 194
median linkage, 87, 92
median transformation, 74
metaobjects, 7
 clustering, 86
methodological artifacts, 178
metric distance defined, 46
microbial communities, 4, 120
midpoints of cover classes, 13
Minchin's R, 28-29
minimum stress, 125
 NMS, 128
Minkowski metric, 47
mirror image, ordination, 124
misclassification
 CART models, 224
 clustering, 82
 table, CART, 230
misclassified items
 DA, 208
 finding with DA, 205
missing data, 58-59
 amount, 58
 CART, 224
 corrective steps, 58-59
 pattern, 58
 regression, 59
models
 habitat, 11-12
 measurement, SEM, 234
 structural, 222
modification indices, 235, 245
molecular markers, 3-4
 bivariate scatterplots, 120
 frequency distribution, 120
 linearity, 120
 PCA, 120
monothetic clustering, 80
monotonic transformations, 67

CCA, 176
 defined, 67
monotonicity, 125
 defined, 126
 departure from, 126
 NMS, 126
Monte Carlo test
 CCA, 172
 example, 175
 eigenvalues, 119
 indicator species analysis, 199
 example, 201
 Mantel test, 211-217
 NMS, 129, 133, 136
 ordination, 179
morphology, 4
MRBP, 193-194
 balanced design, 194
 bivariate correlation, 195
 distance measure, 194
 options, 194
MRPP, 44, 79, 85, 184-185, **188-197**, 216, 223
 agreement statistic, A, 191
 alignment of blocks, 194
 assumptions, 188-189
 avoiding assumptions of DA, 205
 basis for pruning dendrogram, 227
 blocked, 185-186, **193-195**
 chance-corrected within-group
  agreement, 190
 compared to Mantel test, 217
 compared to ANOSIM, 195
 compared to Mantel test, 185, 211, 216
 comparison of distance measures, 191
 complex designs, 188
 distance measures, 189
 effect size, 191
 examples, 191-193
  blocking, 195
  comparing distance measures, 192
  simple, 191-193
 frequency distribution of delta, 190
 group weight, 189
 homogeneity of variance, 197
 how it works, 189-191
 median alignment, 194
 multiple comparisons, 192
 nonmetric, 191, 193
 power, 188-189
 randomization, 190
 rank-transformed, 191, 193
 rare species, 76
 sample size, 190-191, 193
 test statistic, 190
 visualize with overlay, 189
 weighting groups, 189
 what to report, 191
 when to use it, 188

MRT, 83, 177
 assumptions, 177
 compared to CCA and RDA, 177
 distance-based, 83
multidimensional scaling, 102-103, 112, 125-142
 isomap, 139
 nonmetric, 44, 51, 54, 125-142
multiple comparisons, MRPP, 192
multiple regression, 12, 60, 108, 120-121, 216, 238
 calculation of R^2, 239
 CCA, 164, 166-169
 example, 173
 compared to SEM, 241
 detecting outliers, 61
 difficulty of interpretation, 239
 matrix algebra, 259
 ordination axes and environment, 237
 partial correlation, 237-238
 RDA, 177
 relationship to SEM, 234
 R^2, CCA, 169
 with ordination, 108
multiplication of matrices, 7-11, 258
multi-response permutation
 procedures (see MRPP), 188-197
multistage sampling, 22
multivariate analysis, 1, 8, 10, 38
 vs. univariate, 182
multivariate analysis of variance
 (see MANOVA), 184
multivariate experiments, 182-187
 four approaches to analysis, 183
 tools and examples, 184
 what to report, 184
multivariate normality, 38, 116
 DA, 205
 distribution, 41, 184
 MRPP, 188
 PCA, 115
 testing, 115
multivariate regression trees, 83, 177
 assumptions, 177
 compared to CCA and RDA, 177
multivariate responses
 structural equation modeling, 183
 to experiments, 183
mycorrhizae, 4, 37

N

N routine, 84
N squared routine, 84
natural experiments, 182
nearest neighbor linkage, 87, 92

Index

nearest neighbors
    isomap, 139
    LLE, 139
negative association, causes, 37
negative exponential function, 37
nest sites, prediction, 209
nested designs, 80, **218-221**
    compared to factorial, 218-220
    MANOVA, 219
    NPMANOVA, 185, 197, 218, 221
    F-ratio, 218
    spatially nested, 218
    species richness, 218
    univariate examples, 218
    univariate vs. multivariate, 218
nested models, SEM, **235**, 247
nested sampling, 22
niche-space analysis, 3-4
NMDS, 125
NMS (nonmetric multidimensional
        scaling), 125-142
noise
    effect on CCA, 169
    filtering, 76
    from rare species, 73, 75
noise and information, 76
nominal variables, 3, 8
nonadditive effects, 224
noncentered PCA, 121
    variance explained, 121
noncombinatorial strategies, 86
noncombinatorial strategy
    defined, 88
non-Euclidean distance
    group comparison, 188
nonhierarchical clustering, 84
nonlinear structures
    extracting, 137
    PCA, 115
    species relationships, 39-41, 43
    species responses, 226
        CART, 226
    with ordination, 108
nonmetric MRPP, 191, 193
    NMS as companion, 193
    variance represented, 112
nonmetric multidimensional
        scaling, 44, 51, 54, 65,
        103, **125-142**, 144, 157,
        163, 176
    advantages, 125
    autopilot, 130, 135
    axis numbering, 131
    basic procedure, 126
    calibration scores, 140
    choosing the best solution, 131-136
    compared to CA, 152-153, 157, 161
    compared to CCA, 164, 175-176
    compared to DCA, 159, 161-162
    compared to PCA, 152-153, 157
    constrained, 137
    correlations, 136
    dimensionality, 130-132, 136
        example, 130
    disadvantages, 125
    dissimilarity measure, 126
    distance measure, 125-126
    Euclidean distance in k space, 126
    examples, 130, 137-138, 157, 161-162, 175-176
    extrapolation limit, 140-142
    flag for poor fit, 140-142
    general procedure, 135
    global minimum, 128
    global vs. local, 137
    gradient vector, 127
        magnitude, 128
    how it works, 125-128
    instability, 128, 136
        example, 136
    interpretability, 132
    iterations, 128
        defined, 126
    landscape analogy, 128
    large data sets, 125
    local, 137
    local minima, 128-129
    migration of points, 138
    minimum stress, 128
    monotonicity, 125-126
    Monte Carlo test, 129-131, 133, 136
    normalized configuration, 126
    number of axes, 130-132, 136
    overfitting, 134
    overlays, 136
    predicting scores, 139
    predictive mode, 139-142
        example, 141-142
    preliminary runs, 136
    randomization test, 176
    ranked distances, 135
    rare species, 76
    regression on principal components, 241
    relativization, 135
    rotation, 132
    run, defined, 126
    scores, 126, 139
    scores for new data, 139
        flag for poor fit, 140
    scree plot, 130-131, 136
    select the best solutions, 131
    simulated annealing, 128
    small data sets, 133
    species scores, 151
    stability, 128, 134
        examples, 134-136
    stable solution, example, 135
    starting configuration, 126, 129
    steepest-descent, 127
    step size, 128
    stress
        defined, 126
        formulae, 126-127
        rules of thumb, 132
    standardized, 127
    thoroughness, 130
    unstable solutions, 129
        example, 135
    use with other ordinations, 129
    variations, 137
    what to report, 136
    when to use it, 125
    with multiple regression, 137
nonmetrics, 46
nonparametric MANOVA, 185
nonparametric statistics, 3
    assumptions, 189
nonreplicated experiments, 182, 185
normal analysis, 5-7, 105
normal distribution, 43
normality, 13, 38, 67
    bivariate, 38, 42
    DA, 205
    improving for proportion data, 69
    kurtosis, 115
    MRPP, 188
    PCA, 115
    skewness, 115
normalizing, 73
NPMANOVA, 44, 83, 185, 188, **196-197**, 223
    assumptions, 197
    distance measures, 196, 218
    F-ratio, 197
    homogeneity of variance, 197
    nested design, 197, 218, 221
    residual sum of squares, 196
nucleotide, 4
null hypothesis
    CCA, 172
    independence of rankings, 179
    no difference among groups, 191
    ordination, 179
null models
    knight's tour, 179
    no correlation structure, 179
    ordination, 113, 179
    presence data, 179
    random numbers, 179
    randomization tests, 179
    species ranking, 179

O

objective function, 81-82, 86
    example, 89
objects, ecological, 4
oblique rotation, 124
observer effects, 20, 178
one-way classification
    MRPP, 184
    NPMANOVA, 197
operational definition
    by ordination, 104

ordered table, 97-100
   2-D gradient problem, 101
ORDIFLEX, 144
ordinal variables, 3, 8
ordination, 102-181
   abundance scatterplot, 107
   accuracy, 178, 181
   artifacts, 178
   as data reduction, 104
   axis reversals, 122
   axis scores
      correlations with, 106-107
      multiple regression, 237
      regressed against PCA, 240
   biplot, 109-110
   bootstrap, 180
   Bray-Curtis, 103, 143-148
   CA, 152-158
   canonical correlation, 176
   canonical correspondence analysis, 164-177
   CCA, 164-177
   coefficient of determination, 111
   consistency, 178, 181
   constrained, 102
   correlation with axis scores, 106
      p-value, 107
   correlation with species, 107
   correspondence analysis, 152-158
   DA, 205-210
   DCA, 159-163
   defined, 102
   detrended correspondence analysis, 159-163
   diagram, 104
   discriminant analysis, 205-210
   distance-based RDA, 177
   environmental, 9, 102
   evaluation by external criteria, 112
   failure, 112
   for operational definition, 104
   history, 103
   how many axes, 105
   indicator species, 204
   interpretability, 112
   interpretation, 105, 178
   introduction, 102-113
   isomap, 137, 139
   isometric feature mapping, 137, 139
   joint plot, 108-110
   LLE, 137, **139**
   locally linear embedding, 137
   maximum variance, 118
   mirror image, 124
   NMS, 125-142
   NMDS, 125
   noncentered PCA, 121
   nonmetric multidimensional scaling, 125-142
   normal analysis, 105
   null hypothesis, 179
   null model, 113
   overlays, 105-110
   PCA, 114-121
   PCoA, 120
   polar, 143
   principal components, 114-121
   principal coordinates, 120
   principal curves, 139
   quality, 110, 178-181
   RA, 152-158
   randomization test, 112-113, 179-180
   rank correlation with, 107
   RDA, 121, **177**
   reciprocal averaging, 152-158
   redundancy analysis, 121, **177**
   relating variables to, 105
   relation to second matrix, 112
   reliability, 178-181
      accuracy, 179
      bootstrap, 180
      consistency, 179
      strength of pattern, 179
      variance represented, 178
   residuals, 104
   rotation, 108, **122-124**
   scores
      multiple regression, 108
      regression on environment, 237
      use in SEM, 236
   sources of variation, 178
   space
      Euclidean distance, 126
   statistical significance, 105
   strength of pattern, 178-179
   symbol sizes, 106
   timeline, 103
   transpose analysis, 105
   variance represented, 111-112, 178
   weighted averaging, 149-151
   what to tell an audience, 104
orthogonal rotation, 122-123
outliers, **59-61**, 78-79, 147
   bivariate example, 60
   Bray-Curtis ordination, 143, 145-147
   causes, 59
   check for, 78-79
   chi square distance, 59
   comparison with null model, 113
   correlation, 116
   describe with Bray-Curtis, 146
      example, 147
   describing, 60
   detecting, 50, 60, 147
   effect on correlation, 107
   effect on distance measure, 45, 193
   effect on randomization test, 113
   effect on stress, 133
   effect on variance represented, 112, 179
   force as endpoint, 147
   in sequence of adjustments, 78
   multivariate, 60-61, 147
      example, 61, 147
   ordination interpretability, 112
   PCA, 116
   reducing influence, 60
   regression for detecting, 61
   sensitivity of distance measures, 54
   univariate, 59
overfitting
   CART, 226
   NMS, 134
   SEM, 235
overidentification, SEM, 235
overlap, 47
   species abundances, 48
overlays, 79, **106-110**, 110, 189
   biplot, 109
   categorical variables, 106
   grid, 110
   joint plot, 108
   NMS, 136
   symbol sizes, 106

P

paired treatments, 186
paired sample
   designs, 22, 185-186, 194
   MRPP, 184
   t-test, 186
parallel bootstrapping, 180
   PCA, 119
partial correlation, 237-239
   bias in regression, 243
   defined, 238
   derivation, 238
   example, 239
   species richness example, 238
partial least squares, 246, 255
partial Mantel statistic, 216
partitioning a matrix, 85
   defined, 257
path analysis, 241
   vs. SEM, 242
path coefficient defined, 237
path diagram, 235, 249
   initial model, 249
path length, average, 84
path model, **235**, 237-241, 255
   calculation of R^2, 239
   compared to multiple regression, 238
   generated by simulation models, 256
   SEM, 234
PC (principal curves), 139
PCA (*see* principal components analysis) 114-121
PCoA (principal coordinates analysis), 120
Pearson correlation, 107, 116
   assumption of independence, 211
   Mantel test, 216
   permutation test, 195

pellet groups, 18
percent cover, 3, 13-14
    point intercept, 18
percent of variation
    distance matrix, 111
    PCA, 111
permanent
    sample units, 22
    plots, 17, 21-22, 73, 186
permutation test, 184
    CCA, 172
    degrees of freedom, 189
    Mantel test, 211-217
    MRPP, 188-197
    NPMANOVA, 196
    Pearson correlation, 195
    rows and columns, 212
phenetic analysis, 3
Picea mariana, 36
Pielou's *J*, 32, 34
pigeonholes, 1
pin frames, 18
placement of sample units, 17
plane-corrected distance, 24
plots, 16-19, 22-23
    number, 19
    shape, 18
    single large, 23
    size, 22, 28
    tradeoffs, 22-23
PLS (partial least squares), 255
point intercept sampling, 18
point sampling, 18, 23
point-centered quarter, 18
Poisson distribution, 13
polar ordination, 143-148
polymodal responses, 35
polynomial, roots of, 116, 259
polythetic clustering, 80
population, 17
    sample, 13
positive association, causes, 37
power, 3, 180
    experimental studies, 182
    MRPP, 188-189
    multivariate approach, 184
    paired designs, 186
    predictive, 12
    transformation, 67-68
precision, 19, 21, 23
predictions, long term, 2
PRELIS, 250
presence-absence, 3, 10, 15, 29-31, 33, 37-38, 45-46, 51, 67, 70, 78, 99, 108, 179
    association analysis, 83
    correlation, 108
    faithfulness in groups, 199
    frequency, 14
    TWINSPAN, 99
    transformation, 67-68

prevalence index, 149
prime mark, 7-8
    defined, 257
principal components analysis, 10, 35, 44, 77, 79, 103, **114-121**, 176-177
    basics, 116
    bootstrapping, 119
    characteristic equation, 116
    compared to CA, 153, 156-157
    compared to DA, 208
    compared to factor analysis, 120
    compared to NMS, 153, 157
    compared to regression, 114
    compared to SEM, 236, 240
    constrained, 121
    correlation structure, 116
    demographic variables, 120
    eigenvalue, 116-118
    eigenvector, 117-119
    environmental variables, 240
    example, 117-119, 153
    factor analysis, 120
    geometric analog, 117
    horseshoe pattern, 152
    linearity, 116
    loading matrix, 117
    molecular markers, 120
    noncentered, 121
    normality, 115
    objective, 116
    outliers, 116
    parallel bootstrapping, 119
    regression against NMS, 241
    rotation analogy, 117
    sample size, 115
    scaling eigenvectors, 118
    scores, 117
    significance, 119
    step by step, 116
    total variance, 116
    traits, 208
    variance explained, 116
    when to use it, 115
principal coordinates analysis, 120
principal curves, 139
prior probabilities, DA, 206-207
Procrustes rotation, 122-123, 180
proportion coefficients, 47-49
    example, 56
proportion data, 13, 69, 79
    in TWINSPAN, 99
    transformation, 69
proportion of variance
    ordination, 111
pruning
    CART, example, 229
pseudoreplication, 22, 189
    nested design, 221
pseudospecies, 97, 99
    cut levels, 97, 99
pseudo-turnover, 22

Pythagorean distance, 46
Pythagorean theorem, 47

Q

Q route, 5, 7, 120
Q_b method 185, 188, 195
    distance measures, 195
QSK distance, 47-49
quadrat
    number, 19
    shape, 18
    single large, 23
    size, 22, 28
    tradeoffs, 22-23
quadratic distortions in CA, 152
quadratic equation, 118
quality assurance, 19
quantitative symmetric
    dissimilarity, 48
quarter hypersphere, 46, 49
quartimax rotation, 122

R

R route, 5, 7
RA, 77, 103, 152, 164
radiation, direct solar, 24
random numbers, 17, 164
random sampling, 17
randomization test, 215
    CCA, 172, 179
    example, 175
    effect of outlier, 113
    effect of relativization, 113
    Mantel test, 211-217
    MRPP, 179, 188-197
    NMS, 176
        limitations, 133
        small data sets, 133
    ordination, 112-113, 172, 176, 179-180
    parallel bootstrapping, 180
    precision, 172
    Q_b method, 185, 195
randomized block design, 44, 193
range
    data, 78-79
    ecological, 150
    environmental variables, 9, 38
    proportion data, 69
    SEM, 250
    species response, 29
    transformations, 68
rank correlation
    between matrices, 216
    for Wilson's method, 181
    species ranks, 181
    species-environment, 168
    with ordination, 107

rank transformation, 74
   distance matrix, 125, 193, 216
ranked variables, 3, 74
   ties, 74, 193
rare species, 6, 11, 18, 23, 26, 31, 67, 75-77
   bias from removal, 76
   causing noise, 73
   CCA, 166
   change or equalize influence, 67
   chi-square distance, 155
   contribution to outliers, 59
   deleting, 75-78
   effect on diversity measures, 26
   effect on stress, NMS, 133
   example of removal, 76
   habitat models, 11-12, 222-224
   indicator value, 199
   influence of, 6, 73
   information function, 74
   noise, 75
   removal, 76
      for species clustering, 90
   stress, NMS, 133
rate of change, 29
RDA, 121, 164-165, **177**, 183
   basic steps, 177
   compared to CCA, 177
   distance-based, 177
   geographic coordinates, 213
   response variables, 177
   scores, 177
reciprocal averaging (see correspondence analysis) 103, 144, **152-158**, 165
reduced-rank regression, 120
reduction, data, **1**, 3, 6, 10, 23, 104, 111
   goals, 3, 104
   ordination, 104
   quality of, 111
redundancy, 51
   species, 75
redundancy analysis, 121, 177
   maximum, 177
reference points, 143
   in experimental design, 147
reflections, ordination, 129
regression
   CCA, 164, 166, 169
      example, 175
   coefficients
      CCA, 166
      partial, 239
      standardized partial, 239
   compared to PCA, 114
   compared to SEM, 236, 243
   environment and axes, 237
   line, 115
   logistic, 10, 12, 37, 150, 186, 223
      compared to DA, 205
   multiple, 12, 108, 121, 164, 238

multiple dependent variables, 177
   ordination distances, 111
   reduced-rank, 120
   simultaneous, 177, 241-242
   stepwise, 239
   structured, 241
   trees, see also CART, 83-84, **222-232**
      compared to classification trees, 231
      deviance, 231
      distance-based, 83
      multivariate, **83-84**, 177, 223
      residual variance, 231
   vector fitting, 109
relative abundance
   for indicator species, 198
   indicator species, example, 200
relative density, 15
relative dominance, 15
relative Euclidean distance, 46, 48-**49**, 61, 73
relative frequency, 15
   indicator species analysis, 199
relative Manhattan distance, 48
relative Sørensen distance, 46, 48
relative weights of variables, 12
relativization, 9, 49, 67-68, **70**, 73-**75**, 78-79
   blocked MRPP, 194
   by maximum, 73
   by norm, 73, 79
   by species maximum, 6, 73
   by species totals, 6, 73
   by standard deviate, 79
   by ubiquity, 74
   CCA, 176
   contingency deviate, 75
   double, 74
   effect on distances, 48
   effect on ordination problem, 113
   for cluster analysis, 92
   for species clustering, 90
   general, 73
   information content, 74
   information function of ubiquity, 74
   NMS, 135
   traits, 10
repeated measures, 22, 44, 185, 189
rescaling axes
   DCA, 159, 160
   effect on variance represented, 112
resemblance measures, 6, 45-57
residual distances, 148
   Bray-Curtis ordination, 111, 143, 145, 148
   non-Euclidean, 147
residual sum of squares, NPMANOVA, 197
resources used, 3

response curves (see species response functions), 35-37
response variables, 1, 182, 184
   aggregate, 183
   CART, 223
   direct and indirect effects, 183
   linear relationships, 184
   univariate, 183
restriction fragments, 4
reversals
   clustering, 82
   ordination axes, 122
richness, 20, **25-27**, 30-34
   compare with species-area, 33
   measurement error, 27
root of polynomial, 116
rotation, **122-124**
   analytical, 122
   changes, 122
   eigenanalysis-based, 122
   examples, 123-124
   for interpretation, 132
   graphical, 122
   maximum correspondence, 122-123
   oblique, 124
   of joint plot, 123
   ordination, 108, 129
   orthogonal, 122
   Procrustes, 122, 180
   quartimax, 122
   rigid, 122
   to external variable, 123
   to maximize correspondence, 122-123
   varimax, 122
      example, 123
   visual, 122-123
rows
   matrix, 5
   totals, CV, 70
R-route, 116, 120

S

sample, 13
sample size, 18-19, 23, 26, 32, 133
   adequacy, 19
   and effect size, 191
   effect on diversity, 32
   effect on evenness, 32
   PCA, 115
   species-area curves, 19, 33
sample space
   defined, 6-8, **45**
   distances in, 45
sample unit totals
   reduce effect of, 67
sample units, 3, 6
   averaging, 77
   fixed-area, 18
   number, 18
   paired, 186

Index

permanent, 22
placement, 17
sampling, 13-24
  arbitrary, 17
  distance, 18
  error, 178
  haphazard, 17
  line intercept, 18
  multistage, 22
  nested, 22
  paired, 22
  point, 18, 23
  random, 17
  regular, 17
  stratified random, 17
  subjective, 17
  systematic, 17
  time, 23
  tradeoffs, 22-23
scalar, defined, 257
scaling
  CA scores, 154
  dendrograms, 81-82
  dissimilarity, 31
  eigenvectors, 118
  Hill's, 30
  nested designs, 221
  nonmetric, 44, 54, 125-142
  variables, 9
scaling constant
  CA, 155
  CCA, 166-167, 172
scores, 9
  accuracy, 20
  assign to new sample units, 44
  CA
    1-D example, 156
    scaling, 154
  canonical correlation, 122
  CCA, 166
  centered with unit variance, 166
  DA, 206
  defined, 9
  gradient, 20
  NMS, 44, 139
  NMS, depends on number of axes, 131
  ordination
    accuracy, 20
    regression, 237
    regression on principal components, 240
    use in SEM, 237
  PCA, 117
  weighted averaging, 151
    species, 151
  predicting from NMS, 139
  sample unit, CA, 155
  sites, 7
    CA, 155
  species, 7, 9
    based on indicator species analysis, 204

CA, 155
  weighted averaging, 151
standardized
  CA, 155
  CCA, 166
  weighted averaging, 149-150
z, 74, 79
scree plot, NMS, 131-132, 136
second matrix, 9, 65, 121-122, 214
  CCA, 164
segment length, DCA, 161
SEM, see structural equation modeling, 233-256
semimetrics defined, 46
sensitivity, distance measures, 51
Shannon index, 26, 32, 34
Shannon-Wiener diversity, 26, 32, 34
shared zeros
  PCA, 115
shortest path adjustments, 55
shuffle matrices, Mantel test, 211
similarity, 45-57
  rate of change, 29
  matrices
    congruence, 211
Simpson's index, 26
simulated annealing, 128
simultaneous regression, 177
single linkage, 87, 92
site scores
  CCA, 166
  scaling, CCA, 167
size-class data, 15
size-structured populations, 15
skewness, 59, 78-79, 115
  rule of thumb, 116
slope
  aspect, 22
  correction, 24
  steepness, 24
smoothing, 37, 67-68, 70-71
  Beals, 70-71
    example, 71
  principal curves, 139
  species data, 70
sociological distance, 50
sociological favorability index, 70
soil fungi, 3
solar radiation, 24
solid curves, 4, **35-37**, 39, 41, 43
Sørensen distance, 44, 46-47, **48**, 51-54, 56, 87, 176, 192, 217
  compared to Euclidean, MRPP, 193
  example, 56
  Mantel test, 213
  median alignment incompatible, 194
  rank transformed, 193
  relative, 49

species-area, 33
  use with clustering, 90
  use with MRPP, 189, 192
  use with relativization, 73
sorting strategies, 86
space
  environmental, 7
  sample, 6-8, 45
  species, 6-8, 45
space conserving, 85-87, 92
space contracting, 87
  defined, 88
space distorting
  defined, 88
space expanding, 87
  defined, 88
sparse matrices, 8, 38-43
spatial models
  distance apart, 213
  location, 213
spatial proximity, Mantel test, 213
spatial scale, species-area, 33
spatial statistics, 2
species
  abundance, 1
    skew, 37
  accumulation curve, 33
  area, see species-area curves
  association, 37
  capture, 20, 23
  change, rate of, 27-31, 186-187
  composition, matrix, 8
  cryptic, 23, 27
  data, adjusting, 77
  density, 27
  distributions
    bimodal, 36
    bivariate, 37-39
    skew, 37
  diversity, 2, 25-34
  dominant, 13
  effort curve, 33
  focal, 11-12
  groups, 10
    cluster analysis, 90
  indicator value, 149, 151, 198-204
  inventory, 2
  list, completeness, 22
  maxima, 6
    relativization by, 73
  optima, 6
  performance, 43
  presence, 11-12 (see also presence)
  ranking, consistency, 181
  rare, 11-12, 23, 75-77
  relative weighting, 72
  response functions, 29, 35-37, 39, 43
    bimodal, 36
    complex, 36
    examples, 36-37
    mean range, 29
    weighted averaging, 150

richness, 20, 25, **26-27**, 30-34
   bootstrap, 32
   jackknife estimators, 32
   nested design, 218
   regression example, 238
scores, 7, 9, 30
   CCA example, 175
   on categorical variables, 204
   PCA, 117-119
   scaling, CCA, 167
   variance in rank, 180
   weighted averaging, 151
space, 6-8, 37, 45
   CCA, 169
   corners, 38
   defined, 6, **45**
   distances in, 45
   grouping objects, 80
   multivariate, 38
totals, relativization by, 73
traits 5, 7-10, 242
   analysis with DA, 206, 208
   example matrix, 5
   independence, 10
   PCA, 120
   relativization, 10
   turnover, 27
volumes, 6
   in environmental space, 105
weights, 149-151
   example, 151
species × environment matrix, 9
species-area curves, 19, 27, **32-34**
   error, 33
   islands, 34
   sample size, 33
   Sørensen distance, 33
   subsample, 33
species-environment, CCA
   correlation, 168, 172
   rank correlation, 168
square matrix, defined, 257
square root transformation, 67, 69, 73, 79
squared Euclidean distance, 46, 48
stability criterion, NMS, 128
standard deviation
   species turnover, 30
   species-area, 33
standard error, 19
standardizations
   built-in, 79, 144
   by the norm, 73
   scores, DA, 206
   variables, 9
standardized Euclidean distance, 46
standardized Mantel statistic, 211, 213-216
stands, defined, 17
starting coordinates, NMS, 129
statistics

descriptive, 78-79
   summary, 58
   tree data summary, 16
steepest-descent, 127
step size, NMS, 128
stepwise DA, 206, 210
stratified random sampling, 17
stress
   and number of variables, 133
   defined, 126
   depends on sample size, 132-133
   formula one, 127, 132
   formula two, 127
   minimizing, 125
   negative gradient, 127
   NMS, 126
   normalized, 127
   rare species, 133
   rules of thumb, 132
   scree plot, 136
   seek low, 132
   vs. dimensionality, 130-132
   vs. iteration, NMS, 134-136
structural equation modeling, 233-256
   a priori model, 249
   alternative models, 251
   categorical variables, 246
   compared to path analysis, 242
   compared to PCA, 236, 240-241
   compared to regression, 236-239, 241, 243
   competing models, 235, 247
   confirmatory, **235**, 246, 250
   construct validity, 235
   covariance analysis, 234
   data types, 246
   degrees of freedom, 249
   endogenous variables, **235**, 250
   equivalent models, 247
   evaluation of model fit, 250
   examples, 252-255
   exogenous variables, 235, 250
   exploratory analysis, 235
   how it works, 248-251
   hypothesis evaluation, 234
   identification, **235**, 249
   indicator variables, 234-235
   interfering conditions, 248
   interpretation, 243, 251
   latent variables, 235, 242-243
   linearity, 250
   loadings, **235**, 244
   manifest model, **235**, 241
   measurement model, 234, **235**
   model generating, 251
   model refinement, 246
   modification indices, **235**, 245
   multivariate normality, 250
   nested models, 235
   nonlinear relationships, 250
   ordination, use with, 108, 236
   overfitting, 235, 246

   overidentification, 235
   partial correlation, 238
   path coefficient, defined, 237
   path model
      compared to regression, 238
   philosophy and history, 233
   precision of coefficients, 255
   precondition with PLS, 255
   range of variables, 250
   relationship to regression, 234
   sample size, 246
   simulation models, use with 256
   simultaneous regression, 177, 241
   software, 234
   strictly confirmatory, 251
   structural model, 234, **235**
   terminology, 235
   total effects, **235**, 239
   underidentification, 235, 249
   use of prior knowledge, 241
   validation, 247
   variations, 255-256
   what to report, 251-252
   when to use it, 245-248
structural models, 222
   CART, 222-232
   SEM, 233-256
subjective sampling, 17
subsample, 19, 21-22, 189
   species-area curve, 33-34
   to evaluate consistency, 181
   to evaluate error, 178
subtract matrices, 258
successional change
   beta diversity, 30
   first difference transform, 77
   MRPP, 185
   example, 109
   vectors, 108-110, 186-187
summary statistics, 58, 61
symbol size in ordination, 105-106
symmetrical matrix defined, 257
synthetic variables, 1, 104, 184
systematic sampling, 17

T

tau
   Kendall's 107
   with ordination, 107
taxonomy, 3
   analysis, 4
TETRAD, 234, 251
thoroughness, NMS, 130
Thuja occidentalis, 36
ties, ranks, 193
time
   change through, 77
   differences through, 77
   first difference, 77
   series, 67, 77

Index

tolerance
  CCA, 167-168
  DCA, 160
  TWINSPAN, 100
topographic variables, 22
total effects, SEM, 235, 239
trace, defined, 257
tradeoffs in sampling, 22-23
traits, 5, 7-10, 242
  analysis with DA, 206, 208
  example matrix, 5
  independence, 10
  PCA, 120
  relativization, 10
trajectories, 42, 108-110, 186-187
  empty sample unit, 42
  in species space, 38, 110, 186-187
  predicted, 186-187
  sample unit in species space, 38, 110, 147
transformation, **67-79**
  adjustment to mean, 73
  arcsine, 69
  arcsine-squareroot, 13-14, 69
  Beals smoothing, 70
    example, 71
  binary by median, 68
  binary with respect to mean, 74
  binary with respect to median, 74
  by standard deviation, 74
  by the norm, 73
  centering, 73
  contingency deviate, 75
  cube root, 67
  data, 67-79
  difference between dates, 77
  domains and ranges, 67-68
  effect on NMS, 135
  effect on species, 72
  first difference of time series, 77
    PCA, 119
  for PCA, 115
  fourth root, 67
  log, 37, 67-69
  logit, 69
  monotonic, 67, 78-79
  NMS, 135
  normalizing, 73
  notation, 67
  power, 67-68
  presence-absence, 67-68
  probabilistic, 67
  proportion data, 69
  range, 68
  ranks, 74
  relativization, 67-68, **70-75**, 78
  slope correction, 24
  smoothing, 70
  sociological favorability index, 70
  square root, 67
  ubiquity, 68
transpose analysis, 5-7, 105
  CA, 152

transpose matrix, 5-8
  defined, 257
treatments, 3, 5, 9, 11, 22, 33, 76, 182
  blocked, 193-194
  comparison of vectors, 186
  manipulative, 182
  nonreplicated, 185
  species indicators, 198
  within blocks, 194
tree data, 15-16
tree density, 18
tree diagram (*see also* dendrogram) 80, 86
tree pruning, 82, 88, 99
  based on indicator species, 203
  CART, 226, 229
trees (*see also* dendrogram), 80
  classification, 222-232
  regression, 223
T-RFLP, 4
triangle inequality axiom, 46
t-test
  independent-sample, 186
  paired-sample, 186
turnover, 22, 27-30
  equalizing, 159-161
TWINSPAN, 84, **97-101**, 154
  compared to cluster analysis, 100-101
  differential species, 99
  example, 98, 101
    table, 98
  instability, 100
  quantitative data, 97
  preferential species, 99
  tolerance, 100
  when to use it, 97
two-way indicator species analysis, 84, **97-101**
two-way ordered table, 98

U

ubiquity
  information function, 68, 74
  transformation, 68
  weighting by, 74
uncertainty and information, 27
underidentification, SEM, **235**, 249
unimodal species response, 35, 39
  CART, 226
  CCA, 164, 177
units, differing, 73
univariate vs. multivariate, 182
UPGMA linkage, 92
UTM, 213

V

V/M, 16
validation, 235
  CART, 226
variables, 3-11
  binary, 3, 8 (*see also* presence)
  categorical, 3, 8
    recoding, 3
  conceptual, SEM, 234
  continuous, 3
  discontinuous, 3
  dummy, 3
  indicator, 3
    SEM, 234, 241, 243
  latent, SEM, 234
  measurement, 3
  nominal, 3, 8
  ordinal, 3, 8
  ranked, 3
  relative weights, 12, 189
  response, 1
  scaling, 9
  synthetic, 1, 104, 114, 120, 184
  topographic, 22
  z-transformed, 239
variance
  assumption of homogeneous, 67
  within-site, 23
variance represented, 173, 178-179, 254
  CA, 111
  CCA, 111, 164, 173
    example, 175
  high values, 112
  multiple regression, 239
  NMS, 112
  noncentered PCA, 121
  ordination, 111, 178
  path model, 239
  PCA, 116, 118
  trivial structure, 112
variance-covariance matrix, **116**, 259
variance-regression endpoints, 146
  geometry, 146
varimax rotation, 122
  example, 123
vectors
  before/after data, 110
  biplot, 109-110
    CCA, 175
  change through time, 77, 108-110, 186-187
  column, defined, 257
  compare with MANOVA, 186
  compare with MRPP, 186
  compositional, 186-187
  defined, 257
  fitting, 109

gradient, NMS, 127
hypotheses
    difference in direction, 187
    difference in rate, 187
    overall change, 186
inner product, 258
joint plot, 108
    significance, 109
    species, 109
latent, 259
length, 258
    joint plot, 108
length and direction, 186
overlay, 110
paired samples, 110
rotation, 123
row, defined, 257
successional, 109-110
trajectories, 186
treatment and control, 186
vegetation types
    classification tree, 224
vigor, 120
visual integration, 23
visual rotation, 122

W

WA scores, 166, 169, 171, 173
    correlation with LC scores, 168
    example, 173, 175
Ward's method, 87-88, 92-93
    example, 88
    with Sørensen distance, 92
weighted averaging, 149-151
    CCA, 166
    compared to logistic regression, 150
    example, 150
    how it works, 150
    ordination, 149
        example, 151
        when to use it, 149
    range, 150
    related to CA, 152
    weights, 149-150
    when to use it, 149
weights, 12
    Benthic Quality Index, 149
    CA, 154
        initial species, 156
    CCA, 166
        example, 173
    climax adaptation values, 149
    sample unit, 154
    sites and species, CCA, 171
    species, 149-152, 154
        binary, 151
        design, 151
        example, 174
    weighted averaging, 150
wetland zooplankton index, 149
Wilk's lambda, 207

Wilson's method, consistency and accuracy, 181
Wishart's objective function
    example, 89
within-group agreement
    chance-corrected, 190
within-group variance, minimizing, 80
WPGMC, 92
WZI, 149

Z

z scores, 74, 79
    SEM, 239
        Z statistic, 211
zero truncation problem, 35-36, 42-43, 125
    relieving, 44, 70
zeros
    effect on correlation coefficient, 38, 42
    shared, 38, 51
z-transformed, 239